SOTHIS PRESS

Sothis Press aims to completely reimagine the public domain book with an emphasis on universal access to captivatingly designed print editions. We believe that technology will enable every single book ever published to be perpetually accessible in beautiful print form. Sothis is beauty in ubiquity.

www.sothis.press
1 Waterhouse Square, London, EC1N 2ST

# The Metal Worker Pattern Book

A Practical Treatise on the Art and Science of Pattern Cutting as Applied to Sheet Metal Work

A. O. Kittredge

SOTHIS PRESS

ISBN 978-1-0365-1744-1

*The Metal Worker Pattern Book*
is in the public domain.

This edition is published by
 The company
is registered in England and Wales.
The Company Number is 14900389.

# The Metal Worker

# PATTERN BOOK.

A PRACTICAL TREATISE ON THE ART AND SCIENCE OF PATTERN CUTTING AS APPLIED TO SHEET METAL WORK.

BY

A. O. KITTREDGE.



NEW YORK:
DAVID WILLIAMS, 83 READE STREET.
1882.

# PREFACE.

The demand for a second edition of the "Metal Worker Pattern Book," the author feels, may be taken as conclusive evidence that his work has been found useful by those for whom it was specially intended. It was undertaken in response to a well-defined demand upon the part of Tinners and Cornice-makers for a comprehensive exposition of the principles of pattern cutting as applied to sheet-metal work. Some parts were prepared in direct answer to questions asked by subscribers of *The Metal Worker*, and appeared in the columns of that journal during the time the work was in progress. The most careful attention has been given throughout to the needs of sheet-metal workers, as made known to the author during his editorial connection with *The Metal Worker*, and through his previous practical experience in the trade. The aim has been to present a work serviceable alike to the apprentice boy who can afford but a single instruction book, to the mechanic who desires to add to whatever knowledge of pattern cutting he already possesses, to the student who would master the art by systematic investigation, and to all who need occasional assistance in getting over difficult places.

The work is comprised in five general divisions or chapters. In the first, Definitions and Technicalities are considered. The various mathematical and mechanical terms which it has been found necessary to use in the book, and which are current among mechanics, are explained, and such architectural terms as are commonly employed in cornice shops have been defined. Illustrations have been employed wherever their use has seemed advantageous. Following this is a chapter on Drawing Tools and Materials, prepared to meet the wants of those who commence drawing as a preliminary step to pattern cutting. This in turn is followed by a selection of simple Geometrical Problems, chosen with particular reference to the needs of students of pattern cutting. Various ways of solving the same problem, and the use of different instruments in accomplishing the same object, are presented in order to give the mechanic the widest range of choice in methods. At this stage the theoretical chapter of the work is presented, and is entitled the "Art and Science of Pattern Cutting." It is an attempt to explain the principles underlying all the operations of pattern cutting in such a way as will enable the student to make intelligent application of them, irrespective of formulated rules. Mechanics who already possess a fair degree of ability as pattern cutters, but who are perplexed when unusual combinations arise, may find in this division of the book all they require to render them proficient. Following this is a selection of Pattern Problems, arranged for the most part in classes, so that those of a kind will be found together. Each problem, so far as possible, has been demonstrated independently of all others. This arrangement has been followed to facilitate occasional use of the book by those who do not care to go through it in course. The work is completed by a comprehensive index, which it is believed will be found useful when searching for any required problem. The names of some of the ordinary articles of ware made in tin shops have been included in it, with references to the rules which may be employed in developing their patterns.

The "Metal Worker Pattern Book" was prepared from the mechanic's standpoint. In diction and style it will be found suited to the needs of workmen of the most ordinary attainments. Each proposition is expressed and demonstrated in language which the average reader will have no difficulty in understanding, and which the apprentice can read without becoming confused or discouraged.

83 Reade Street, New York, November, 1882.

# DEFINITIONS AND TECHNICALITIES.

1. A treatise on Pattern Cutting as applied to sheet-metal work, is only an exposition and application of geometrical principles. Any work on geometry, or, more particularly, upon geometrical drawing, presents in a general way all the principles that enter into the art of Pattern Cutting. It remains for us, therefore, in this work to make specific application of those principles, and to describe them in a way that will be readily understood by mechanics who have not had the advantages of a mathematical education. While in each problem and demonstration we shall be careful to avoid, as far as possible, the use of technical terms and words not in common use among mechanics, the necessity for precise language in describing geometrical figures, together with the fact that the every-day vocabulary of the workshop is not sufficiently comprehensive to enable us to restrict ourselves entirely to it, compels us to employ some terms not in general use which it is proper we should define and explain at the outset. In this connection it may not be out of place to remark that the advantage to the mechanic of accurate language is so considerable that every student of this book will be justified in giving careful attention to the introductory chapter, for the purpose of increasing and improving his vocabulary, as well as for the sake of being able to readily comprehend the demonstrations in the pages following. For this reason we have extended the list of terms to be defined somewhat beyond the strict requirements of this book. We have made it include nearly all of the terms belonging to plane geometry and those peculiar to pattern cutting, and we have added a few architectural terms suggested by the problems relating to cornice work. We introduce the terms and definitions in the way of a familiar talk, rather than in the set form of a glossary, because we believe the former will be of greater advantage to the mechanic. By reference to the index, which is arranged alphabetically, any term can be readily found.

*Fig 1.—A Straight Line.*

2. *Geometry* is that branch of mathematics which investigates the relations, properties and measurements of solids, surfaces, lines and angles.

3. *Sheet-Metal Pattern Cutting* is founded upon those principles of geometry which relate to the surfaces of solids. Although articles made from sheet metal are hollow, being only shells, they are all considered in the process of pattern cutting as though they were solids. Thus the pattern for a cone is called the envelope of a cone, as though it were a casing stripped from a solid cone.

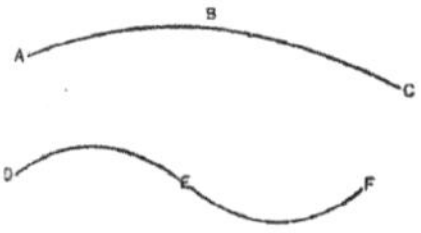


*Fig. 2.—Curved Lines.*

4. A *Point* is that which has place or position without magnitude, as the intersection of two lines or the center of a circle; it is therefore frequently represented to the eye by a small dot.

5. A *Line* is measured by length merely, and may be straight or curved.

6. A *Straight Line*, or, as it is sometimes called, a *right line*, is the shortest line that can be drawn between two given points. Straight lines are generally designated by letters or figures at their extremities, as A B, Fig. 1.

7. A *Curve* is a bending without angles.

8. A *Curved Line* is one which changes its direction at every point, or one of which no portion, however small, is straight. It is therefore longer than a straight line connecting the same points. Curved lines are designated by letters or figures at their extremities and at intermediate points. (Fig. 2.)

9. A *Given Point* or a *Given Line* expresses a point or line of fixed position or dimension.

10. *Parallel Lines* are those which have no inclination to each other, being everywhere equidistant, as A B and A' B' in Fig. 3, which can never meet though produced to infinity. C D and C' D' are also parallel lines, being arcs of circles which have a common center.

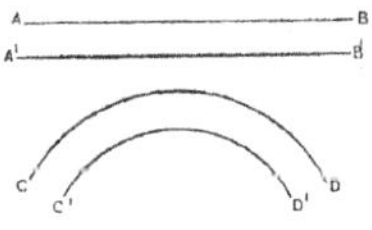


*Fig. 3.—Parallel Lines.*

11. *Horizontal Lines* are such as are parallel to the horizon, or level.

12. A Horizontal Line in a drawing is indicated by a line across the paper, as A B in Fig. 4; or, in other words, by a line drawn from left to right in front of the draftsman.

13. *Vertical Lines* are such as are parallel to the position of a plumb-line suspended freely in a still atmosphere.

14. A Vertical Line in a drawing is represented by a line drawn up and down the paper, or at right angles to a horizontal line, as E C in Fig. 4.

15. *Inclined Lines* occupy an intermediate between horizontal and vertical lines, as C D, Fig. 4. Two lines which converge toward each other, and which, if produced, would meet or intersect, are said to incline to each other.

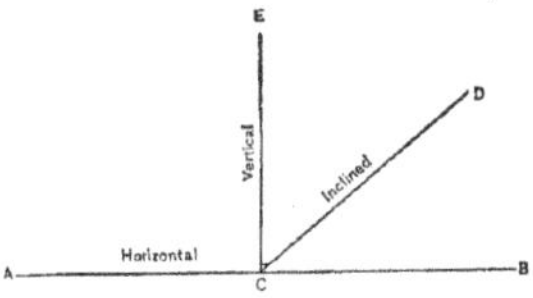


*Fig. 4.—Names of Lines by Direction.*

16. *Perpendicular Lines.*—Lines are perpendicular to each other when the angles on either side of the point of junction are equal. Vertical and horizontal lines are always perpendicular to each other, but perpendicular lines are not always vertical and horizontal. They may be at any inclination to the horizon, provided that the angles on either side of the point of intersection are equal. In Fig. 5, C F, D H and E G are said to be perpendicular to A B. Also in Fig. 6, C D and E F are perpendicular to A B.

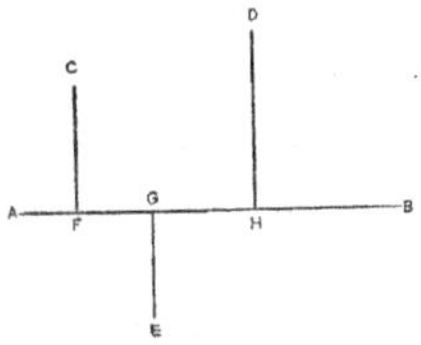


*Fig. 5.—Perpendicular Lines.*

17. An *Angle* is the opening between two straight lines which meet one another. An angle is commonly designated by three letters, and the letter designating the point in which the straight lines containing the angle meet, is put between the other two letters.

18. A *Right Angle.*—When a straight line meets another straight line, so as to make the adjacent angles equal to each other, each angle is called a right angle, and the straight lines are said to be perpendicular to each other. (See C B E and C B D, Fig. 7.)

19. An *Acute Angle* is an angle less than a right angle, as A B D or A B C, Fig. 7.

20. An *Obtuse Angle* is an angle greater than a right angle, as A B E, Fig. 7.

21. A *Surface* is that which has length and breadth without thickness.

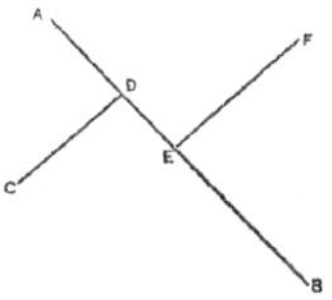


*Fig. 6.—Perpendicular Lines.*

22. A *Plane* is a surface such that if any two of its points be joined by a straight line, such line will be wholly in the surface. Every surface which is not a plane surface, or composed of plane surfaces, is a *curved surface.*

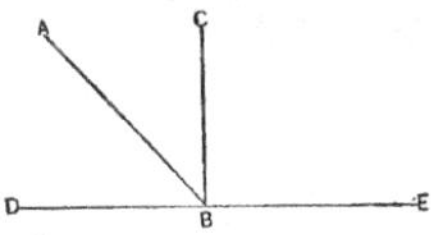


*Fig. 7.—Angles.*

C B E, C B D, right angles. A B D, A B C, acute angles. A B E, an obtuse angle.

23. A *Plane Figure* is a portion of a plane terminated on all sides by lines either straight or curved.

24. *Rectalinear Figures.*—When surfaces are bounded by straight lines they are said to be rectalinear. (See Figs. 8, 16, 21, &c.)

25. *Polygon* is the general name applied to all rectalinear figures, but is commonly applied to those having more than four sides. A *regular polygon* is one in which the sides are equal.

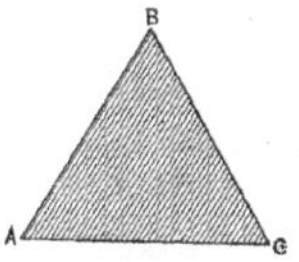


*Fig. 8.—An Equilateral Triangle.*

26. A *Triangle* is a flat surface bounded by three straight lines. (Figs. 8, 9, 10, 11, 13, &c.)

27. An *Equilateral Triangle* is one in which the three sides are equal. (Fig. 8.)

28. An *Isosceles Triangle* is one in which two sides are equal. (Fig. 9.)
29. A *Scalene Triangle* is one in which all the sides are of different lengths. (Fig. 10.)
30. A *Right-Angled Triangle* is one in which one of the angles is a right angle. (Fig. 11.)
31. An *Acute-Angled Triangle* is one which has its three angles acute. (Fig. 12.)

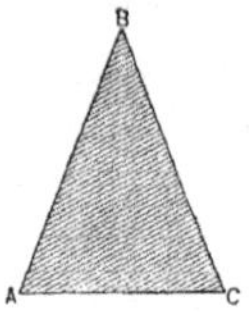


*Fig.* 9.—*An Isosceles Triangle.*

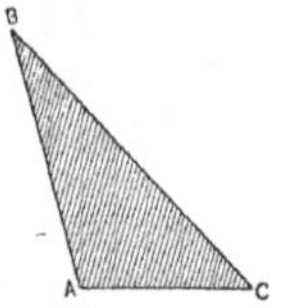


*Fig.* 10.—*A Scalene Triangle.*

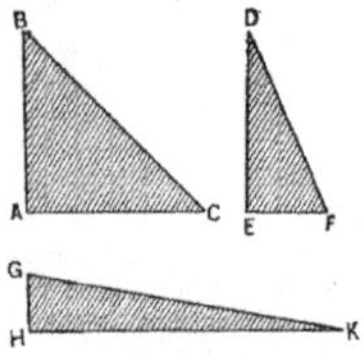


*Fig.* 11.—*Right-Angled Triangles.*

32. An *Obtuse-Angled Triangle* is one which has an obtuse angle. (Fig. 13.)
33. A *Hypothenuse* is the longest side in a right-angled triangle, or the side opposite the right angle. (Fig. 15.)
34. The *Apex* of a triangle is its upper extremity. (Fig. 14.) Also called vertex.

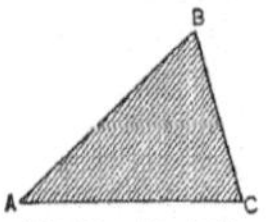


*Fig.* 12.—*An Acute-Angled Triangle.*

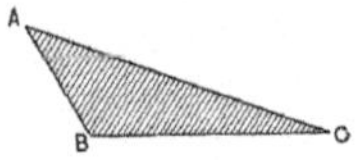


*Fig.* 13.—*An Obtuse-Angled Triangle.*

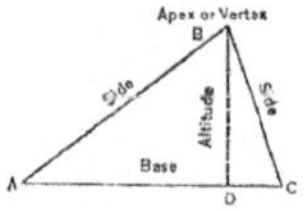


*Fig.* 14.—*Names of the Parts in a Triangle.*

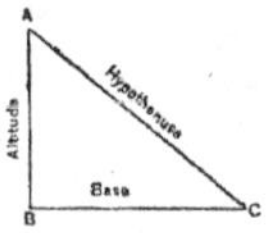


*Fig.* 15.—*Names of the Parts in a Right-Angled Triangle.*

35. The *Base* of a triangle is the line at the bottom. (Figs. 14 and 15.)
36. The *Sides* of a triangle are the including lines. (Fig. 14.)
37. The *Vertex* is the point in any figure opposite to and furthest from the base. The vertex of an angle is the point in which the sides of the angle meet. (Fig. 14.)
38. The *Altitude* of a triangle is the length of a perpendicular let fall from its vertex to its base, as B D, Fig. 14.
39. A *Quadrilateral* figure is a surface bounded by four straight lines.

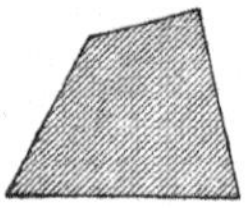

*Fig.* 16.—*A Trapezium.*

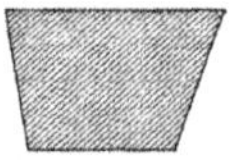

*Fig.* 17.—*A Trapezoid.*

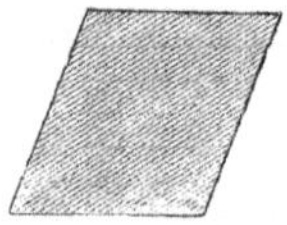

*Fig.* 18.—*A Rhomboid.*

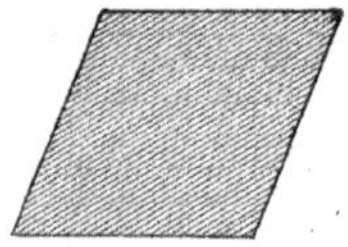

*Fig.* 19.—*A Rhombus or Lozenge.*

There are three kinds of quadrilaterals:
40. The *Trapezium*, which has no two of its sides parallel. (Fig. 16.)
41. The *Trapezoid*, which has only two of its sides parallel. (Fig. 17.)
42. The *Parallelogram*, which has its opposite sides parallel.
There are four varieties of parallelograms:
43. The *Rhomb*, *Rhombus* or *Lozenge*, in which the several sides are equal, and whose opposite sides are parallel, and in which two angles are obtuse and two acute. (Fig. 19.)
44. The *Rhomboid*, which has only the opposite sides equal, the length and width being different. (Fig. 18.)

45. The *Rectangle*, which is an equiangular parallelogram. (Fig. 20.)
46. The *Square*, which is both equilateral and equiangular. (Fig. 21.)
47. A *Pentagon* is a plane figure of five sides. (Fig. 22.)
48. A *Hexagon* is a plane figure of six sides. (Fig. 23.)
49. A *Heptagon* is a plane figure of seven sides. (Fig. 24.)

*Fig.* 20.—*An Equiangular Parallelogram, called a Rectangle.*

*Fig.* 21.—*A Square.*

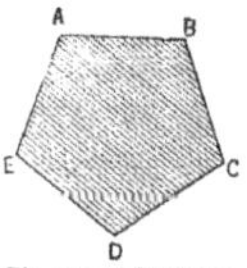


*Fig.* 22.—*A Pentagon*

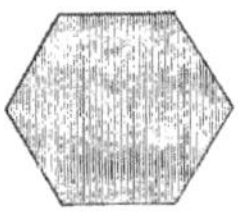

*Fig.* 23.—*A Hexagon.*

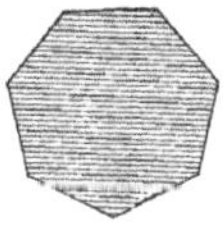

*Fig.* 24.—*A Heptagon.*

50. An *Octagon* is a plane figure of eight sides. (Fig. 25.)
51. A *Nonagon* is a plane figure of nine sides. (Fig. 26.)
52. A *Decagon* is a plane figure of ten sides. (Fig. 27.)
53. A *Dodecagon* is a plane figure of twelve sides. (Fig. 28.)
54. The *Peremeter* of a polygon is the broken line that bounds it, as A B C D E, Fig. 22.
55. A *Diagonal* is a straight line joining two opposite angles of a figure, as A B and C D, Fig. 29.

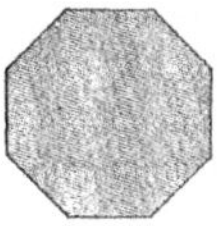

*Fig.* 25.—*An Octagon.*

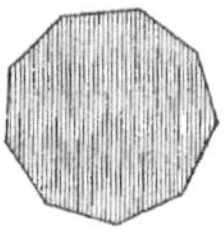

*Fig.* 26.—*A Nonagon.*

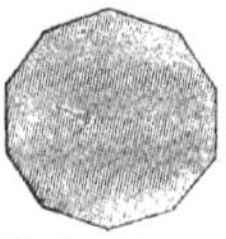

*Fig.* 27.—*A Decagon.*

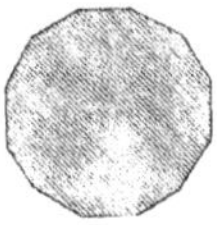

*Fig.* 28.—*A Dodecagon.*

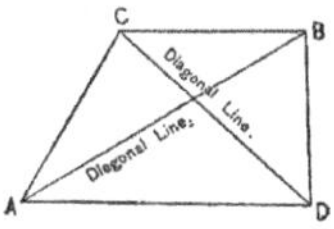


*Fig.* 29.—*Diagonals.*

56. A *Circle* is a plane figure contained by one curved line, everywhere equidistant from its center. (Fig. 30.) The term circle is also used to designate the boundary line. (See also Circumference.)
57. The *Circumference* of a circle is the boundary line of the figure. (Fig. 30.)
58. The *Center* of a circle is a point within the circumference equally distant from every point in it, as A, Fig. 30.
59. The *Radius* of a circle is a line drawn from the center to any point in the circumference, as A B, Fig. 30. The plural of radius is *radii.*

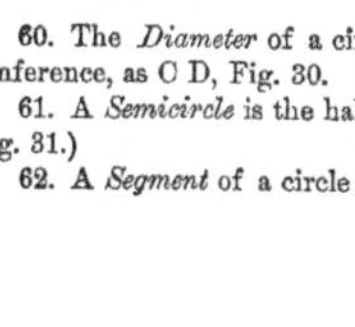


*Fig.* 30.—*A Circle.*

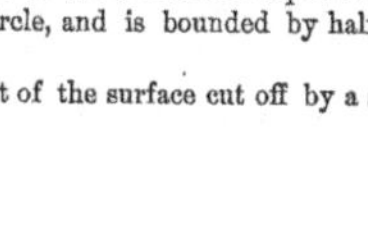


*Fig.* 31.—*A Semicircle.*

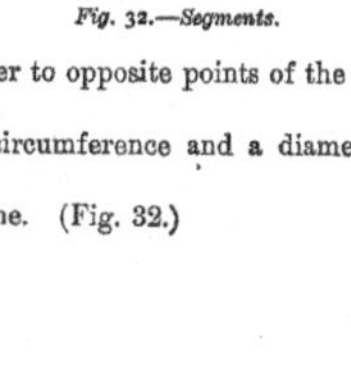


*Fig.* 32.—*Segments.*

60. The *Diameter* of a circle is any straight line drawn through the center to opposite points of the circumference, as C D, Fig. 30. The length of the diameter is equal to two radii.
61. A *Semicircle* is the half of a circle, and is bounded by half of the circumference and a diameter. (Fig. 31.)
62. A *Segment* of a circle is any part of the surface cut off by a straight line. (Fig. 32.)

63. An *Arc* of a circle is any part of the circumference, as A B E and C F D, Fig. 33.

64. A *Sector* of a circle is the space contained between two radii and the arc which they intercept, as A C B and D C E, Fig. 34.

65. A *Quadrant* is a sector whose area is equal to one-fourth of the circle. (B A C, Fig. 35.) In a quadrant the two radii are at right angles.

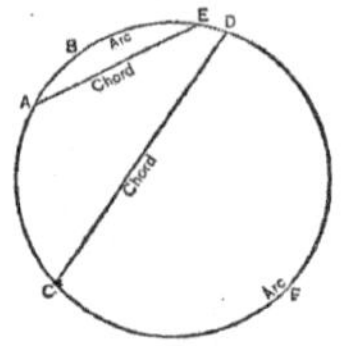


*Fig.* 33.—*Arcs and Chords.*

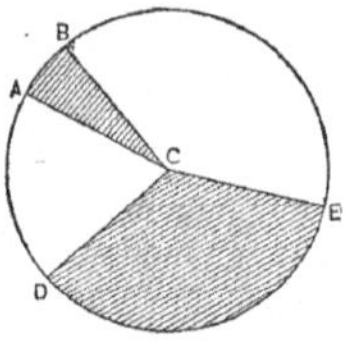


*Fig.* 34.—*Sectors.*

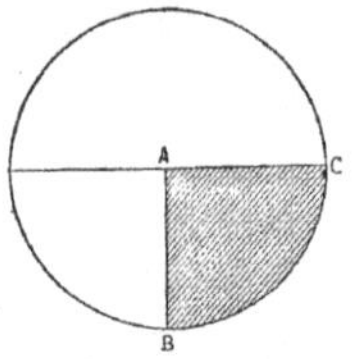


*Fig.* 35.—*A Quadrant.*

66. A *Chord* is a straight line joining the extremities of an arc, as A E and C D, Fig. 33.

67. A *Tangent* to a circle or other curve is a straight line which touches it at only one point, as E D and A C, Fig. 36.

68. Circles are *concentric* when described from the same center. (Fig. 37.)

69. Circles are *eccentric* when described from different centers. (Fig. 38.)

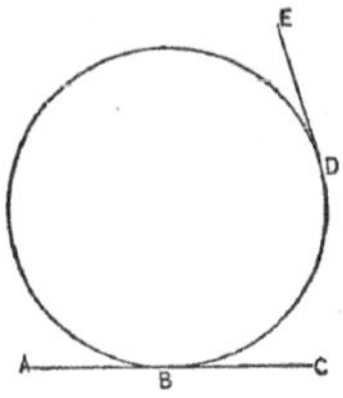


*Fig.* 36.—*Tangents.*

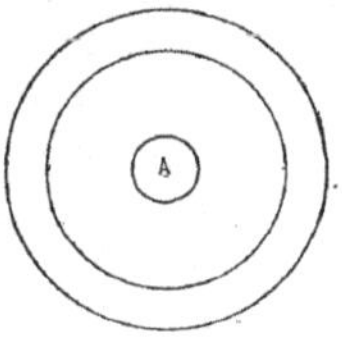


*Fig.* 37.—*Concentric Circles.*

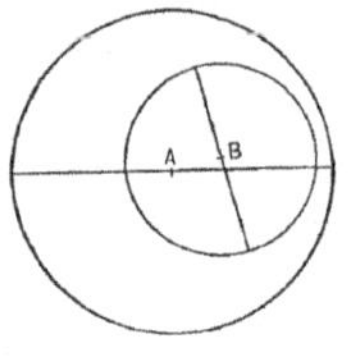


*Fig.* 38.—*Eccentric Circles.*

70. Triangular figures and those with a greater number of sides are *inscribed in*, or *circumscribed by*, circles when the vertices of all their angles are in the circumference. (Fig. 39.)

71. A circle is *inscribed* in a straight-sided figure when it is tangent to all sides. (Fig. 40.) All regular polygons may be inscribed in circles, and circles may be inscribed in the polygons; hence the facility with which polygons may be constructed.

72. A *Degree.*—The circumference of a circle is considered as divided into 360 equal parts, called *degrees* (marked °). Each degree is divided into 60 *minutes* (marked ′); and each minute into 60 *seconds* (marked ″). Thus if the circle be large or small the number of divisions is always the same, a degree being equal to $\frac{1}{360}$th part of the whole circumference; the semicircle is equal to 180° and the quadrant to 90°. The radii drawn from the center of a circle to the extremities of a quadrant are always at right angles with each other; a right angle is therefore called an angle of 90° (A E B, Fig. 41). If we bisect a right angle by a straight line, it divides the arc of the quadrant also into two equal parts, each being equal to one-eighth of the whole circumference, or 45°, (A E F and F E B, Fig. 41); if the right angle were divided into three equal parts by straight lines, it would divide the arc into three equal parts, each containing 30° (A E G, G E H, H E B, Fig. 41). Thus the degrees

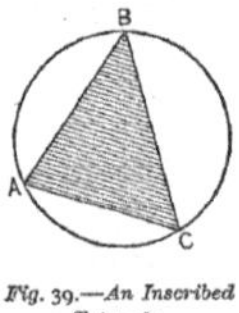


*Fig.* 39.—*An Inscribed Triangle.*

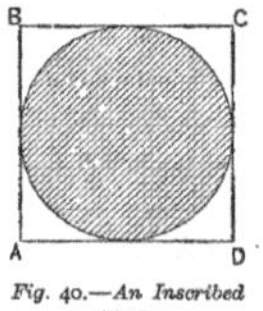


*Fig.* 40.—*An Inscribed Circle.*

of the circle are used to measure angles, and when we speak of an angle of any number of degrees, it is understood that if a circle with any length of radius be struck with one foot of the compasses in the angular point, the sides of the angle will intercept a portion of the circle equal to the number of degrees given. Thus the angle A E H, Fig. 41, is an angle of 60°.

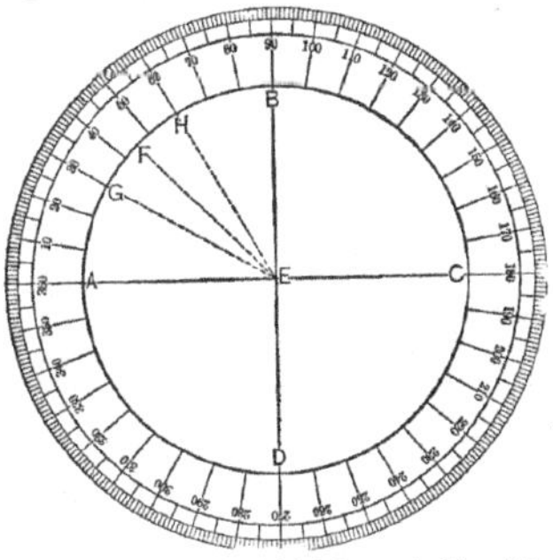


*Fig. 41.—The Circle Divided into Degrees, for Measuring Angles.*

73. In the measurement of angles by the circumference of the circle, and in the various mathematical calculations based thereon, use is made of certain lines, always bearing a fixed relationship to the radius of the circle and to each other, which gives rise to a number of terms, some of which, at least, it is desirable for the pattern cutter to understand.

74. The *Complement* of an arc or of an angle is the difference between that arc or angle and a quadrant. In Fig. 42, A D B is the complement of B D C, and *vice versa.*

75. The *Supplement* of an arc or of an angle is the difference between that arc or angle and a semicircle. In Fig. 43, B D C is the supplement of A D B, and *vice versa.*

76. A *Tangent* has already been defined as a straight line drawn without a circle, touching it at only one point. (See Fig. 36.) The *Tangent of an Angle,* or of an arc, is a line which touches the arc at one extremity. In Fig. 44, C B is the tangent of the arc E C, or of the angle E A C. Every tangent is perpendicular to a radius at the point it touches. Thus, B C is perpendicular to A C.

77. A *Secant* is a straight line drawn from the center of a circle, cutting its circumference and prolonged to meet a tangent. (A B, Fig. 44.)

78. The *Co-Tangent* of an arc is the tangent of the complement. (F G, Fig. 45.)

79. The *Sine* of an arc is a straight line drawn from one extremity perpendicular to a radius drawn to the other extremity of the arc. (H B, Fig. 45.)

80. The *Co-Sine* of an arc is the sine of the complement of that arc. (H K, Fig. 45.)

81. The *Versed Sine* of an arc is that part of the radius intercepted between the sine and the circumference. (A B, Fig. 45.)

82. The *Co-Secant* of an arc or angle is the secant of the complement of that arc or angle, as F C, Fig. 45.

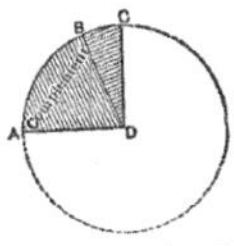


*Fig. 42.—Complement.*

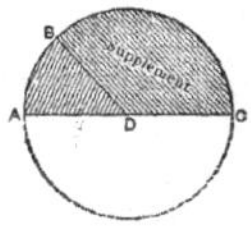


*Fig. 43.—Supplement.*

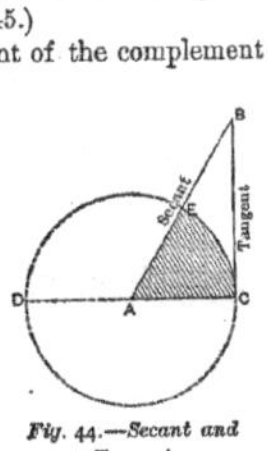


*Fig. 44.—Secant and Tangent.*

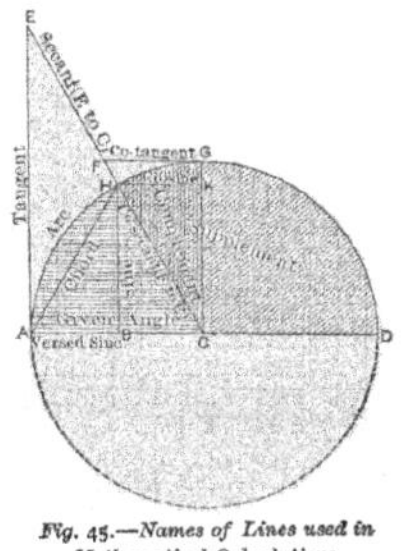


*Fig. 45.—Names of Lines used in Mathematical Calculations.*

83. In Fig. 45 are shown all the various lines and divisions appertaining to a given angle. A C H represents the angle; H C G is the complement of that angle, and H C D is the supplement. The names of the several parts are given in the diagram, and also have been defined and described above.

84. An *Ellipse* is an oval-shaped curve (Fig. 46), from any point in which, if straight lines be drawn to two fixed points within the curve, their sum will be always the same. These two points are called *foci* (F and H). The line A B, passing through the foci, is called the *transverse axis.* The line E G, perpendicular to the cen-

ter of the transverse axis, and extending from one side of the figure to the other, is called the *conjugate axis.* There are various other definitions of the ellipse besides the one given here, dependent upon the means employed for projecting it, which will be fully explained at the proper place among the problems.

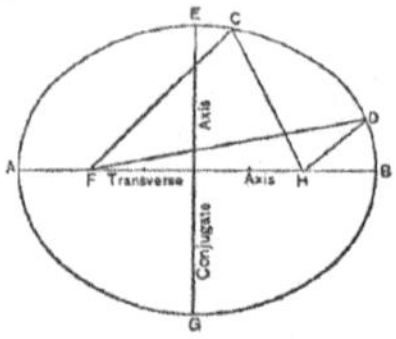


*Fig. 46.—An Ellipse.*

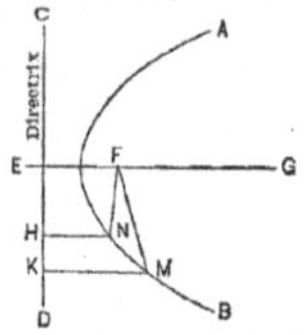


*Fig. 47.—A Parabola.*

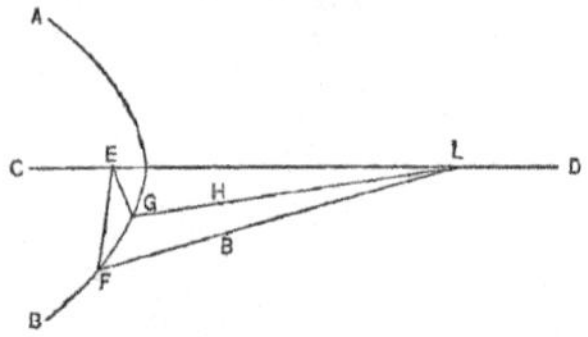


*Fig. 48.—A Hyperbola.*

85. A *Parabola* (A B, Fig. 47) is a curve in which any point is equally distant from a certain fixed point and a straight line. The fixed point (F) is called the *focus*, and the straight line (C D) the *directrix.* In this figure any point, as N or M, is equally distant from F and the same point in C D, as H or K.

86. A *Hyperbola* (A B, Fig. 48) is a curve from any point in which, if two straight lines be drawn to two fixed points, their difference shall always be the same. Thus, the difference between E G and G L is H L, and the difference between E F and F L is B L. H L and B L are equal. The two fixed points, E and L, are called *foci.*

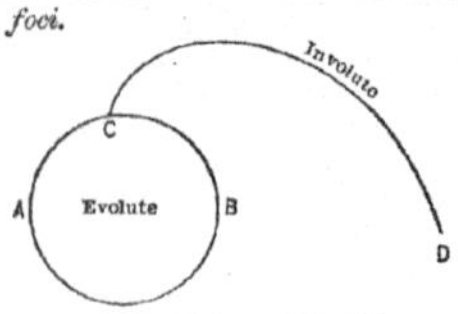


*Fig. 49.—Evolute and Involute.*

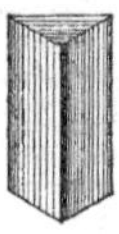

*Fig. 50.—A Triangular Prism.*

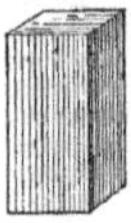

*Fig. 51.—A Quadrangular Prism.*

87. An *Evolute* is a circle or other curve from which another curve, called the *involute* or *evolutent,* is described by the aid of a thread gradually unwound from it. (Fig. 49.)

88. An *Involute* or *Evolutent* is a curve traced by the end of a string wound upon another curve or unwound from it. (Fig. 49.)

89. A *Solid* has length, breadth and thickness.

90. A *Prism* is a solid of which the ends are equal, similar and parallel straight-sided figures, and of which the other sides are parallelograms.

91. A *Triangular Prism* is one whose bases or ends are triangles. (Fig. 50.)

92. A *Quadrangular Prism* is one whose bases or ends are quadrilaterals. (Fig. 51.)

*Fig. 52.—A Pentagonal Prism.*

*Fig. 53.—A Hexagonal Prism.*

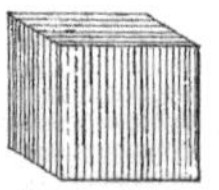

*Fig. 54.—A Cube.*

*Fig. 55.—A Cylinder.*

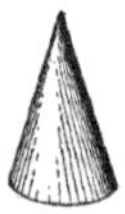

*Fig. 56.—A Cone.*

93. A *Pentagonal Prism* is one whose bases or ends are pentagons. (Fig. 52.)

94. A *Hexagonal Prism* is one whose bases or ends are hexagons. (Fig. 53.)

95. A *Cube* is a prism of which all the sides are square. (Fig. 54.)

96. A *Cylinder* is a round solid of uniform thickness, of which the ends are equal and parallel circles. (Fig. 55.)

97. A *Cone* is a round solid with a circle for its base, and tapering uniformly to a point at the top. (Fig. 56.)

98. A *Right Cone* is one in which the perpendicular let fall from the vertex upon the base passes through the center of the base. This perpendicular is then called the *axis* of the cone. (Fig. 57.)

99. An *Oblique Cone* or *Scalene Cone* is one of which the axis is inclined to the plane of its base, and of which the sides are unequal. (Fig. 58.)

100. A *Truncated Cone* is one whose vertex is cut off by a plane parallel to its base. (Fig. 59.) This figure is also called a *frustum* of a cone. (See Figs. 73 and 74 and accompanying definitions.)

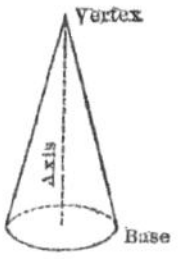

*Fig.* 57.—*A Right Cone.*

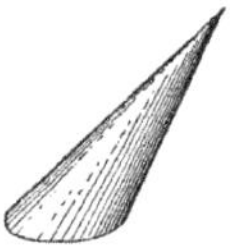
*Fig.* 58.—*An Oblique or Scalene Cone.*

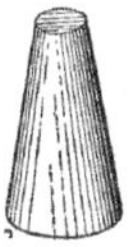
*Fig.* 59.—*A Truncated Cone.*

*Fig.* 60.—*A Sphere or Globe.*

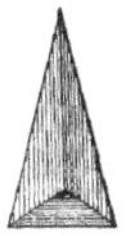
*Fig.* 61.—*A Triangular Pyramid.*

101. A *Sphere* or *Globe* is a solid bounded by a uniformly curved surface, any point of which is equally distant from the center, a point within the sphere. (Fig. 60.)

102. A *Pyramid* is a solid having a straight-sided base and triangular sides terminating in one point or *vertex.* Pyramids are distinguished as *triangular, quadrangular, pentagonal, hexagonal,* etc., according as the base has three sides, four sides, five sides, six sides, etc. (Figs. 61, 62 and 63.)

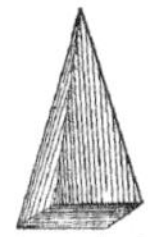
*Fig.* 62.—*A Quadrangular Pyramid.*

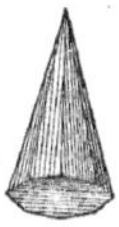
*Fig.* 63.—*An Octagonal Pyramid.*

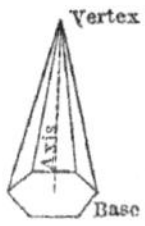

*Fig.* 64.—*A Right Pyramid.*

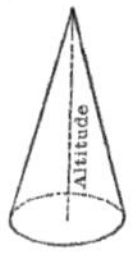

*Fig.* 65.—*Altitude of a Cone.*

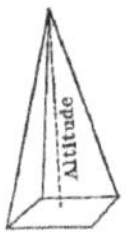

*Fig.* 66.—*Altitude of a Pyramid.*

103. A *Right Pyramid* is one whose base is a regular polygon, and in which the perpendicular let fall from the vertex upon the base passes through the center of the base. This perpendicular is then called the *axis* of the pyramid. (Fig. 64.)

104. The *Altitude* of a pyramid or cone is the length of the perpendicular let fall from the vertex on the plane of the base. The altitude of a prism or cylinder is the distance between its two bases or ends, and is measured by a line drawn from a point in one base perpendicular to the plane of the other. (Figs. 65, 66, 67 and 68.)

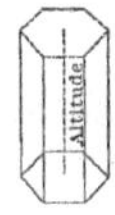

*Fig.* 67.—*Altitude of a Prism.*

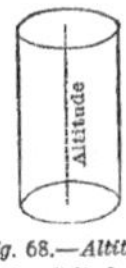

*Fig.* 68.—*Altitude of a Cylinder.*

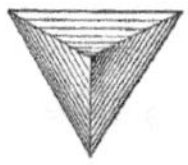
*Fig.* 69.—*A Tetrahedron.*

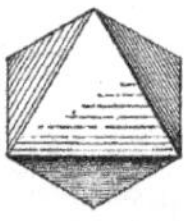
*Fig.* 70.—*An Octahedron.*

105. Besides the solids already described, there are others to which designating names have been applied.

106. A *Tetrahedron* is a solid bounded by four equilateral triangles. (Fig. 69.)

107. A *Hexahedron* is a solid bounded by six squares. The common name for this solid is *cube.* (See definition under cube, Fig. 54.)

108. The *Octahedron* is a solid bounded by eight equilateral triangles. (Fig. 70.)

109. The *Dodecahedron* is a solid bounded by twelve pentagons.

110. The *Icosahedron* is a solid bounded by twenty equilateral triangles.

111. An *Axis* is a straight line, real or imaginary, passing through a body on which it revolves, or may be supposed to revolve. The axis of a circle is any straight line passing through the center. The axis of a cylinder is the straight line joining the centers of the two ends. (Figs. 57 and 64.)

112. An *Envelope* of a solid is that which envelopes, encases or surrounds it, as the envelope of a cone.

113. Solids are said to *penetrate* each other when they are so fitted together as to appear to pass through each other. Hence we have the term *penetration of solids*. (Fig. 71.)

*Fig. 71.—Penetration of Solids.*

*Fig. 72.—Intersection of Solids.*

*Fig. 73.—Frustum of a Scalene Cone.*

*Fig. 74.—Frustum of a Pyramid.*

114. *Intersection of Solids* is a term meaning substantially the same as penetration of solids, and is used to describe the condition of solids which are so joined and fitted to each other as to appear to pass through each other. (Fig. 72.)

115. When a solid, as, for example, a cone, is cut through transversely by a plane parallel or inclined to the base, the part next the base is called a *frustum of the solid*. Hence we have the terms *frustum of a cone*, *frustum of a pyramid*, etc. (Figs. 59, 73 and 74.)

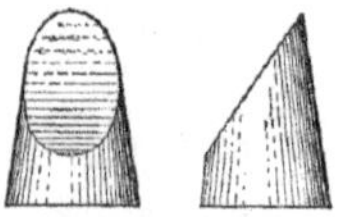

*Fig. 75.—Conical Ungula.*

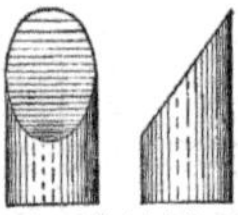

*Fig. 76.—Cylindrical Ungula.*

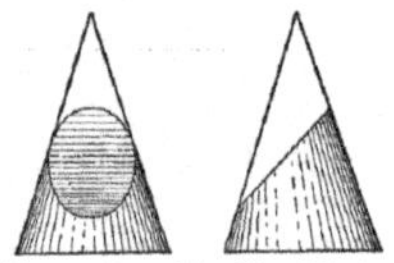

*Fig. 77.—Cone cut by a Plane obliquely through its opposite sides.*

116. When a section of a solid of revolution, as, for example, a cylinder or a cone, is cut off by a plane oblique to the base, it is called an *ungula*. (Figs. 75 and 76.)

117. A *Conic Section* is a curved line formed by the intersection of a cone and a plane.

118. When a cone is cut by a plane obliquely through its opposite sides, the resulting figure is an *ellipse*. (Fig. 77.)

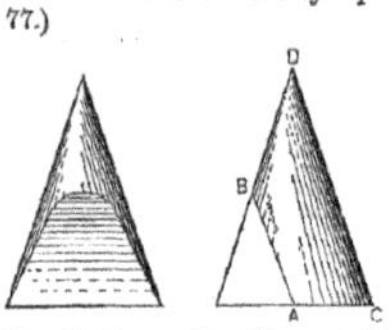


*Fig. 78.—Cone cut by a Plane parallel to one of its sides.*

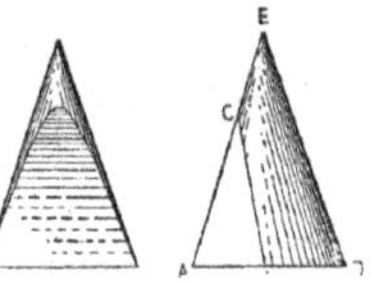


*Fig. 79.—Cone cut by a Plane which makes an angle with the base greater than the angle formed by the side.*

*Fig. 80.—A Concave Surface.*

*Fig. 81.—A Convex Surface.*

119. When a cone is cut by a plane parallel to one of its sides, the resulting figure is a *parabola*. Thus in Fig. 78, the cutting plane A B is parallel to the side of the cone C D.

120. When the cutting plane makes a greater angle with the base than the side of the cone makes, the resulting figure is a *hyperbola*. Thus in Fig. 79, the angle A B C is greater than the angle A D E.

121. The parabola and hyperbola resemble each other, both being incomplete figures, with arms extending indefinitely. The ellipse is a complete figure, but of varying proportions, as the cutting plane is inclined more or less.

122. Means of producing these several figures has been illustrated in Figs. 46, 47 and 48. See also the accompanying definitions. Further remarks concerning the ellipse will be found among the problems.

123. *Concave* means hollowed and curved or rounded, said of the interior of an arched surface or curved line in opposition to convex. (Fig. 80.)

124. A *Convex* surface is one that is regularly protuberant or bulging, when viewed from without. The opposite of convex is concave. (Fig. 81.)

125. *Diamond* is the name applied to a geometrical figure consisting of four equal straight lines and having two of the interior angles acute and two obtuse; a rhombus; a lozenge. (Fig. 18.)

126. The term *Cornice* is ordinarily used to designate any molded projection which finishes or crowns the part to which it is affixed. Hence common usage accepts the term cornice in the sense of an entire entablature, while by strict definition it is restricted to the upper division of the entablature as that word is understood in classical architecture. (Fig. 82.)

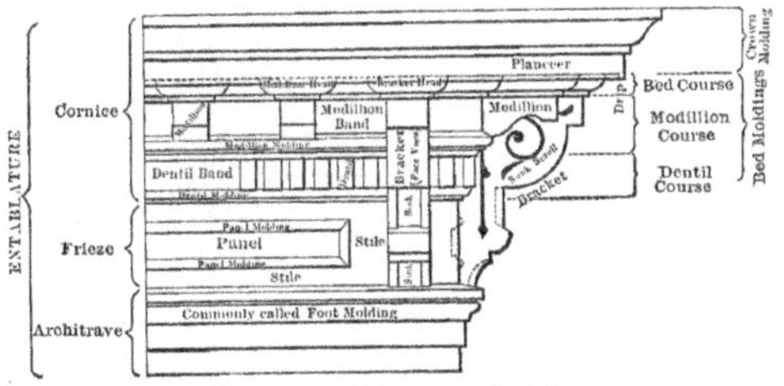


*Fig. 82.—The Entablature and its Parts.*

127. An *Entablature* consists of three parts, the cornice, the frieze and the architrave, as illustrated by the accompanying engraving (Fig. 82).

128. The *Frieze*, the middle division of the entablature (Fig. 82), is sometimes treated very plainly and sometimes receives considerable ornamentation, being subdivided into panels or enriched by scrolls, etc. Hence we have the terms *plain frieze*, designating a frieze devoid of ornamentation; *frieze-piece* or *frieze-panel*, designating one of the parts of which a frieze is constructed.

129. The *Architrave* is the third or lower division of the entablature. (Fig. 82.) This term is also used to designate a molding running around the exterior curve of an arch.

130. *Crown Molding* is the term applied to the front or projecting member of a cornice. (Fig. 82.)

131. *Planceer* is the term indicating the ceiling or under side of the cornice. (Fig. 82.)

132. *Soffit* is the term applied to the under side of a projecting molding.

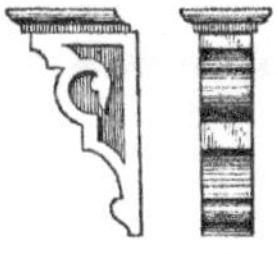

*Fig. 83.—A Cornice Bracket.*

*Fig. 84.—A Cornice Modillion.*

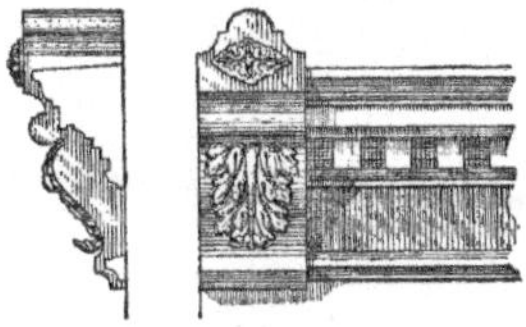

*Fig. 85.—A Lintel Cornice.*

133. A *Drip* is a downward projecting member in a cornice or in a molding, used to throw the water off from the other parts. (Fig. 82.)

134. A *Bracket* (Fig. 83), as used in a sheet-metal cornice, is simply an ornament of the cornice. Brackets were originally used as supports of the parts coming above them. Modern architecture retains the form, but with changes in construction has kept nothing of the original use. (Fig. 82.)

135. *Modillions* are also cornice ornaments, and differ from brackets only in general shape. (Fig. 84.) While a bracket has more depth than projection, modillions have more projection than depth.

136. A *Lintel Cornice* is a cornice covering a lintel or occurring just over an opening. This term is very generally used to designate the cornice used above the first story of stores. (Fig. 85.)

137. A *Deck Cornice* or *Deck Molding* is the cornice or molding used to finish the edge of a flat roof where it joins a steeper portion.

138. A *Sink* is a depression in the face of a piece of work or in a plain surface. (See face of bracket, Fig. 82, and side of bracket, Fig. 83.)

139. A *Truss* is a large terminal bracket in a cornice, projecting sufficiently to receive all the moldings against its side, thus forming a finish to the end of the cornice. (Fig. 86.)

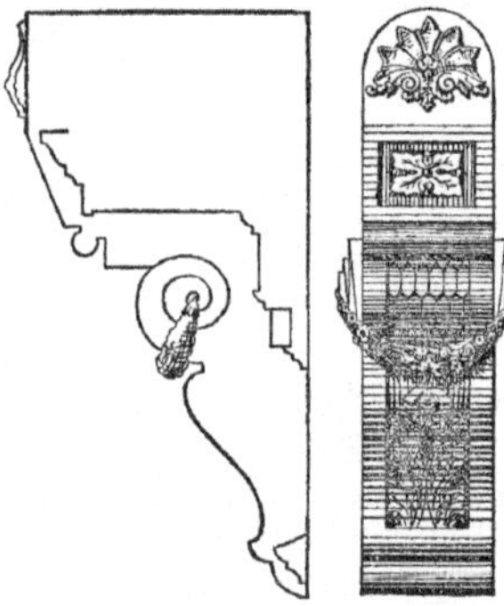

*Fig. 86.—A Cornice Truss.*

140. A *Stop Block* is a block-shaped structure, variously ornamented, which is placed above an ordinary bracket in a cornice, and which projects far enough to receive against its side the various moldings occurring above the bracket, thus forming an end finish to a cornice. (Fig. 87.)

141. A *Head Block* is a structure in general shape and appearance similar to a stop block, but which, unlike the latter, is placed outside of the various moldings, and whose sides finish against their face, forming an ornament to the crown molding.

142. A *Corbel* is a modified form of bracket. It is used to terminate the lower parts of window caps, and also forms the support for the lower end of arches, etc., in gothic forms.

143. A *Molding* is an assemblage of forms projecting beyond the wall, column, etc., to which it is affixed.

144. The *Bed Moldings* of a cornice are those moldings forming the lower division of the cornice, and which are made up of the bed course, modillion course and dentil course. (Fig. 82.)

145. The *Bed Course* is the upper division of the bed moldings, the part with which the bracket heads and modillion heads ordinarily correspond, and against which they miter. (Fig. 82.)

146. The *Modillion Course* of a cornice embraces those moldings which are immediately back of and below the modillions. It is subdivided into the *modillion molding* and the *modillion band.* (Fig. 82.)

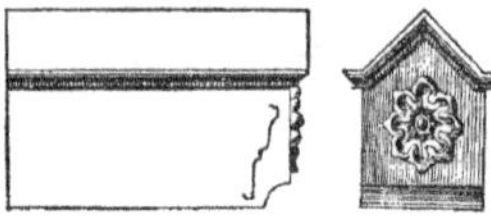

*Fig. 87.—A Cornice Stop-Block.*

147. The *Dentil Course* of a cornice embraces those moldings to which the dentils are attached as ornaments, and consists of the *dentil molding* and *dentil band.* (Fig. 82.)

148. *Foot Molding* is the common term used to designate the lower molding in a cornice. It is frequently in this connection used in the sense of architrave. (Fig. 82.)

149. *Curved Moldings* are those moldings whose plan or elevation is a curve.

150. A *Bracket Molding,* also called *bracket head,* is the molding around the upper part of a bracket, and which generally members with the bed molding, against which it finishes. (Fig. 82.)

151. A *Horizontal Molding* is one whose course is in a horizontal direction.

152. A *Vertical Molding* is one whose course is vertical, or at right angles to the horizon.

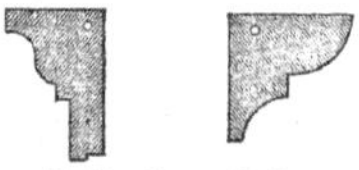

*Fig. 88.—Stays or Profiles.*

153. An *Inclined Molding* is one whose course is intermediate between vertical and horizontal.

154. A *Gable Molding* is an inclined molding which is used in the finish of a gable.

155. A *Ridge Molding* is a molding used to cap or finish a ridge. It is also called a *ridge capping,* or simply *ridging.*

156. A *Hip Molding* is a molding used to protect and finish the hips or angles of a roof. It is very frequently included in the more general term *ridging.*

157. The *Face* of a molding is its outer surface when placed in the position it is intended to occupy.

158. The *Stay* of a molding is its shape or profile cut in sheet metal. (Fig. 88.)

159. *Rake Moldings* are those which are inclined, as in a gable or pediment; but since to miter a rake molding or an inclined molding, under certain conditions, necessitates a change or modification of profile in one or the other of the moldings, *to rake* has come to mean to make such change of profile.

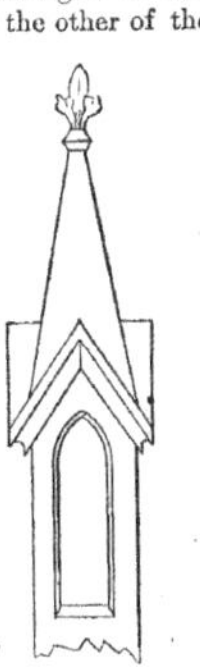

*Fig. 89.—A Pinnacle.*

160. A *Raked Molding*, therefore, is a term describing a molding of which the profile is a modification of some other profile.

161. A *Raked Profile* or *Raked Stay* describes the profile or stay which has been derived from another profile or stay, by certain established rules, in a process like that of mitering a horizontal and inclined molding together.

162. The *Normal Profile* or *Normal Stay* is the original profile or stay from which the raked profile or stay has been derived.

163. The term *Miter* designates a joint in a molding, or between two pieces not moldings, at any angle.

164. A *Butt Miter* is the term applied to the cut made upon the end of a molding to fit it against another molding or against a surface.

165. A *Gable Miter* is the name applied to the miter either at the peak or at the foot of a gable molding.

166. A *Rake Miter* is a miter between two moldings, one of which has undergone a modification of profile to admit of the joint being made.

167. *Square Miter* is the common term for a joint at right angles, or at 90°.

168. An *Octagon Miter* is a miter joint between two sides of a regular octagon, or between any two pieces at an angle of 135°.

169. An *Inside Miter* indicates a joint at an interior or re-entrant angle.

170. An *Outside Miter* is a joint at an exterior angle.

171. A *Miter Piece* is one of the pieces next the proper cut made upon it, between which a miter joint is to be made.

172. A *Complete Miter* is the structure formed by the union of two pieces of molding by means of a miter joint.

173. A *Fillet* is a little square member, and is of frequent occurrence in moldings.

174. A *Flange* is a projecting edge by which a piece is strengthened or fastened to anything.

175. A *Pinnacle* is a slender turret or part of a building elevated above the main building. (Fig. 89.)

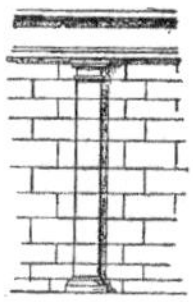

*Fig. 90.—A Pilaster.*

176. A *Pilaster* is a square column, usually set within a wall, and projecting part of its diameter. (Fig. 90.)

177. A *Pediment* is a triangular ornamental facing of a portico, or a similar decoration over doors, windows, etc. The name is also applied to arched ornaments of a like kind. (Figs. 91 and 92.)

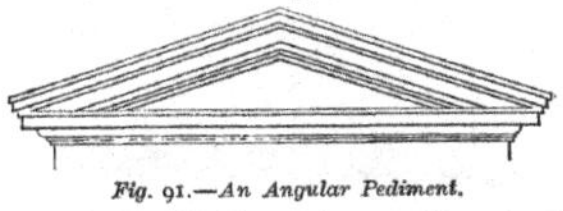

*Fig. 91.—An Angular Pediment.*

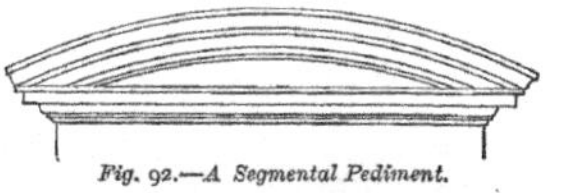

*Fig. 92.—A Segmental Pediment.*

178. A *Broken Pediment* is one, either in the form of a gable or arch, which is cut away in its central portion for the purpose of ornamentation. (Fig. 93.)

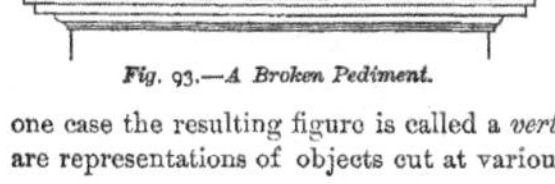

*Fig. 93.—A Broken Pediment.*

179. An *Elevation* is a geometrical projection of a building or other object on a plane perpendicular to the horizon. (Fig. 94.)

180. A *Plan* is the representation of the parts as they would appear if cut upon a horizontal line. (Fig. 95.)

181. A *Section* is a view of the object as it would appear if cut in two at a given vertical or horizontal plane. (Fig. 96.) In the one case the resulting figure is called a *vertical section*, and in the other a *horizontal section*. Oblique sections are representations of objects cut at various angles.

182. A *Perspective* is a representation of a building or other object upon a plane surface, presenting the same appearance that the object itself would present if viewed from a particular point. (Fig. 97.)

183. A *Draft* is a figure described on paper.

184. A *Drawing* is a representation on a plane surface, by means of lines, or by means of lines and shades, of the appearance or figure of objects.

185. A *Detail Drawing* is a drawing commonly full size, for the use of mechanics in constructing work.

186. A *Working Drawing* is the same as a detail drawing.

187. A *Scale Drawing* is one made to some scale less than full size.

*Fig. 94.—Elevation of House.*

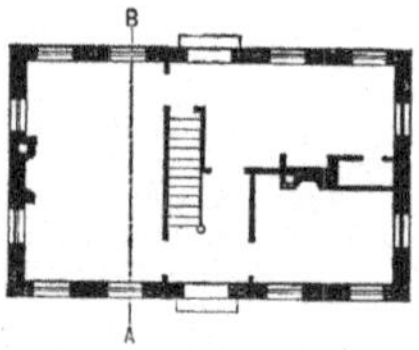


*Fig. 95.—Plan of House.*

188. *Incised Work* is a style of ornamentation consisting of fine members and irregular lines, sunken or cut into a plane surface.

189. A *Hip* is the external angle formed by the meeting of two sloping sides or skirts of a roof which have their wall plates running in different directions.

190. A *Gable* is the vertical triangular end of a house or other building, from the cornice or eaves to the top.

191. A *Problem* is a question proposed for solution. This term also describes anything which is required to be done, as to bisect a line.

192. A *Proposition* is that which is offered for consideration or adoption—a statement in terms, either of a truth to be demonstrated or of an operation to be performed.

193. An *Hypothesis* is something not proved, but assumed for the sake of argument; a supposition.

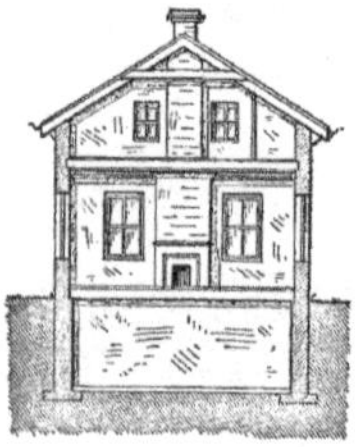

*Fig. 96.—Section of House on Line A B.*

*Fig. 97.—Perspective View of House.*

194. A *Demonstration* is a course of reasoning showing that a certain result is the necessary consequence of assumed premises.

195. A *Premise* is something previously stated or assumed as ground for further argument.

196. A *Basis* of anything is its groundwork or first principle.

197. A *Conclusion* is the final decision or determination.

198. To *Develop* a pattern is to define its shape and boundaries by a series of progressive steps.

199. The *Development* of a surface is the process of changing a given surface into another form of equivalent area or value.

200. To *Project* a figure is to construct, by means of lines, etc., on paper, a representation of the figure as it would appear from a given point of sight.

201. *Ratio* is the relation which one quantity or magnitude has to another of the same kind, as expressed by the quotient of the second divided by the first. Thus the ratio of 4 to 8 is expressed by $\frac{8}{4}$ or 2.

202. The *Area* of a figure is its superficial contents, or the surface contained within its boundary lines.

203. To *Raise*, means to form or to shape by hammering or stamping.

204. *Bisect*, signifies to divide into two equal parts, or, in other words, to cut in the middle.

205. *Prolong*, means to continue in the same direction; to draw still further.

206. *Indefinitely*, signifies without a limit. To prolong a line indefinitely means to draw it without regard to a limit of length.

# DRAWING TOOLS AND MATERIALS.

207. In the following description of the appliances, tools and materials likely to be of service to the pattern cutter, we purposely omit all mention of some special tools, although in general use, because they are not likely to be of service in the class of work in which the reader is supposed to be most interested. We limit our description, therefore, to articles of general use to the pattern cutter, rather than extend them into a general treatise upon drawing tools and materials. We shall only treat exhaustively such topics as are of special value to the pattern cutter. All others will be discussed only so far as they are likely to interest the special class for which the book is prepared. Those who are interested in drawing tools and materials upon a broader basis than here presented, are respectfully referred to special treatises on drawing and to the catalogues of manufacturers and dealers in drawing materials and drawing instruments.

208. *Drafting Tables.*—A drafting table suitable for a jobbing shop should be about five feet in length and three to four feet in width. It is better to have a table somewhat too large, than to have one so small that it is frequently inadequate for work that comes in. In hight the table should be such that the draftsman, as he stands up, may not be compelled to stoop to his work. While for some reasons it is desirable that the table should be fixed upon a strong frame and legs, for convenience such tables are generally made portable. Two horses are used for supports and a movable drawing board for the top. A shallow drawer is hung by cleats fastened to the under side, and is arranged for pulling either way. Sometimes horizontal pieces are fastened to the legs of the horses, and a shelf or shelves are formed by laying boards upon them. In Fig. 98 we show such a table as we have just described. When properly made, using heavy rather than light material, such a table is quite solid and substantial, yet when not in use can be packed away into a very small space.

*Fig. 98.—Drafting Table.*

209. For cornice makers' use, a table similar in all respects to the one we have described and illustrated (Fig. 98), except in size, is well adapted. Its dimensions, considering the extremes of work that are likely to arise, should be twelve to fourteen feet in length by about five feet in breadth. Three horses are necessary, and two drawers may be suspended. With very large work, one draftsman or pattern cutter will require the whole table, but for ordinary work, such as window caps, cornices, etc., two men can work at it without interfering with or incommoding each other.

210. Various woods may be used for drawing tables, but white pine is the cheapest and best for the purpose. Inch and one-half to two-inch stuff will be found economical, as it allows for frequent redressing—made necessary by pricking in the process of pattern cutting. Narrow stuff, tongued and grooved together or joined by glue, is preferable to wide plank, as being less liable to warp. Rods run through the table edgeways, as shown in Fig. 98, are desirable for drawing the parts together and holding in one compact piece. The nut and washer are sunken into the edge of the table, a socket wrench being used to operate them.

211. Each drafting table should be an accurate rectangle. Every corner should be a right angle, and the opposite sides should be parallel. The edges should be exactly straight throughout their length. Methods of testing drafting tables and drafting boards, with reference to these points, will be found on the opposite page. The usual way of adjusting a table or board to make it accurate, is to plane off its edges as required. But this is a task less simple and easy than it appears. It requires the nicest skill and accuracy to render it at all satisfactory. When it is remembered that no matter how well seasoned the lumber employed may be, the table will be affected by even slight changes in the atmosphere, it is apparent that dressing off the edges with a plane, under certain circumstances, might be constantly required. Hence, in some of the best shops, an adjustable metal strip is fastened to the edge of the table in such a manner that, by simply turning a few screws, any variation in the table may be compensated. This metallic edge is variously constructed. One of the simplest forms is described as follows: The edge of the table on all sides is cut away so as to allow a bar of steel, say one-eighth or one-sixteenth of an inch thick and about an inch wide, to lie in the cutting, so that its surface is even with the face of the table, with one edge projecting somewhat beyond the edge of the table. Slotted holes are made in the table, through which bolts with countersunk heads are passed for holding the steel strips. A washer and nut are used on the under side of the table. The adjustment required is very slight, so that this arrangement works very well, although other and more accurate methods, and more expensive also, are in use. Any plan similar to this will be found very useful. Iron instead of steel, if planed accurately, can be made to answer a good purpose. The edge of the metal projecting slightly, as we have described, is well adapted for receiving the the head of the T square, rendering the use of that instrument more satisfactory than when it is used against the plane edge of the table, even if equally accurate.

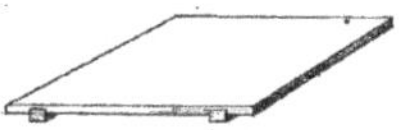

*Fig. 99.—Drawing Board, with Ledges.*

212. *Drawing Boards.*—For a pattern cutter's use, the principal difference between a drafting table and a drawing board is in the size. The same general requirements in point of accuracy, etc., are necessary in each. We have indicated convenient sizes of tables for various uses in our remarks under drafting tables, but to point out sizes of boards for different purposes is not so easy a matter, their application being far more extended and their use more general. A drawing board may be made of any required size, from the smallest for which such an article is adapted, up to the extreme limit consistent with convenience in handling. In the larger sizes the general features of construction noted under drafting tables are entirely applicable, save that thinner material should be used in order to reduce the weight. In small sizes there is the choice between several different modes of construction. We shall describe but two or three of them, remarking that boards of almost any required construction can be purchased ordinarily of dealers in drawing tools and materials at lower prices than they can be made. However, it is very convenient, in many cases, to have boards made to order, and therefore detailed descriptions of good constructions are desirable. Any carpenter or cabinet maker should be able to make such boards as we present.

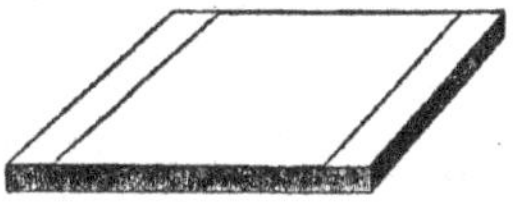

*Fig. 100.—Drawing Board, with Clamped Edges.*

213. In Fig. 99 is shown a very common form of drawing board, being a pine-wood top with hard-wood ledges. The ledges are put on by means of a tapering dovetail, and are so arranged that while allowing entire freedom for seasoning, so that there is no danger of cracking the board, they may be driven tight as required. Where it is desirable to use screws in the ledges, they are passed through slotted holes furnished with a metallic bushing.

214. In Fig. 100 is shown a still simpler form of board, which is adapted only for the smallest sizes. The edges are clamped by hard-wood strips, as shown in the engraving. By using strips of wood thicker than the board, keeping their upper surfaces flush with the surface of the board, this style is sometimes constructed so as to have the advantage of ledges on the under side equivalent to those shown in Fig. 99.

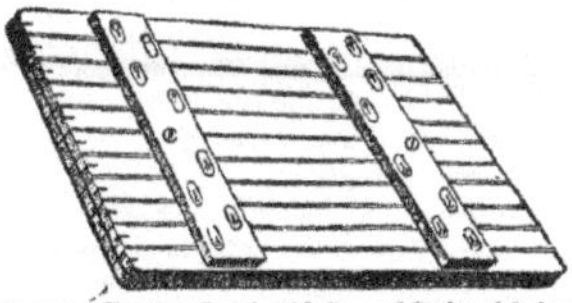

*Fig. 101.—Drawing Board, with Grooved Back and Ledges put on with Slotted Holes.*

215. Fig. 101 shows a construction of a board which, while being somewhat more expensive than others, is undoubtedly much better. It is made of pine wood, glued up to the required width. A pair of hard-wood

ledges are screwed to the back, the screws passing through the ledges in oblong slots with brass bushings, which fit closely under the heads and yet allow the screws to move freely when drawn by the shrinkage of the board. To give the ledges power to resist the tendency of the surface to warp, a series of grooves are sunk in half the thickness of the board over the entire back. These grooves take the transverse strength out of the wood, to allow it to be controlled by the ledges, leaving at the same time the longitudinal strength of the wood nearly unimpaired. To make the two working edges perfectly smooth, allowing an easy movement with the T-square, a strip of hard wood is let into the end of the board. This strip is afterward sawn apart at about every inch, to admit of contraction. In the construction of such boards, additional advantage is obtained by putting the heart side of each piece of wood to the surface.

216. Boards of the several kinds described above use the paper fastened to them, either by means of tacks or by gluing. Boards are sometimes made with a hard-wood frame, the board proper fitting into it as a panel. It is fastened into the frame by means of buttons. The paper is spread over the board, the frame passed down around it, carrying the edge of the paper with it, and when in proper position the buttons are turned into their places. Such boards are not well adapted to practical use. It is more difficult to stretch the paper by means of them than it would seem by the description. A considerable waste of paper is involved, and the necessary play of the parts, to allow room for the paper between them, sometimes leads to inaccuracies in drawings. Stretching paper by gluing or by the use of thumb tacks is found far more satisfactory. For such drafting as the pattern cutter is called upon to perform, thumb tacks are used almost exclusively. For architectural drawings and for drawings of machinery, usually made on white paper, gluing is preferable.

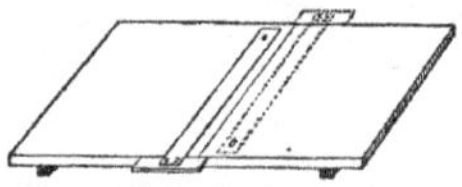

*Fig. 102.—Testing the Opposite Sides of a Drawing Board.*

217. *Testing Drawing Boards and Tables.*—The great desideratum in a drawing table or board is its accuracy. It should be an accurate rectangle, in order to facilitate work that is to be done upon it. If each angle is a right angle—if its opposite sides are exactly parallel—the T-square may be used at will from any portion of it with satisfactory results. If the board is accurate the drawing will be accurate. If the board is not accurate the drawing can only be made accurate at the cost of extra trouble and care. While it is easy to get a board approximately correct by ordinary means, one or two simple tests, which we shall describe, serve to point out inaccuracies for correction which by ordinary means would pass unnoticed. We assume at the outset that we have a T-square and an ordinary two-foot steel square that are exactly correct.

218. Having made the opposite sides and ends of the board as nearly accurate as possible, place the head of the T-square against one side, as shown in Fig. 102, and with a hard pencil sharpened to a chisel edge, or with the blade of a knife, scribe a fine line across the board. Then carrying the T-square to the opposite side of the board, as shown by the dotted lines, bring the edge of the blade to within a short distance of the line just described and draw another. If the two lines are found upon measurement to be exactly parallel, it is satisfactory proof that the opposite edges of the board are parallel at the points tested. Instead of drawing the lines a short distance apart, they may be drawn at the same point; then instead of measuring, it will be necessary simply to see that they exactly coincide throughout their length. Repeat this operation at frequent intervals along the edges of the board, both at the sides and ends. Remove any small inaccuracies by means of a file, or fine sand paper folded over a block of wood. Careful work in this manner will produce very satisfactory results.

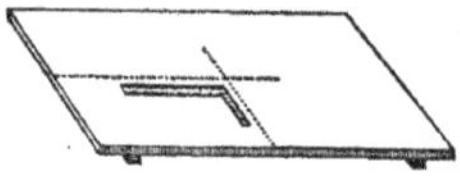

*Fig. 103.—Testing the Corner of a Drawing Board.*

219. A means of testing a board with reference to the accuracy of the corners, is shown in Fig. 103. A carpenter's try-square or an ordinary steel square used upon the corners, does not ordinarily reach far enough in either direction to satisfactorily determine that the adjacent end and side are perpendicular to each other; hence it is desirable to obtain some kind of a test with reference to this point from the central portions of the edges. With the head of the T-square placed against one side of the board draw a fine line, as indicated by the dots in the engraving, and from one end draw a second line in the same manner. If the side and end are at right angles, the two lines will correspond with the arms of a square when placed as shown in the engraving. Repeat this operation for each of the corners.

220. The two methods above described for testing drawing boards, especially when used together, cannot fail to enable any one to obtain a board as nearly accurate as it is possible to make things accurate by ordinary

mechanical means and of the materials used. Modifications of the methods here given, and based upon the same principles, will suggest themselves to any one who will give the matter careful thought.

221. *Straight-Edges.*—In connection with every set of drawing instruments there should be one or more straight-edges. If nothing but pencil or pen lines are to be made upon paper, hard wood or hard rubber as a material will answer very well; but if lines are to be drawn upon metal, steel is the only satisfactory material. The length of the straight-edge must be determined by the work to be done, but a safe rule is to have it somewhat near the length of the table or board. Of course this is out of the question in cornice work, where tables are frequently upward of twelve feet in length. In such cases the size of the material to be cut determines this matter. If iron 96 inches long is used, the straight-edge, for convenience, should be not less than 8½ feet. If shorter iron is regularly used, a shorter straight-edge will answer. In cornice work, two and even three different lengths are found advantageous. The longest we have just described; a second might be about four feet in length and made proportionately lighter, while the smallest might be two feet and also still lighter than the four-foot size. For the latter, however, the long arm of the common steel square serves a good purpose.

*Fig.* 104.—*Straight-Edge.*

222. For tinners' use in general jobbing shops, a three-foot straight-edge in many cases, and a four-foot one in a few instances, will be found quite convenient. Some mechanics desire their straight-edges graduated, the same as a steel square, into inches and fractions. We see no special advantage in this; it adds considerably to the cost, without rendering the tool more useful.

223. A hole should be provided in one end of the straight-edge for hanging up. It should always be suspended when not in use, as in that position it is not liable to receive injury from any cause whatever.

224. It is almost superfluous to add that straight-edges must be entirely accurate, for if inaccurate they would belie their name. A simple and convenient method of testing straight-edges is to place two of them together by their edges, or a single one against the edge of a square, as shown in Fig. 107, and see if light passes between them. If no space is to be observed between the edges, it is satisfactory evidence that they are as straight as they can be made by ordinary appliances. In addition to having the edges straight, it is also necessary to have the two sides parallel.

*Fig.* 105.—*Fixed-Head* T-*Square.*

225. T-*Squares.*—With this instrument, as with almost all drawing instruments, there is the choice of various qualities, sizes and kinds, and selection must be made with reference to the kind of work that is to be performed. Whatever quality may be chosen, the desirable features of a T-square are strict accuracy in all respects, a thin blade and one that will lie close to the paper when in use. For most purposes a fixed head, as shown in Fig. 105, is preferable. For drawings in which a great number of parallel oblique lines are required, and particularly where a small size T-square can be used, a swivel head, as shown in Fig. 106, is sometimes desirable. The objectionable feature about a swivel head is the difficulty of obtaining positive adjustment. When made in the ordinary manner, and depending upon the friction of the nut of a small bolt for holding the head in place, it is almost impossible to obtain a bearing that can be depended upon during even a simple operation. In practice it is found to be far less trouble to work from a straight-edge—properly placed across the board and weighted down or otherwise held in place—by means of a triangle or set-square. Greater accuracy is also assured by this plan.

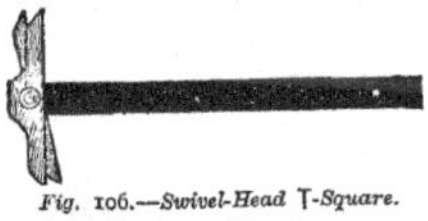

*Fig.* 106.—*Swivel-Head* T-*Square.*

226. In point of materials, probably a T-square constructed with walnut head and maple blade is as likely to give good satisfaction as any. This kind is the cheapest, and is generally considered the best for practical purposes. A good article, but of a higher price, consists of a walnut head with a hard-wood blade, lined with some other kind of wood. Still another variety has a mahogany blade lined with ebony. T-squares, constructed with cast-iron head—open work finished by japanning—with a nickel-plated steel blade, are also to be had from dealers. For accuracy probably these are the best, but they are several times more expensive than the simple wooden material first above described.

227. T-squares are also made with a hard rubber blade, of which Fig. 106 is an illustration. The liability to fracture, however, by dropping necessitates the greatest care in use; otherwise hard rubber makes a very desirable article, and is the favorite material with many draftsmen.

228. In point of size, T-squares should be selected with reference to the use to be made of them. Gener-

ally, the blade should be a very little less in length than the width of the table or board upon which it is to be used. Where a large board or a table is used, it will be found to be economy to have two instruments, one large one and one small one, the former being used for the principal lines in laying off the work, while the latter is used in miter cutting and wherever the diagram can be made near enough the edge of the table to admit of its employment.

229. *The Steel Square.*—One of the most useful tools in connection with the pattern cutter's outfit is an ordinary steel square. The divisions upon it concern him much less than its quality in the way of accuracy. He seldom requires other divisions than inches and eighths of an inch; therefore in selection the principal point to be considered is that of accuracy. The finish, however, is a matter not to be overlooked. Since a nickel-plated square costs but a trifling advance upon the plain article, it is cheaper in the long run to have the plated tool.

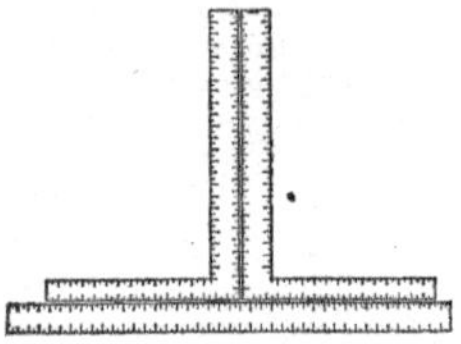

*Fig.* 107.—*Testing the Exterior Angle of a Steel Square.*

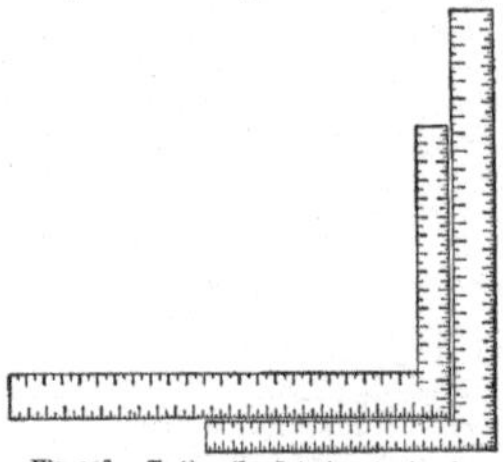

*Fig.* 108.—*Testing the Interior Angle of a Steel Square.*

230. A convenient method of testing the correctness of the outside of a square, and one which can be used at the time and place of purchase, is illustrated in Fig. 107. Two squares are placed against each other and against a straight-edge, or against the arm of a third square. If the edges exactly coincide throughout, the squares may be considered correct.

231. Having procured a square which is accurate upon the outside, the correctness of the inside of another square may be proven, as shown in Fig. 108. Place one square within the other, as shown. If the edges fit together tightly and uniformly throughout, the square may be considered entirely satisfactory.

232. An accurate square is especially desirable, as it affords the readiest means of testing the T-square and the drawing table or board, as elsewhere described. The greatest care should be given, therefore, to the selection of a square. For all ordinary purposes the two-foot size is most desirable. In some cases the one-foot size is better suited. Many pattern cutters on cornice work like to have both sizes at their command, making use of them interchangeably, according to the nature of the work to be done.

*Fig.* 109.—*Open Hard Rubber Triangle, or Set-Square,* 45, 45 *and* 90 *degrees.*

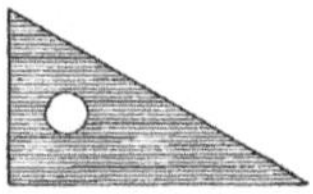

*Fig.* 110.—*Hard Wood Triangle, or Set-Square,* 30, 60 *and* 90 *degrees.*

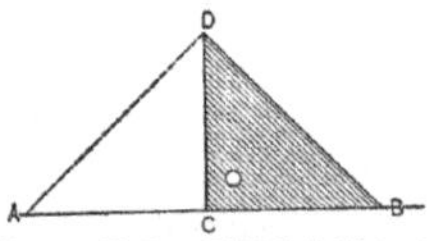


*Fig.* 111.—*Testing a Right-Angled Triangle, or Set-Square.*

233. *Triangles, or Set Squares.*—In the selection of triangles, the draftsman has the choice in material between pear wood; mahogany, ebony lined; hard rubber; German silver; and steel, silver or nickel plated. In style he has the choice between open work, of the general form shown in Fig. 109, and solid, of the general form of Fig. 110. In shape, the two general kinds which are adapted to the pattern cutter's use we have shown in Figs. 109 and 110, the latter being commonly described as 30, 60 and 90 degrees, and the former as 45, 45 and 90 degrees. The special uses of each of these two articles are shown among the problems. In size, the pattern

cutter requires large rather than small. If he can have two sizes of each, the smaller might measure from 4 to 6 inches on the side, and the larger from 10 to 12 inches; but if only a single size is to be had, one having dimensions intermediate to those named will be found most serviceable.

234. The value of a triangle, for whatever purpose used, consists of its accuracy. Particularly is this to be said of the right angle, which is used more than either of the others. A method of testing the accuracy of the right angle is shown in Fig. 111. Draw the line A B with an accurate ruler or straight-edge. Place the right angle of a triangle near the center of this line, and make one of the edges coincide with the line, and then against the other edge draw the line D C. Turn the triangle on this perpendicular line, bringing it into the position indicated by D C A. If it is found that the sides agree with A C and C D, it is proof that the angle is a right angle and that the sides are straight.

235. Besides the kinds of triangles we have described above, a fair article can be made by the mechanic from sheet zinc or a heavy piece of tin. Care need only be taken in cutting to obtain the greatest possible accuracy. For many of the purposes for which a large size 45 degree triangle would be used, the steel square is available; but as the line of the hypothenuse is lacking in it, it cannot be considered a substitute.

236. *Compasses and Dividers.*—The term compasses is applied to those tools, of various sizes and descriptions, which hold a pencil or pen in one leg, while dividers are those tools which, while of the same general form as compasses, have both legs in the shape of fixed points. They derive their name from their obvious use, that of spacing or dividing. A special form of dividers—used exclusively for setting off spaces, as in the divisions of a profile line—is called spacers, as illustrated and described below.

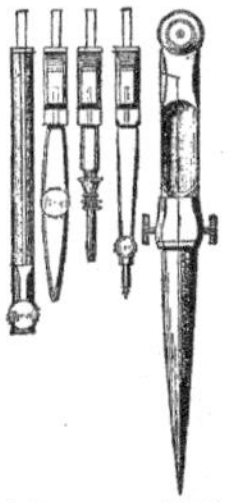

*Fig. 112.—Compasses, with Needle Point, Pencil Point, Pen and Lengthening Bar.*

*Fig. 113.—Plain Dividers.*

*Fig. 114.—Hair-Spring Dividers.*

*Fig. 115.—Steel Spring Spacers.*

237. A pair of compasses consists of the parts as shown in Fig. 112, being the instrument proper with detachable points, and extras comprising a needle point, a pencil point, a pen and a lengthening bar, all as shown to the left. In selection, care should be given to the workmanship; notice whether the parts fit together neatly and without lost motion, and whether the joint works tightly and yet without too great friction. A good German silver instrument, although quite expensive at the outset, will be found the cheapest in the end. A pencil point of the kind shown in our engraving is to be preferred over the old style which clamps a common pencil to the leg. The latter is not nearly so convenient and is far less accurate.

238. Of dividers there are two general kinds, the plain dividers, as shown in Fig. 113, and the hair-spring dividers, as shown in Fig. 114. The latter differ from the former simply in the fact of having a fine spring and a joint in one leg, the movement being controlled by the screw shown at the right. In this way, after the instrument has been set approximately to the distance desired, by means of the screw the adjustable leg is moved, as may be required, either in or out, thus making the greatest accuracy of spacing possible. Both instruments are found desirable in an ordinary set of tools. The plain dividers will naturally be used for larger and less particular work, while the hair-spring dividers will be used in the finer parts. It frequently happens that two pairs of dividers, set to different spaces, are convenient to have at the same time. Then the possession of these two articles is especially desirable.

239. A pair of spacers, shown in Fig. 115, is almost indispensable in a pattern cutter's outfit. He will find advantageous use for this tool, even though possessing both pairs of dividers described above. In size it should

be a little less than that of the dividers. The points should be needle-like in their fineness, and should be capable of adjustment to within a very small distance of each other. It is sometimes desirable to divide a given profile into spaces of an eighth of an inch. The spacers should be capable of this, as well as adapted to spaces of three-quarters of an inch, without being too loose. As will be seen from the engraving, this instrument is arranged for minute variations in adjustment. It has a marked advantage over the hair-spring dividers, in that the legs are controlled by the spring and screw direct; in the latter but one leg is affected by the spring, leaving only the friction of the joint to keep the legs in one constant position relative to each other.

240. *Beam Compasses and Trammels.*—In Fig. 116 we show a set of beam compasses, together with a portion of the rod or beam on which they are used. The latter, as will be seen by the section drawn to one side (A), is in the general shape of a T. This form has considerable strength and rigidity, while at the same time it is not clumsy or heavy. Beam compasses are provided with extra points, for pencil and ink work, as shown. While the general adjustment is effected by means of the clamp against the wood, minute variations are made by the screw shifting one of the points, as shown. This instrument is quite delicate, and when in good order is very accurate. It should be used only for fine work on paper, and never for scribing on metal.

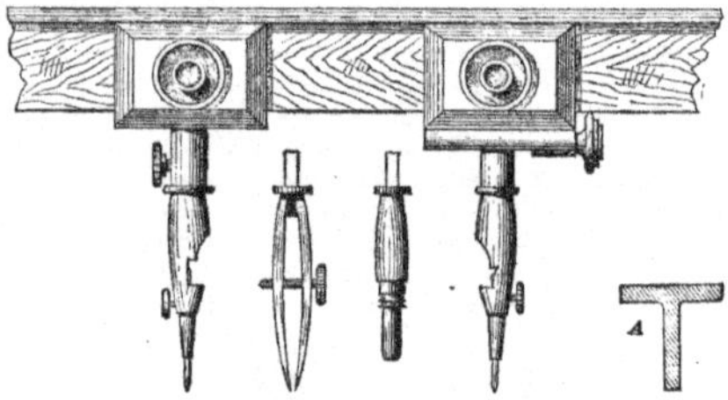


*Fig. 116.—Beam Compasses.*

241. A coarser instrument, and one especially designed for use upon metal, is shown in Fig. 117, and is called a trammel. It is to be remarked in this connection that the name trammel, by common usage, is applied to this instrument and also to a device for drawing ellipses, which will be found described at another place. There are various forms of this instrument, all being the same in principle; our engraving shows one that is in quite common use. A heavier stick is used with it than with the beam compasses, and no other adjustment is provided than that which is afforded by clamping against the stick. In Fig. 117 a carrier at the side is shown, in which a pencil may be placed. Some trammels are arranged in such a manner that either of the points may be detached and a pencil substituted. In others it is intended that the pencil shall be placed in a side carrier without removing the point. In Fig. 118 we show the form of trammel just described, arranged for using a pencil.

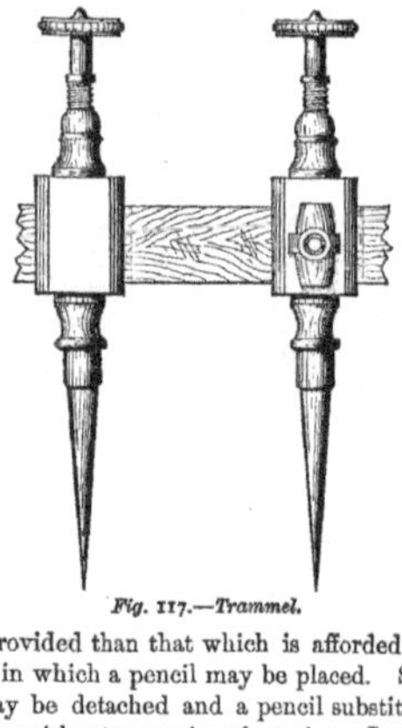

*Fig. 117.—Trammel.*

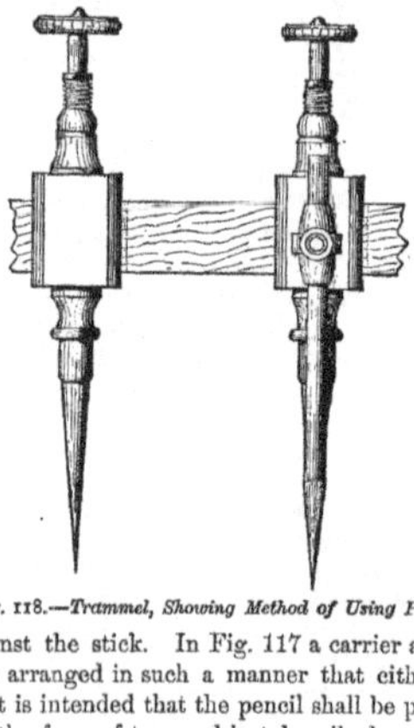

*Fig. 118.—Trammel, Showing Method of Using Pencil.*

242. A trammel, by careful management, can be made to describe very accurate curves, and hence can be used in place of the beam compasses in many instances. For all coarse work it is to be preferred to the beam compasses. It is useful for all short sweeps upon sheets of metal, but for very long sweeps a strip of sheet iron or a piece of wire will be found of more practical service than even this tool.

243. The length of rods for both beam compasses and trammels, up to certain limits, is determined by the nature of the work to be done. The extreme length is determined by the strength and rigidity of the rod itself. It is usually convenient to have two rods for each instrument, one about 3½ or 4 feet in length and the other considerably longer—as long as the strength of material will admit. In the case of the trammel, by means of a simple clamping device, or, in lieu of better, by use of common wrapping twine, the rods may be spliced when unusual length is required; but, as remarked before, a strip of sheet iron or a piece of fine wire forms a better radius, under such circumstances, than the rod.

244. *The Protractor* is an instrument for laying down and measuring angles upon paper. The instrument, when by itself, consists of a semicircle of thin metal or horn, as represented in Fig. 119, the circumference of which is divided into 180 equal parts or degrees. The principles upon which the protractor is constructed and used are clearly explained in the chapter of definitions, under the head "Degree." The methods of employing it in the construction of geometrical figures are shown in the proper place among the problems. For purposes of accuracy, a large protractor is to be preferred to a small size, because in the former fractions of a degree are indicated.

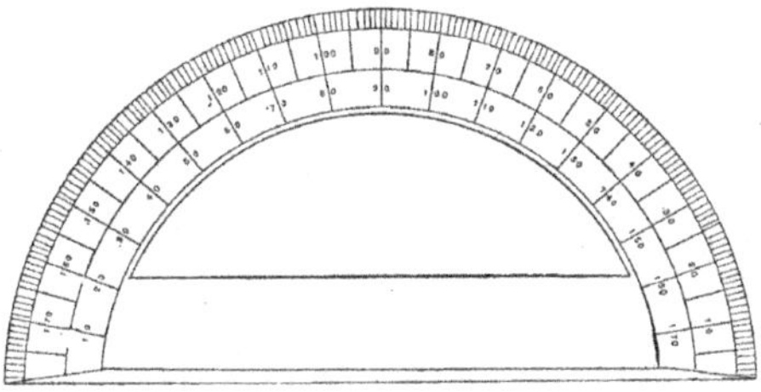

*Fig. 119.—A Semicircular Protractor.*

245. While a number of geometrical problems are conveniently solved by the use of this instrument, it is not one that is specially adapted to the pattern cutter's use. All the problems which are solved by it are capable of other accurate and expeditious methods, which, in most cases, are preferable. It is one of the instruments, however, included in almost every case of instruments sold, and the student will find it advantageous to become thoroughly familiar with it, whether in practice he employs it or not.

246. Besides the semicircular form of the protractor shown in Fig. 119, corresponding lines and divisions to those upon it are sometimes put upon some of the varieties of scales in use, allusion to which will be found in our remarks upon scales.

247. *Scales.*—Many of the drawings from which the pattern cutter works—that is, from which he gets dimensions, &c.—are what are called scale drawings, being some specified fraction of the full size of the object represented. Architects' elevations and floor plans are very generally made either ⅛ or ¼ inch to the foot, or, in other words, 1-96 or 1-48 full size. Scale details are also employed quite extensively by architects, scales in very common use for the purpose being 1½ inches to the foot and 3 inches to the foot, or, in other words, ⅛ and ¼ full size respectively. It is essential that the pattern cutter should be familiar with the various scales in common use, that he may be able to work from any of them on demand. Several of the scales are easily read by means of the common rule, as, for example, 3 inches to the foot, in which each quarter inch on the rule becomes one inch of the scale; also, 1½ inches to the foot, in which each eighth of an inch on the rule becomes an inch of the scale; and, likewise, ¾ inch to the foot, in which each sixteenth of an inch on the rule becomes an inch of the scale. However, other scales besides these are occasionally required, which are not easily read by the common rule, and sometimes special scales are used which are not shown on the instruments especially calculated for the purpose. Accordingly, it is sometimes necessary for the pattern cutter to construct his own scale.

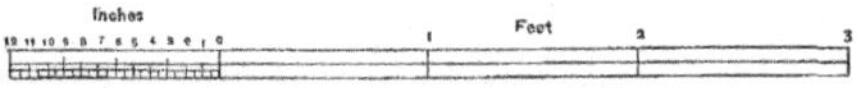


*Fig. 120.—Plain Scale.—1 inch to the foot.*

248. The method of constructing a scale of 1 inch to the foot is illustrated in Fig. 120, in which the divi-

sions are made by feet, inches and half inches. In constructing such scales, it is usual to set off one division at the left, as shown, for division into inches and fractions of an inch. In using the scale, measurements are made to commence with the second division. When the number of feet has been found in this way, the instrument is shifted from left to right until the nearest division of feet comes opposite the end of the space measured; the feet are read by the number thus found, while a glance at the other end of the rule shows how many inches constitute the fraction of the foot.

249. Besides scales of the kind just described, which are termed plain divided scales, there are in common use what are known as diagonal scales, an illustration of one of which we show in Fig. 121. The scale represented is $1\frac{1}{2}$ inches to the foot. The left-hand unit of division has been divided by means of the vertical lines into 12 equal parts, representing inches. In width the scale is made to equal 8 of these parts, and the intermediate parallel lines are drawn. Next the diagonal lines are drawn. By a moment's inspection it will be seen that, by means of these diagonal lines, one-eighth of an inch and multiples thereof are shown on the several horizontal lines. If we have a distance equal to the space from A to B, as marked on the scale, we read it (first at the right for feet) 2 feet, (then in the left for inches by means of the vertical lines figured both at top and bottom) 6 inches (and last by means of the diagonal line, figured at the end of the scale, for fractions) and three-eighths. The top and bottom lines of the scale measure feet and inches only. The other lines measure feet, inches and fractions of an inch, each horizontal line having its own peculiar fraction, as shown. Such scales are frequently quite useful, and, as the reader will see, may be constructed by any one to any unit of measurement, and divided by diagonal lines into any desired fractions.

*Fig. 121.—Diagonal Scale.—1½ inches to the foot.*

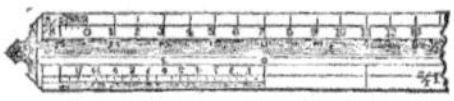
*Fig. 122.—Triangular Boxwood Scale.*

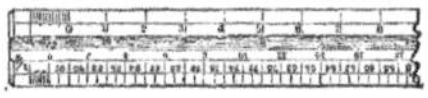
*Fig. 123.—Flat Boxwood Scale.*

250. A scale in common use, and known as the triangular scale, is shown in Fig. 122. The shape of this scale, which is indicated by the name, and which is also shown in the cut, presents three sides for division. By dividing these through the center lengthways by a groove, as shown, six spaces for divisions are obtained, and by running the scales in pairs—that is, taking two scales, one of which is twice the size of the other—and commencing with the unit at opposite ends, the number of scales which may be put upon one of these instruments is increased to twelve. This article, which may be had in either boxwood, ivory or plated metal, and of 6, 12, 18 or 24 inches in length, is probably the most desirable for general use of any sold.

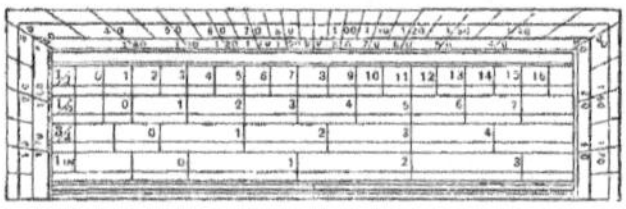
*Fig. 124.—Flat Scale, with Dimensions of the Circle on the Margins.*

251. In Fig. 123 we show what is known as a flat scale, and which is also manufactured in both boxwood and ivory. Less scales or divisions can be put upon it than upon the triangular scale, yet for certain purposes it is to be preferred to the latter. There are less divisions to perplex the eye in hunting out just what is required, and, accordingly, there is less liability to error in its use. However, the limited number of scales which it contains greatly restricts its usefulness.

252. In Fig. 124 we show another form of the flat scale, one in quite common use in the past, but now virtually discarded in favor of more convenient dimensions and shapes. This scale combines with the various divisions of an inch the divisions of the protractor, as shown around the margin. The fact that the divisions of an inch for purposes of a scale are located in the middle of the instrument, away from the edge, which makes it necessary to step off all spaces for measurement with the dividers, renders the article awkward for use. The arrangement of the divisions of the circle, as shown on the margins, is less satisfactory for use than the same thing upon the circular protractor.

253. *Lead Pencils.*—Various qualities of pencils are sold, some at much lower prices than others, but, all things considered, in this as in other cases, the best are the cheapest. Of leading brands, which are likely to give both draftsman and mechanic entire satisfaction, there may be mentioned Faber's, the American, and Dixon's. The former are perhaps the best known, having been before the public for the longest time, and accordingly we base our remarks concerning hardness, etc., upon them, as being likely to be more generally understood than if we referred to newer and less generally known pencils, although equally good. Both Faber's and the American, in the ordinary grades, employ numbers, 1, 2, 3, etc., to indicate hardness of lead, No. 1 being the softest, and No. 5 being the hardest in common use. A finer grade of pencils manufactured by Faber, known as poligrades, is marked by letters, commencing at the softest with B B, and ending at the hardest with H H H H H H. The Dixon pencils are graded to correspond with the qualities in greatest demand of the older manufacturers, but are marked upon a system peculiar to themselves.

*Fig. 125.—Pencil Sharpened to Chisel Point.*

254. Of either make of pencil the draftsman has the choice of round or hexagon shape, in all except the finest grades, the latter being made exclusively hexagon. The same quality of lead is said to be put in each, the only difference being in the shape of the wood and the finish. The round costs from 10 to 15 cents per dozen less than the hexagon. The poligrades in price are about double the common pencils, and save for exceptionally fine work, which will be mentioned further on, are no better for the purposes of drawing and pattern cutting than the ordinary kind. Besides pencils with fixed leads, which we have been describing, there are several styles of pencils with movable leads. They are of various lengths and prices. Some are made of wood, hexagon in shape, finished and polished the same as an ordinary pencil, the point being of plated metal and the top surmounted by an ivory cap; some are of hard rubber; some are made of ivory. Leads of various qualities, and of different degrees of hardness, may be bought for any of them. While, no doubt, such articles are a trifle more ornamental than common pencils with fixed leads, we think that all gained in this direction is sacrificed in utility. The ordinary pencil is not only cheaper, but it is better for all practical purposes.

*Fig. 126.—Pencil Sharpened to Round Point.*

*Fig. 127.—Drawing Pen.*

255. Whatever kind of pencil the draftsman or mechanic uses, he will require different numbers for different purposes. For working drawings, full-sized details, etc., on manila paper, a No. 3 is quite satisfactory. Some like a little harder lead, and therefore prefer a No. 4. For lettering and writing in connection with drawings upon manila or ordinary detail paper, a No. 2 is usually chosen. For fine lines, as in developing a miter, in which the greatest possible accuracy is required, a No. 5 is very generally used, although many pattern cutters prefer the finer grade for this purpose and use a H H H H H of the poligrades.

256. The quality and accuracy of drawings depend, in a considerable measure, upon the manner in which pencils are sharpened. A pencil used for making straight lines, as, for instance, the measuring lines in miter cutting, and also in the dropping of points in pattern cutting generally, should be sharpened to a chisel edge, as illustrated in Fig 125. Pencils for making dots, for marking points and for general work away from the edges of the T-square, triangle, etc., should be sharpened to a round point, as shown in Fig. 126. It facilitates work, and it is quite economical to have several pencils at command, sharpened in different ways for different purposes. Where for any reason only one pencil of a kind can be had, both ends may be sharpened, one to a chisel edge and the other to a point.

257. For keeping a good point upon a pencil, a piece of fine sand paper or emery paper, glued upon a piece of wood, will be found very serviceable. A flat file, mill-saw cut, is also useful for the same purpose. Sharpen the pencil with a knife, so far as the wood part is concerned, and then shape the lead as required upon the file or sand paper.

258. *Drawing Pens.*—Although most of the pattern cutter's work is done by use of the pencil, there occasionally arise circumstances under which the use of ink is desirable. Tracings of parts of drawings are frequently required which can be better made with ink than with pencil. The pattern cutter, by the very force of

circumstances, gradually assumes the functions and duties of a draftsman, dependent altogether upon his skill in the management of tools, and his acquirement of knowledge concerning the draftsman's art. Therefore our remarks concerning drawing instruments would be quite incomplete with no mention of ink-using tools and the management of ink itself.

259. The drawing pen, as illustrated in Fig. 127, is used for drawing straight lines. Attachments with corresponding members, to which the following remarks may also be applied, have been shown in connection with both compasses and beam compasses for drawing curved lines. The drawing pen consists of two blades with steel points, fixed to a handle. The blades are so bent that a sufficient cavity is left between them for ink when the ends of the points meet close together or nearly so. The blades are set with the points more or less nearly together, by means of the screw shown in the engraving, so as to draw lines of any required thickness. One of the blades is provided with a joint, so that, by taking out the screw, the blades may be completely opened and the points readily cleaned after use. The ink is put between the blades by means of a common pen, or sometimes by a small hair brush. In using the pen, it should be slightly inclined in the direction of the line to be drawn, and care must be taken that both points touch the paper. The drawing pen should be kept close to the ruler or straight-edge during the whole operation of drawing a line.

260. To keep the blades of his pens clean is the first duty of a draftsman who is to make a good piece of work. Pieces of blotting, or unsized paper or cotton velvet, or even the sleeve of a coat, should always be at hand when a drawing is being inked. When a small piece of blotting paper is folded twice, so as to present a corner, it may be passed between the blades of the pen now and then, as the ink is liable to deposit at the point and obstruct the passage. To do this the screw must be loosened. The same purpose may be accomplished, in a measure, by drawing the pen over a piece of velvet, or even over the surface of thick blotting paper. When the pen is done with for the occasion, it should be thoroughly cleaned at the nibs. This will preserve its edges and prevent rusting. If the draftsman is careless in this particular, the ink will soon corrode the points to such an extent that it will be impossible to draw fine lines.

261. Pens will gradually wear away, and in course of time they require dressing. To dress up the tips of the blades of a pen, since they are generally worn unequally by customary usage, is a matter of some nicety. A small oil stone is most convenient for use in the operation. The points should be screwed into contact in the first place, and passed along the stone, turning upon the point in a directly perpendicular plane until they acquire an identical profile. Next they are to be unscrewed and examined to ascertain the parts of unequal thickness around the nib. The blades are then to be laid separately upon their backs upon the stone, and rubbed down at the points until they are brought up to an edge of uniform fineness. It is well to screw them together again and pass them over the stone once or twice more to bring up any fault, to retouch them also at the outer and inner side of each blade to remove barbs or frasing, and finally to draw them across the palm of the hand.

Inferior. Good.

*Figs. 128 and 129.—India Ink of Different Qualities.*

262. *India Ink.*—For tracings, and for some kinds of drawings, which the pattern cutter is obliged to make occasionally, India ink is much better than the pencil, which is used for the greater part of his work. Care is to be exercised in the selection of ink, as poor grades are sold as well as good ones. Some little skill is required in dissolving or mixing it for use.

263. India ink is sold in cakes or sticks, of a variety of shapes. It is prepared for use by the process technically known as rubbing, which consists of dissolving a portion of it by rubbing it upon the surface of a glass, or of a porcelain slab or dish, in a very small quantity of water.

264. As to the quality of the ink, upon general principles it may be determined by the price. The common size sticks are about 3 inches long. Inferior grades of this size can be bought at 40 cents, 50 cents and 60 cents per stick, while good quality is worth \$1.50 to \$2 per stick, and the very best, still higher figures. However, except in the hands of a responsible and experienced dealer, this method of judging is hardly satisfactory.

To a certain extent ink may be judged by the brands upon it, although in the case of the higher qualities the brands frequently change, so that this test may not be infallible. A common brand of ordinary quality (about 50 cents per stick at present prices) is shown by our engraving, Fig. 128, full size. In shape the stick is oval, and is known as the "Lion's Head." An article of good quality for general use, and which is also adapted to fine work, is shown full size in Fig. 129. This stick is nearly square in shape, and at present price is worth $2. There is a great deal more ink in this stick than in the one first described, while its quality renders its use so much preferable to the other that it may safely be considered the cheaper of the two. These two brands have been selected for our illustrations because they are commonly known to the trade, and because they represent the two extremes between which the draftsman ordinarily chooses. There are other brands of about the same grade as each of these, and also those of intermediate and still better quality.

265. The quality of India ink is quite apparent the moment it is used. The best is entirely free from grit and sediment, is not musky, and has a soft feel when wetted and smoothed. The color of the lines may also be used as a test of quality. With a poor ink it is impossible to make a black line. It will be brown or irregular in color. With poor ink the line will also present an irregular edge, as though broken or ragged, while an ink of satisfactory quality will produce a clean line, whether drawn very fine or quite coarse.

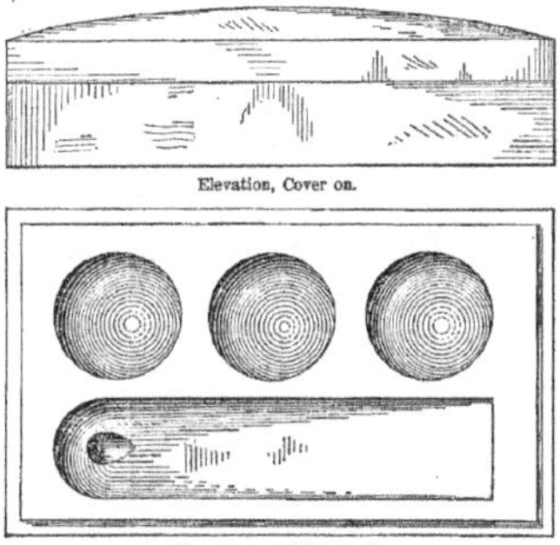

Elevation, Cover on.

Plan, View with Cover off.

*Fig. 130.—India Ink Slab with Cover.*

266. In rubbing down ink ready for using, it should be made just so thick as to run freely from the pen. The degree can be determined at first by trial, but after a time it will be recognized by the appearance of the ink in the dish. The rubbing of a stick of ink in water tends to crack and break away the surface at the points. To prevent this, the stick may be shifted in the hand at intervals while being rubbed, thus rounding the surface. For the same reason, it is not advisable to bear very hard upon the stick while rubbing, as the mixture is otherwise more evenly made and the enamel of the pallet is less liable to be worn off. When drawings are being made which require the use of ink for some time, a considerable quantity of it should be rubbed down at one time, as the water continually evaporates. By having quite a quantity prepared, it will remain longer in fit condition for use. As evaporation takes place the ink may be thinned from time to time, as required, by the addition of more water.

267. Various shaped cups, slabs and dishes are in use for mixing and containing India ink. In many respects they are like those used for mixing and holding water colors. Indeed, in many cases the same articles are employed. Our engraving (Fig. 130) shows what is termed an India ink slab, with three holes and one slant. This article is in common use among draftsmen and serves a satisfactory purpose. In order to retard evaporation, a kind of saucers, in sets, is frequently used, so constructed that one piece will form a cover to the other, and which are known in the trade as cabinet sets or cabinet saucers. They are from 2½ to 3½ inches in diameter and come six in a set. In the absence of ware especially designed for the purpose, India ink can be satisfactorily mixed in and used from an ordinary saucer or plate of small size, or even on a piece of glass. The articles made especially for it, however, are convenient, and in facilitating the care and economical use of the ink are well worth the small price they cost.

*Fig. 131.—Thumb Tacks or Drawing Pins.*

268. *Thumb Tacks* or *Drawing Pins*, both names being in common use, are made of a variety of sizes, ranging from those with heads one-quarter of an inch in diameter up to eleven-sixteenths of an inch in diameter. They are likewise to be had of various grades and qualities. The best for general use are those of German silver, about three eighths to five-eighths of an inch in diameter, and with steel points screwed in and riveted. Those which have the points riveted only, are of the second quality. The heads should be flat, to allow the T-square to pass over them readily. In the annexed cut, Fig. 131, we show an assortment of sizes. Those which are beveled upon their upper edges are preferable to those which are beveled underneath.

269. *A Box of Instruments.*—In Fig. 132 we show a box of instruments of medium grade, as made up and sold by the trade generally. While it contains some pieces that the pattern cutter has no use for, it also contains the principal tools he requires, all put together in compact shape, and in a convenient manner for keeping the instruments clean and in good order. The tray of the box lifts out, there being a space underneath it in which may be placed odd tools, pencils, etc. We do not recommend this particular box of instruments to the pattern cutter, nor, for that matter, any other. We introduce it as showing of what a box of tools ordinarily consists, and as indicating the advantages of a case in which to keep whatever tools the mechanic may possess. Tools may be selected, as required, of most of the large dealers in drawing instruments. A case or box fitted for their reception, neatly lined and with proper spaces, may be obtained at a small additional cost. We believe it to be to the advantage of the pattern cutter to buy his instruments odd—that is, not to buy a case as ordinarily made up. By buying in single pieces he will get only what he requires for use, and will probably secure quite as good quality in the tools. After he has made his selection, a box properly fitted and lined should be provided for them. A little skill and ingenuity upon the part of the mechanic will enable him to make his own instrument case. Wood, as a material, is to be preferred to metal, although there is less objection to the latter if the spaces for the instruments are properly padded and lined so that the tools need not come in contact with the metal of the box. Velvet is probably the best material for the lining.

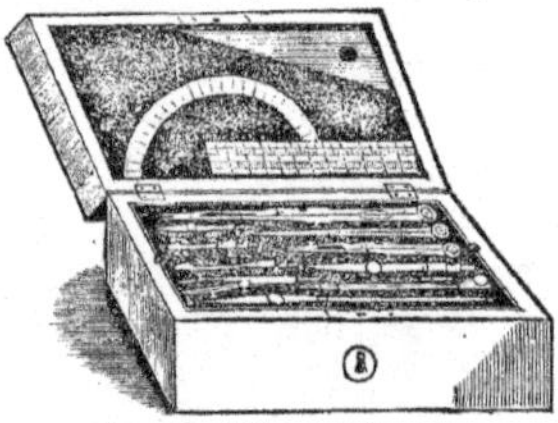

*Fig. 132.—A Box of Instruments.*

270. *India Rubber.*—A good rubber with which to erase erroneous lines is indispensable in the pattern cutter's outfit. The several pencil manufacturers have put their brands upon rubber as well as upon pencils, and satisfactory quality can be had from any of them. In size, a large piece, since it continually becomes less by use, is more economical than a small piece. The shape is somewhat a matter of choice. Flat cakes are perhaps the most used. The same quality can be obtained in diamond or lozenge shape, and in short square sticks or blocks. A very soft rubber is not so well adapted to erasing on detail paper as the harder varieties, but is to be preferred for use in fine drawings on good quality paper. Erasers put up in wooden holders are not economical for use upon rough paper, as they wear out too fast in such work.

271. Besides the cakes and blocks of rubber described above, rubber is fastened to pencils by a number of devices. There is the plain rubber cap; the rubber let into the pencil, something as the lead is put in, and the rubber held in a metallic case, which also forms a shield for the point of the pencil when carried in the pocket. Rubber in all of these shapes is very useful and convenient, but considering the small quantity that can be got into any one of them, it should not be depended upon for other than occasional use where very small erasures are to be made. The larger piece, in the form of the cake first described, should be used for general work, and the piece in connection with the pencil used only as supplementary to it.

272. *Paper.*—The principal paper that the pattern cutter has anything to do with is known as brown detail paper, or manila detail paper. It can be bought of almost any width, from 30 inches up to 54 inches, in rolls of 50 to 100 pounds each. It is ordinarily sold in the roll by the pound, but can be bought at retail by the yard, although at a higher figure. There are different thicknesses of the same quality. Some dealers indicate them by arbitrary marks, as XX, XXX, XXXX; others by numbers 1, 2, 3; and still others as thin, medium and thick. The most desirable paper for the pattern cutter's use is one which combines several good qualities. It should be just as thin as is consistent with strength. A thick paper, like a stiff card, breaks when folded or bent short, and is, therefore, objectionable. The paper should be very strong and tough, as the requirements in use are quite severe. The surface should be very even and smooth, yet not so glossy as to be unsuited to the use of hard pencils. It should be hard, rather than soft, and should be of such a texture as to withstand repeated erasures in the same spot without damage to the surface.

273. White drawing paper, which the pattern cutter has occasionally to use in connection with his work, can be had of almost every conceivable grade and in a variety of sizes. The very best quality, and the kinds suited for the finest drawings, come in sheets exclusively, although the cheaper kinds are also made in the shape of sheets as well as in rolls. White drawing paper in rolls can be bought of different widths, ranging from 36 to 54 inches, and from a very thin grade up to a very heavy article, and of various surfaces. It is sold by the pound, rolls ranging from 30 to 40 pounds each, and also at retail by the yard.

274. Drawing paper in sheets is sold by the quire, and at retail by the single sheet. The sizes are generally indicated by names which have been applied to them. The following are some of the terms in common use, with the dimensions which they represent placed opposite:

| | | | | | |
|---|---|---|---|---|---|
| Cap | 13 x 17 | Super Royal | 19 x 27 | Columbier | 23 x 35 |
| Demy | 15 x 20 | Imperial | 22 x 30 | Double Elephant | 27 x 40 |
| Medium | 17 x 22 | Elephant | 23 x 28 | Antiquarian | 31 x 53 |
| Royal | 19 x 24 | Atlas | 26 x 34 | Emperor | 48 x 68 |

Still another set of terms is used in designating French drawing papers. Different qualities of paper, both as regards thickness, texture and surface, can be had of any of the sizes above named.

275. The pattern cutter has frequent use for tracing paper, and a good article, one which combines strength, transparency and suitable surface, is very desirable. Tracing paper is sold both in sheets, in size to correspond to the drawing papers above described, and in rolls, to correspond in width to the roll drawing paper. It is usually priced by the quire and by the roll, although single sheets or single yards are to be obtained at retail. The rolls, according to the kinds, contain from 20 to 30 yards. We cannot offer any other good rule for selection of suitable quality than inspection and actual trial. There are various manufacturers of this article, but it is usually sold upon its merits, rather than by any brand or trade-mark. Tracing cloth, or tracing linen, is used in place of tracing paper where great strength and durability are required. This article comes exclusively in rolls, ranging in width from 18 to 42 inches. There are generally 24 yards to the roll, and prices are made according to the width, or, in other words, according to the superficial contents of the roll. Two grades are usually sold, the first being glazed on both sides and suitable only for ink work, and the second on but one side, the other being left dull, rendering it suitable for pencil marks. Upon general principles, pencil marks are not satisfactory upon cloth, even upon the quality specially prepared with reference to them. It is but a very little more labor or expense to use ink, and a much more presentable and usable drawing is made. Tracing paper may be used satisfactorily with either pencil or pen.

# GEOMETRICAL PROBLEMS.

276. Very much of the pattern cutter's skill depends upon his knowledge of fundamental geometrical principles. He should know how to lay off an octagon or a pentagon, or any required figure or angle, as well as how to cut a miter to fit the given angle after the figure is drawn. He should know how to draw a plan and elevation of an oval flaring dish, as well as how to develop the patterns for it after the drawing is given him. It is designed that this book shall be complete in itself—that it shall fully illustrate the science and art of pattern cutting in all its phases. It does not presuppose a knowledge of geometry upon the part of the student, but undertakes to supply all that it is necessary for him to learn in acquiring a knowledge of pattern cutting, from the definition of simple terms, up to the cutting of the most intricate and complex patterns. In the preceding chapters we have described at some length the various instruments and tools which the pattern cutter is likely to use, and have, along with other terms, defined and illustrated various figures and shapes in which his work is likely to occur. It remains for us, therefore, to illustrate the use of these instruments and tools, and show methods of constructing the various figures commonly occurring in pattern cutting, before commencing the demonstration of practical problems. There are also various expedients for shortening and simplifying what would otherwise prove long and tedious operations, at which it will be well to glance in passing. In the arrangement of the problems in this chapter, it has been found difficult to follow any one logical system throughout. Several schemes of order for the problems have suggested themselves, each of which, for certain parts, has appeared better than the others. Accordingly, to the critical reader, the arrangement as here presented may appear defective in some particulars, or, at least, show inconsistencies. But the student is reminded that the intent of the book is to afford not only a complete exposition of the art of pattern cutting, but also to serve as a ready reference book for answering vexed questions. It attempts not only to present the subject, from beginning to end, in an arrangement that will be acceptable to those who desire to make the book a regular and systematic study, but also to exhibit each individual principle and problem in a complete and independent form, so that when any one item is referred to it shall be found self-explanatory, and therefore ready for use, without tedious search through problems in other portions of the book in order to fully comprehend it. Accordingly, the use of the index is recommended to all who desire to pursue any order of study different from the arrangement we have followed. Since each rule and demonstration, so far as possible, is made independent of all other rules and demonstrations, the student, by referring to the several pages indicated by the topic heads as given in the index, can obtain an exhaustive presentation of any phase of the subject upon any system of classification he chooses to follow. We deem no further explanation necessary for the somewhat arbitrary arrangement we have found it desirable to follow in carrying out the special purposes of the book.

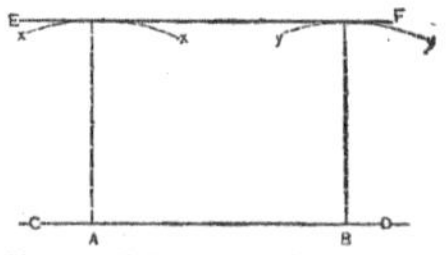

*Fig. 133.—To Draw a Straight Line Parallel to a Given Straight Line, and at a Given Distance from it, Using the Compasses and a Straight-Edge.*

277. *To Draw a Straight Line Parallel to a Given Line, and at a Given Distance from it, Using the Compasses and a Straight-Edge.*—In Fig. 133, let C D be the given line, parallel to which it is desired to draw another straight line. Take any two points, as A and B, in the given line as centers, and, with a radius equal to the given distance, describe the arcs $x$ $x$ and $y$ $y$. Draw a line touching these arcs, as shown by E F. Then E F will be parallel to C D.

278. *To Draw a Line Parallel to Another by the Use of Triangles or Set-Squares.*—In Fig. 134, let A B be the line parallel to which it is desired to draw another. Place one of two triangles or set-squares, F', against it, as indicated by the dotted lines. While holding F' firmly in this position, bring the second triangle, E, against one of its other sides, as shown. Then, holding the second triangle firmly in place, slide the first away from the given line, keeping the edges of the two triangles in contact, as shown in the figure. Against the same edge of the first triangle that was placed against the given line draw a second line, as shown by C D. Then C D will be parallel to A B. In drawing parallel lines by this method, it is found advantageous to place the longest edges of the triangles against each other, and to so place the two instruments that the movement of one triangle against the other shall be in a direction oblique to the lines to be drawn. Greater accuracy is attainable in this way than is possible otherwise.

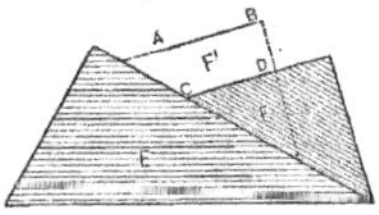


*Fig. 134.—To Draw a Line Parallel to Another by the Use of Triangles or Set-Squares.*

279. *To Erect a Perpendicular at a Given Point in a Straight Line by Means of the Compasses and Straight-Edge.*—In Fig. 135, let A B represent the given straight line, at the point C in which it is required to erect a perpendicular. From C set off on each side equal distances of any convenient space, as shown by D and B. With D and B as centers, and with any radius longer than the distance from each of these points to C, strike arcs, as shown by *x x* and *y y*. From the point at which these arcs intersect, E, draw a line to the point C, as shown. Then E C will be perpendicular to A B.

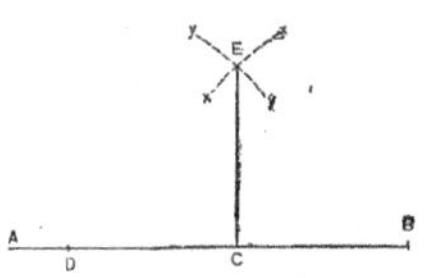


*Fig. 135.—To Erect a Perpendicular at a Given Point in a Straight Line by Means of the Compasses and Straight-Edge.*

280. *To Erect a Perpendicular at or near the End of a Given Straight Line by Means of the Compasses and Straight-Edge.—First Method.*—In Fig. 136, let A B be the given straight line, to which, at the point P, situated near the end, it is required to erect a perpendicular. Take any point (C) outside of the line A B. With C as center, and with a radius equal to the distance from C to P, strike the arc, as shown, cutting the given line A B in the point P, and also in another point, as at E. From E, through the center C, draw the line E F, cutting the arc, as shown at F. Then from the point F, thus determined, draw a line to P, as shown. The line F P is perpendicular to A B.

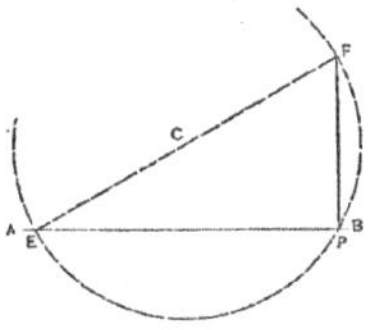


*Fig. 136.—To Erect a Perpendicular at or near the End of a Given Straight Line by Means of the Compasses and Straight-Edge.—First Method.*

281. *To Erect a Perpendicular at or near the End of a Given Straight Line by Means of the Compasses and Straight-Edge.—Second Method.*—In Fig. 137, let B A be the given straight line, to which, at the point P, it is required to erect a perpendicular. From the point P, with a radius equal to three parts, by any scale, describe an arc, as indicated by *x x*. From the same point, with a radius equal to four parts, cut the line B A in the point C. From the point C, with a radius equal to five parts, intersect the arc first drawn by the arc *y y*. From the point of intersection D draw the line D P. Then D P will be perpendicular to B A.

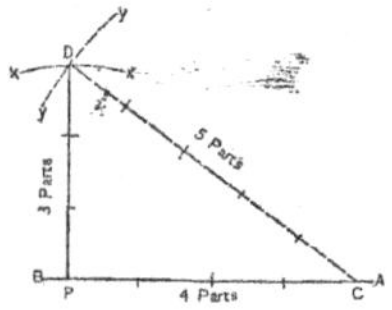


*Fig. 137.—To Erect a Perpendicular at or near the End of a Given Straight Line by Means of the Compasses and Straight-Edge.—Second Method.*

282. *To Draw a Line Perpendicular to Another Line by the Use of Triangles or Set-Squares.*—In Fig. 138, let C D be the given line, perpendicular to which it is required to draw another line. Place one side of a triangle, B, against the given line, as shown. Bring another triangle, A, or any straight edge, against the long side of the triangle B, as shown. Then move the triangle B along the straight-edge or triangle A, as indicated by the dotted lines, until the opposite side of B crosses the line C D at the required point. When against it, draw the line E F, as shown. Then E F is perpendicular to C D. It is evident that this rule is adapted to drawing perpendiculars at any point in the given line, whether central or located near the end. Its use will be found especially convenient for erecting perpendiculars to lines which run oblique to the sides of the drawing board.

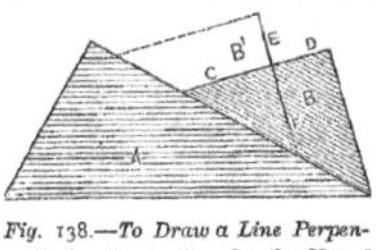


*Fig. 138.—To Draw a Line Perpendicular to Another by the Use of Triangles or Set-Squares.*

283. *To Divide a Given Straight Line into Two Equal Parts, with the Compasses, by Means of Arcs.*—In Fig. 139, let it be required to divide the straight line A B into two equal parts. From the extremes A and B as centers, and with any radius greater than one-half of A B, describe the arcs *d f* and *a e*, intersecting each other on opposite sides of the given line A B. A line drawn through these points, as shown by G H, will bisect the line A B, or, in other words, divide it into two equal parts.

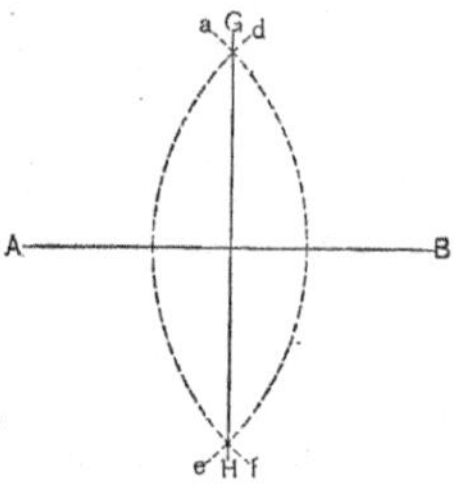


*Fig. 139.—To Divide a Given Straight Line into Two Equal Parts, with the Compasses, by Means of Arcs.*

284. *To Divide a Straight Line into Two Equal Parts by the Use of a pair of Dividers.*—In Fig. 140, it is required to divide the line A B into two equal parts, or to find its middle point C. Open the dividers to as near half of the given line as possible by the eye. Place one point of the dividers on one end of the line, as at A. Bring the other point of the dividers to the line, as at C, and turn on this point, carrying the first around to D. Should the point D coincide with the other end of the line, we have the division required. But should the point D fall within (or without) the end of the line, divide this deficit (or excess) by the eye into two equal parts, and extend (or contract) the opening of the dividers to this point and apply them again as at first. Thus, finding that the point D falls within the end of the line, we know our first division is too short. We therefore divide the deficit D B by the eye, as shown by E, and increase the space of the dividers to the amount of one of these divisions. Then, commencing again at A, we step off as before, and finding that upon turning the dividers upon the point F the point coincides with the end of the line B, we know that F is the middle point in the line. In some cases it may be necessary to repeat this operation several times before the exact center is obtained. The smaller the space to be divided, the more accurate is the spacing of it by the eye.

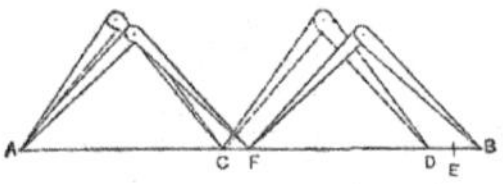


*Fig. 140.—To Divide a Line into Two Equal Parts by the Dividers.*

285. *To Divide a Straight Line into Two Equal Parts by the Use of a Triangle or Set-Square.*—In Fig. 141, let A B be a given straight line. Place a T-square or some straight edge parallel to A B. Then bring one of the right-angled sides of a set-square against it, and slide it along until its long side, or hypothenuse, meets one end of the line, as A. Scribe along the long side of the triangle indefinitely. Reverse the position of the set-square, as shown by the dotted lines, bringing its long side against the end, B, of the given straight line, and in like manner scribe along its long side. Next slide the set-square along until its vertical side meets the intersection of the two lines scribed, as shown at C, from which point drop a perpendicular to the line A B, cutting it at D. Then D will be equidistant from the two extremities A and B.

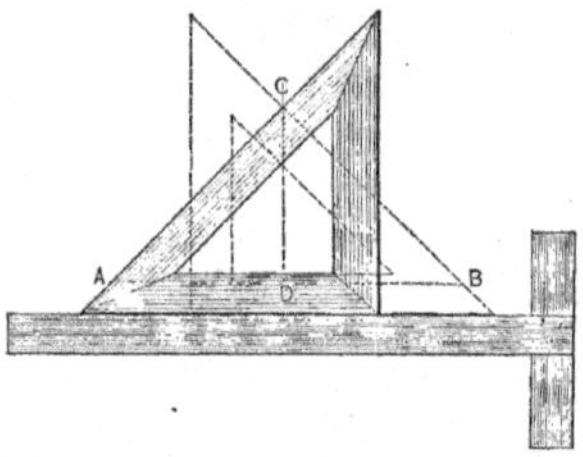


*Fig. 141.—To Divide a Straight Line into Two Equal Parts by the Use of a Triangle or Set-Square.*

286. *To Divide a Given Straight Line into Any Number of Equal Parts.*—In Fig. 142, let A B be a given straight line to be divided into equal parts, in this case eight. From one extremity of this line, as at A, draw a line, as either A C or A D, oblique to A B. Set the dividers to any convenient space, and step off the oblique line, as A C, eight divisions, as shown by *a b c d*, etc. From the last of the points, *h*, thus obtained, draw a line to the end of the given line, as shown by $h$ $h^2$. Parallel to this line draw other lines, from each of the other points to the given line. The divisions thus obtained, indicated in the engraving by $a^2$ $b^2$ $c^2$, etc., will be the desired spaces in the given line. It is evident by this rule that it is immaterial, except as a matter of convenience, at what space the dividers are set. The object of the second oblique line in the engraving is to illustrate this. Upon A C the dividers were set so as to produce spaces shorter than those required in the given line A B, while in A D the spaces were made longer than those

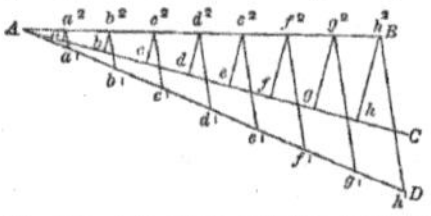


*Fig. 142.—To Divide a Given Straight Line into Any Number of Equal Parts.*

required in the given line. By connecting the extremes, as shown by the lines $h\ h^2$ and $h^1\ h^2$, and drawing lines from the points in each line parallel to these lines respectively, it will be seen that the same divisions are obtained in the given line A B.

287. *A Scale by which to Divide a Straight Line into Any Number of Equal Parts.*—It frequently happens in pattern cutting that it is more convenient to transfer the length of a given line to a slip of paper, and by laying the paper across a scale, as shown in Fig. 143, mark the required dimensions upon it, and afterward transfer them to the given line, than to divide the line itself by one of the methods explained for that purpose. It also occasionally occurs that it is desirable to divide lines of different lengths into the same number of equal parts, or the same lengths of lines into different numbers of equal parts. Such a scale as is shown in Fig. 143 is adapted to all of these purposes. The scale may be ruled upon a piece of paper or upon a sheet of metal, as is preferred. The lines may be all of one color, or two or more colors may be alternated, in order to facilitate counting the lines or following them by the eye across the sheet. In size, the scale is to be adapted to the special purposes for which it is intended to be used. For cornice makers' use it should not be less than 18 inches in width, and might with advantage be as wide as the widest sheet of metal commonly worked. The length should be proportioned to the width, to adapt it to the use of strips diagonally, as shown in the engraving. The size of the spaces into which it is to be divided also depends altogether upon the character of the work in connection with which it is to be used. For cornice makers' purposes, the divisions might be made from a half inch up to an inch in width. By the contrast of two colors in ruling the lines, one scale may be adapted to both coarse and fine work. For instance, if the lines are ruled a half inch apart, in colors alternating red and blue, in fine work all the lines in a given space may be used, while in large work, in which the dimensions are not required to be so small, either all the red or all the blue lines may be used, to the exclusion of those of the other color. We have indicated approximately the size desirable in such a scale for cornice makers' use. When designed for other purposes, the size must be made suitable. In Fig. 143, let it be required to divide the line A B into thirty equal parts. Transfer the length A B to a slip of paper, as shown by $A^1\ B^1$, and placing $A^1$ against the first line of the scale, carry $B^1$ to the thirtieth line. Then mark divisions upon the strip of paper opposite each of the several lines it crosses, as shown. Let it be required to divide the same length, A B, into fifteen equal parts by the scale. Transfer the length A B to a straight strip of paper, as shown by $A^2\ B^2$. Place $A^2$ against the first line and carry $B^2$ against the fifteenth line, as shown. Then mark divisions upon the strip of paper opposite each line of the scale, as shown. A problem of frequent occurrence in pattern cutting is to divide the circumference of a circle into a given number of equal parts. By first obtaining a straight line equal to the circumference of the circle, the division may be readily performed by means of this scale. Several rules for obtaining a straight line approximately equal to the circumference of a circle are given in their appropriate place.

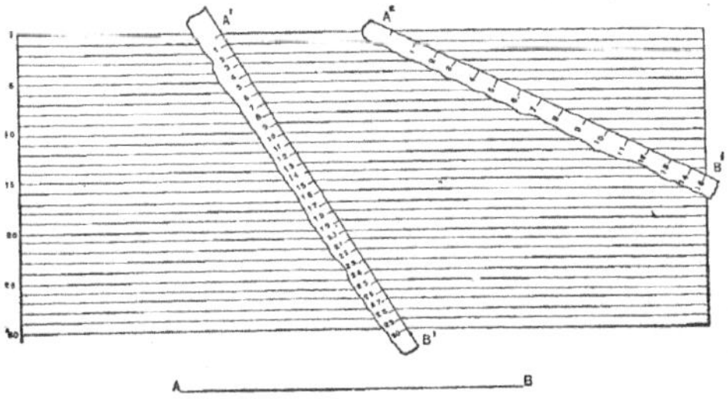

Fig. 143.—*A Scale by which to Divide a Straight Line into Any Number of Equal Parts.*

288. *To Divide a Given Angle into Two Equal Parts.*—In Fig. 144, let A C B represent any angle, through the center of which it is required to draw a straight line. From the vertex, or point C, as center, with any convenient radius, strike the arc D E. From D and E as centers, with any radius greater than one-half the length of the arc D E, strike short arcs intersecting at G, as shown. Through the point of intersection, G, draw a line to the vertex of the angle, as shown by F C. Then F C will divide the angle into two equal parts.

*Fig. 144.—To Divide a Given Angle into Two Equal Parts.*

289. *To Find the Center from which a Given Arc is Struck.*—In Fig. 145, let A B C represent the given arc, the center from which it was struck being unknown and to be found. From any point near the middle of the arc, as B, with any convenient radius, strike the arc F G, as shown. Then from the points A and C, with the same radius, strike the intersecting arcs I H and E D. Through the points of intersection draw the lines K M and L M, which will meet in M. Thus M is the center from which the given arc was struck. Instead of the points A and C being taken at the extremities of the arc, which would be quite inconvenient in the case of a long arc, the points may be located in any part of the arc which is most convenient. The greater the distance between A and B, and B and C, the greater will be the accuracy of succeeding operations. The essential feature of this rule is to strike an arc from the middle one of the points, and then strike intersecting arcs from the other two points, using the same radius. It is not necessary that the distance from A to B and from B to C shall be exactly the same.

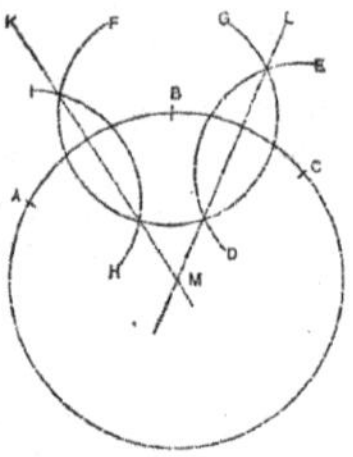


*Fig. 145.—To Find the Center from which a Given Arc is Struck.*

290. *The Chord and Hight of a Segment of a Circle being Given, to Find the Center by which the Arc may be Struck.*—In Fig. 146, let A B represent the chord of a segment or arc of a circle, and D C the rise or hight. It is required to find a center from which an arc, if struck, will pass through the three points A, D and B. Draw A D and B D. Bisect A D, as shown, and prolong the line H L indefinitely. Bisect D B and prolong I M until it cuts H L, produced in the point E. Then E, the point of intersection, will be the center sought. It will be observed that by producing D C, and intersecting it by either H L or I M prolonged, the same point is found. Therefore, if preferred, the bisecting of either A D or D B may be dispensed with. A practical application of this rule occurs quite frequently in cornice work, in the construction of window caps and other similar forms, to fit frames already made. In the conveying of orders from the master builder or carpenter to the cornice worker, it is quite customary to describe the shape of the head of the frames which the caps are to fit by stating that the width is, for example, 36 inches, and that the rise is 4 inches. To draw the shape thus described, proceed as follows: set off A B equal to 36 inches, from the center of which erect a perpendicular, D C, which make equal to 4 inches. Continue D C in the direction of E indefinitely. Draw A D, which bisect, as shown, and draw H L, producing it until it cuts D C prolonged, in the point E. Then with E as center and E D as radius, strike the arc A D B.

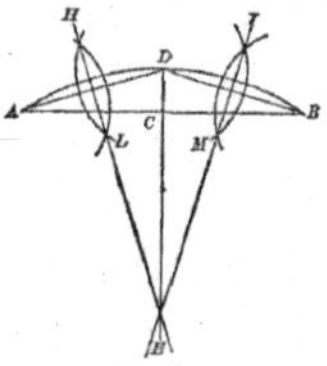


*Fig. 146.—The Chord and Hight of a Segment of a Circle being Given, to Find the Center by which the Arc may be Struck.*

291 *To Find the Center from which a Given Arc is Struck by the Use of the Square.*—In Fig. 147, let A B C be the given arc. Establish the point B at pleasure and draw two chords, as shown by A B and B C. Bisect these chords, obtaining the points E and D. Place the square against the chord B C, as shown in the engraving, bringing the heel against the center point, D, and scribe along the blade indefinitely. Then place the square as shown by the dotted lines, with the heel against the center point, E, of the second chord, and in like manner scribe along the blade, cutting the first line in the point F. Then F will be the center of the circle, of which the arc A B C is a part. This rule will be found very convenient for use in all cases where the radius is less than twenty-four inches in length.

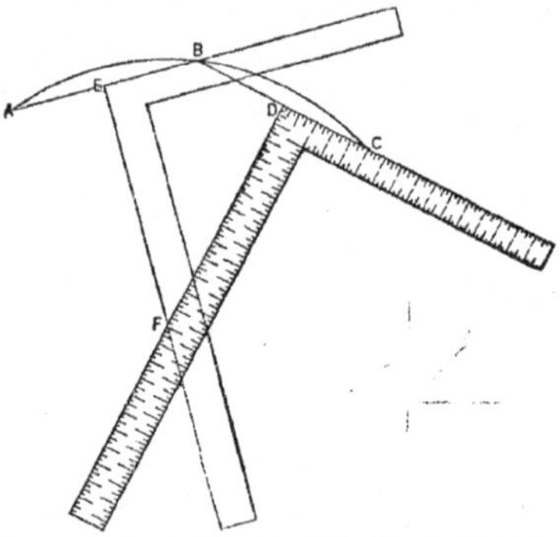


*Fig. 147.—To Find the Center from which a Given Arc is Struck by the Use of the Square.*

292. *To Strike a Segment of a Circle by a Triangular Guide, the Chord and Hight being Given.*—In Fig. 148, let A D be the given chord and B F the given hight. The first step is to determine the shape and size of the triangular guide. Connect A and F, as shown. From F, parallel to the given chord A D, draw F G, making it in length equal to A F, or longer. Then A F G, as shown in the engraving, is the angle of the triangular guide to be used. Construct the guide of any suitable material, making the angle of two of its sides equal to the angle A F G. Drive pins at the points A, F and D. Place the guide as shown. Put a pencil at the point F. Shift the guide in such a manner that the pencil will move toward A, keeping the guide at all times against the pins A and F. Then reversing, shift the guide so that the pencil at the point F will move toward D, keeping the guide during this operation against the pins F and D. By this means the pencil will be made to describe the arc A F D.

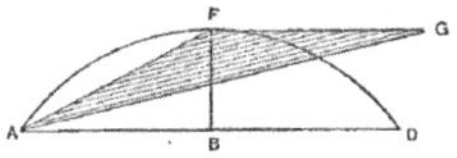

*Fig. 148.—To Strike a Segment of a Circle by a Triangular Guide, the Chord and Hight being Given.*

293. *To Draw a Circle Through any Three Given Points not in a Straight Line.*—In Fig. 149, let A, D and E be any three given points not in a straight line, through which it is required to draw a circle. Connect the given points by drawing the lines A D and D E. Bisect the line A D by F C, drawn perpendicular to it, as shown. Bisect D E by the line G C, also perpendicular to it, as shown. Then the point C, at which these lines meet, is the center of the required circle.

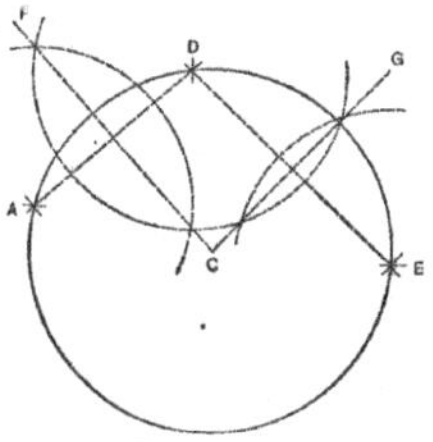

*Fig. 149.—To Draw a Circle Through any Three Given Points not in a Straight Line.*

294. *To Raise a Perpendicular to an Arc of a Circle, without having Recourse to the Center.*—In Fig. 150, let A D B be the arc of a circle to which it is required to erect a perpendicular. With A as center, and with any radius greater than half the length of the given arc, describe the arc *x x*, and with B as center, and with the same radius, describe the arc *y y*, intersecting the arc first struck, as shown. Through the points of intersection draw the line F E. Then F E will be perpendicular to the arc, and if sufficiently produced will reach the center from which the arc A B is drawn.

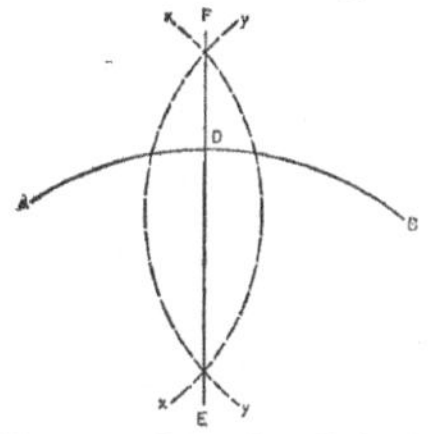

*Fig. 150.—To Raise a Perpendicular to an Arc of a Circle, without having Recourse to the Center.*

295. *To Draw a Tangent to a Circle, or a Portion of a Circle, without having Recourse to the Center.*—In Fig. 151, let A D B be the arc of a circle, to which a tangent is to be drawn at the point D. With D as center, and with any convenient radius, describe the arc A F B, cutting the given arc in the points A A B. Join the points A and B, as shown. From D draw a straight line perpendicular to A B, as shown by D C, and from B erect another perpendicular to A B, as shown by B G. Make B G equal to C D. Draw E H through the points D and G. Then E H will be the required tangent.

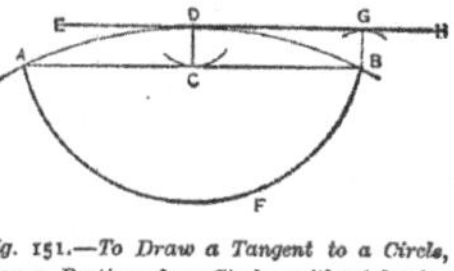

*Fig. 151.—To Draw a Tangent to a Circle, or a Portion of a Circle, without having Recourse to the Center.*

296. *To Draw a Straight Line Equal to the Circumference of a Given Circle.—First Method.*—In Fig. 152, let A D B C be the circle, equal to the circumference of which it is desired to draw a straight line. Draw two diameters, A B and D C, as shown, at right angles. Connect the points A and D. Bisect the line A D and draw E F. To three times the diameter (A B or D C) add the length E F. The result will be very nearly the circumference of the circle. This rule gives a length slightly in excess of the true circumference, the error being about one-sixteenth of an inch in the circumference of a circle the diameter of which is one foot.

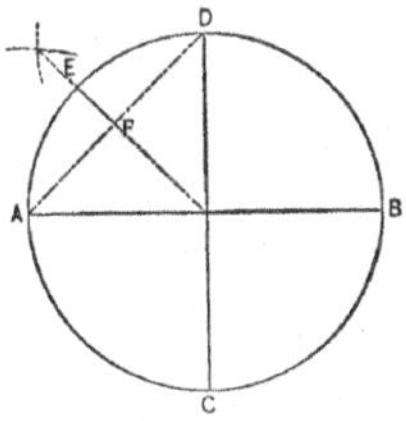

*Fig. 152.—To Draw a Straight Line Equal to the Circumference of a Given Circle.—First Method.*

297. *To Draw a Straight Line Equal to the Circumference of a Given Circle.—Second Method.*—In Fig. 153, let A D B C be the circle, equal to the circumference of which it is required to draw a straight line. Draw any two diameters at right angles, as shown by A B and D C. Divide one of the four arcs, as, for instance, D B, into eleven equal parts, as shown. From 9, the second of these divisions from the point B, let fall a perpendicular to A B, as shown by 9 F. To three times the diameter of the circle (A B or D C) add the length 9 F, and the result will be a very close approximation to the length of the circumference. This rule, upon a diameter of 1 foot, gives a length of about $\frac{3}{50}$ths of an inch in excess of the actual length of the circumference.

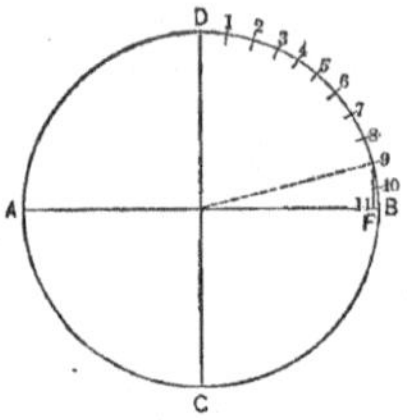

*Fig. 153.—To Draw a Straight Line Equal to the Circumference of a Given Circle.—Second Method.*

298. *To Draw a Straight Line Equal to the Semi-Circumference of a Given Circle.*—In Fig. 154, let A B C represent the semi-circle, equal to the circumference of which it is required to draw a straight line. Divide the semicircle into two equal parts by the line B F. Divide the arc B C into eleven equal parts, as shown by the small figures. From the first division, above the radius F C, drop a perpendicular upon that line, as shown by D E. To three times the radius F C add the distance D E. The result will be the length of the semi-circumference A B C. This rule, in principle, is the same as that presented in Section 295, to which refer for the measure of its accuracy.

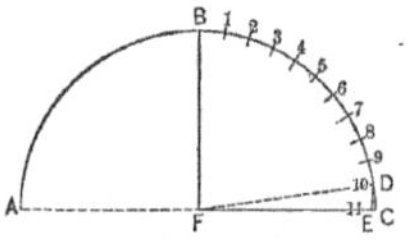

*Fig. 154.—To Draw a Straight Line Equal to the Semi-Circumference of a Given Circle.*

299. *To Draw a Straight Line Equal to the Quarter Circumference of a Given Circle.*—In Fig. 155, A B C represents a quarter circle, equal to the arc A C of which it is required to draw a straight line. Divide the arc into eleven equal parts, as shown, and from the first division above the radius B C drop a perpendicular on to B C, as shown by D C. Bisect this perpendicular, as shown at F. Also bisect the radius B C, as shown at E. Then to three times B E, or one-half of the radius, add the length D F, being half of the perpendicular. The result will be the length of a straight line which shall equal the quarter circumference, or the arc A C. This rule, also, is the same in principle as that given in Section 295, and produces a close approximation to the actual length.

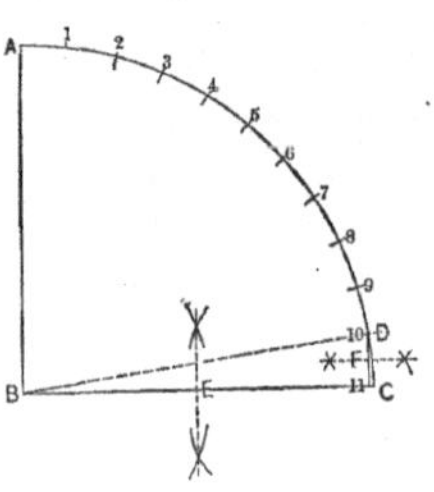

*Fig. 155.—To Draw a Straight Line Equal to the Quarter Circumference of a Given Circle.*

300. *To Draw a Straight Line Equal to any Given Part of a Circle less than a Semi-circle.—First Method, Using the Center.*—In Fig. 156, let A F B represent the given arc, equal to which it is desired to draw a straight line. Draw the chord A B to the arc, which bisect. From the middle point C, through the center of the circle of which the given arc is a part, draw the line J K indefinitely. Divide a radius of the circle, as, for example, D E, into four equal parts, and set off three of those parts from E toward K, as indicated by the small figures. Draw the tangent G H to the arc at the point F, or where J K cuts the arc. From the point L, obtained as just before explained, draw lines through the extremities of the arc, or through A and B, cutting the tangent in the points G and H. The line G H will then be equal to the length of the arc A B. This rule, like others of its class, is only approximately correct, but the variation is so slight as to make its use entirely safe in ordinary mechanical operations.

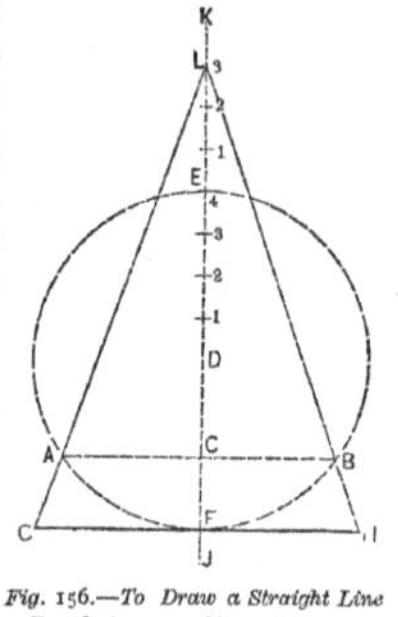

*Fig. 156.—To Draw a Straight Line Equal to any Given Part of a Circle less than a Semicircle.—First Method, Using the Center.*

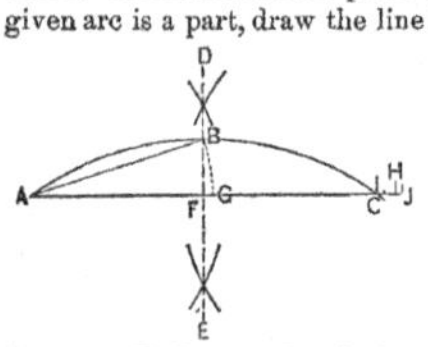

*Fig. 157.—To Draw a Straight Line Equal to any Given Part of a Circle less than a Semicircle.—Second Method, Without Using the Center.*

301. *To Draw a Straight Line Equal to any Given Part of a Circle less than a Semicircle.—Second*

*Method, Without Using the Center.*—In Fig. 157, let A B C represent the given part of a circle, equal to which a straight line is to be drawn. Draw the chord A C, which bisect as shown by D E. Draw A B, which is the chord of half of the given arc. Lay off the length A B twice on the chord A C. The distance will exceed the length A C, as indicated by A G and G H, by a certain distance. Divide this excess, or the space C H, into three equal parts, and increase the length stepped off by twice the chord A B, by the amount of one of these parts, as shown by H D. Then A J will be a straight line, which in length is equal to the arc A B C. This rule, like the one preceding it, is sufficiently accurate for ordinary mechanical operations, but is not absolutely correct.

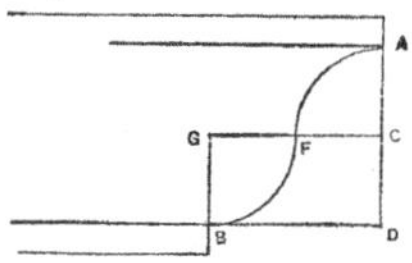

*Fig. 158.—To Draw an Ogee by Means of Two Quarter Circles.*

302. *To Divide an Arc of a Circle into Any Given Number of Equal Parts.*—Several somewhat intricate rules for performing this operation have been devised, but as they are not practical, we have not considered it worth while to present them in this connection. The simplest way of performing this oft-recurring operation in the pattern cutter's work is as follows: Lay off a straight line equal to the arc of the circle by either of the rules already given, by stepping around the arc with the dividers, or by measuring the arc with a strip of metal bent to fit it. Having obtained the straight line by one or the other of these ways, divide it into the required number of equal parts by either of the rules already given for that operation. Take one of the spaces thus obtained in the dividers, or, what is better for the purpose, the spacers, and step around the arc, marking the places where the points of the instrument come.

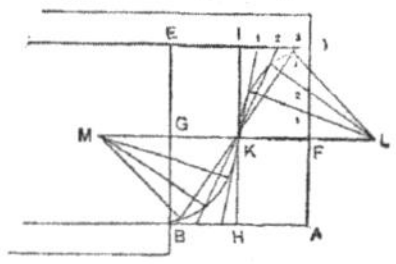

*Fig. 159.—To Draw a Grecian Ogee.*

303. *To Draw an Ogee by Means of Two Quarter Circles.*—In Fig. 158, let A D be the hight of the ogee, which is also equal to D B, the projection. Bisect A D, obtaining the point C, from which, parallel to D B, draw C G. From C as center, and with C A as radius, describe the arc A F. From B erect a perpendicular, cutting C G in the point G. From G as center, and with G F, which is equal to A C, as radius, strike the arc F B, which will complete the ogee.

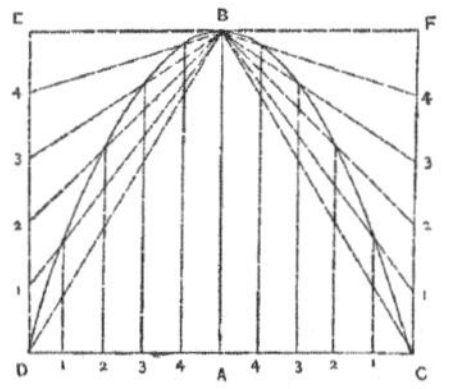

*Fig. 160.—To Draw a Parabola by the Intersection of Lines, its Hight and Base or Ordinate being Given.*

304. *To Draw a Grecian Ogee.*—In Fig. 159, let A D be the hight of the required form and A B the projection. Upon these two sides erect a rectangle, as shown by B A D E. Bisect A D, and through the point F thus obtained draw a line at right angles to A D indefinitely, as shown by M L. Bisect A B by the line H I, which make equal to A D, cutting M L in the point K. Make F L and G M each equal to K F. Divide I D, D F, G B and B H into the same number of equal parts. From the divisions in I D and B H draw lines to K. From L draw lines through the points in D F, to intersect the lines drawn from I D, and from M, through the divisions in G B, draw lines to intersect the lines drawn from B H. A line traced through the points of intersection thus obtained will be the curve sought.

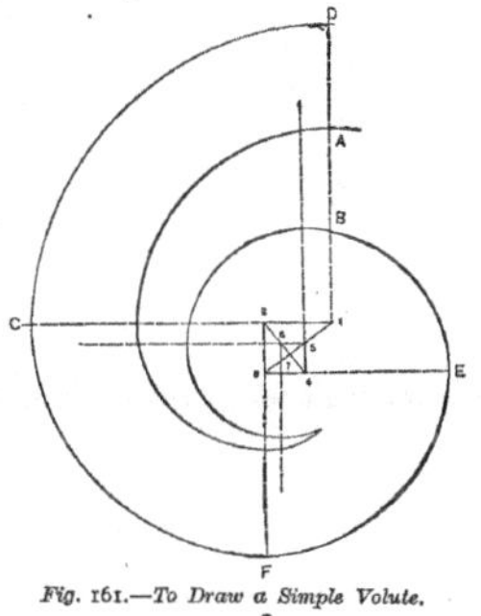

*Fig. 161.—To Draw a Simple Volute.*

305. *To Draw a Parabola by the Intersection of Lines, its Hight and Base or Ordinate being Given.*—In Fig. 160, let A B be the hight and D C the base of the required figure. Draw D E and C F equal to the hight and parallel to it. Divide D E and C F into any convenient number of equal parts. Divide each half of the base into the same number of equal parts, as shown. Draw lines from the points 1 2 3 4 in D E and C F to the point B. Erect perpendiculars to the base D C on each of the points 1 2 3 4. Then a line traced through the points in which these lines intersect will describe one-half of the required figure.

306. *To Draw a Simple Volute.*—Let D A, in Fig. 161, be the width of a scroll or other member for which it is desired to draw a volute termination. Draw the line D 1, in length equal to three times D A, as shown by

D A, A B and B 1. From the point 1 draw 1 2 at right angles to D 1, and in length equal to two-thirds the width of the scroll, or, what is the same, to two-thirds the width of D A. From 2 draw the line 2 3 perpendicular to 1 2, and in length equal to three-quarters of A D. Draw the diagonal line 1 3. From 2 draw a a line perpendicular to 1 3, as shown by 2 4, indefinitely. From 3 draw a line perpendicular to 2 3, producing it until it cuts the line 2 4 in the point 4. From 4 draw a line perpendicular to 3 4, producing it until it meets the line 1 3 in the point 5. In like manner draw 5 6 and 6 7. The points 1, 2, 3, 4, etc., thus obtained are the centers by which the curve of the volute is struck. From 1 as center, and with 1 D as radius, describe the quarter circle D C. Then from 2 as center, and 2 C as radius, describe the quarter circle C F, and so continue until the figure is completed, as shown.

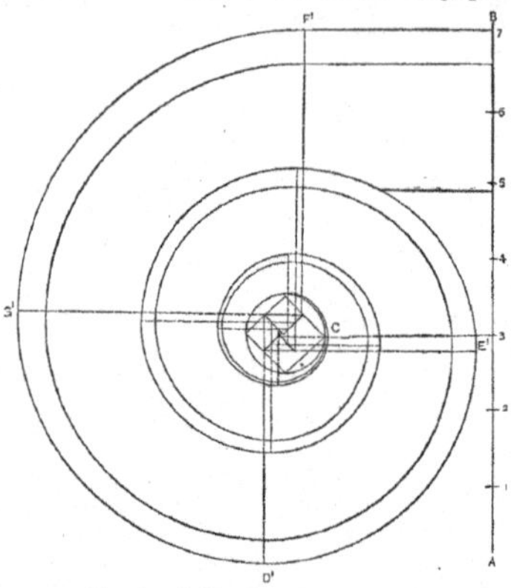


*Fig. 162.—To Describe an Ionic Volute.*

307. *To Describe an Ionic Volute.*—Draw the line A B, Fig. 162, equal to the hight of the required volute, and divide it into seven equal parts. From the third division draw the line 3 C, and from a point at any convenient distance on this line from A B describe a circle, the diameter of which shall equal one of the seven divisions of the line A B. This circle forms the eye of the volute. In order to show its dimensions, etc., it is enlarged in Fig. 163. A square, D E F G, is constructed, and the diagonals G E and F D are drawn. F E is bisected at the point 1, and the line 1 2 is drawn parallel to G E. The line 2 3 is then drawn indefinitely from 2 parallel to F D, cutting G E in the point H. The distance from H to the center of the circle, O, is divided into three equal parts, as shown by H *a* *b* O. The triangle 2 O 1 is formed. On the line O H set off a point, as *c*, at a distance from O equal to one-half of one of the three equal parts into which O H has been divided. From *c* draw the line *c* 3 parallel to 1 O, producing it until it cuts 2 3 in the point 3. From 3 draw the line 3 4 parallel to G E indefinitely. From the point *c* draw a line *c* 4 parallel to 2 O, cutting the line 3 4 in the point 4, completing the triangle *c* 3 4. From 4 draw the line 4 5 parallel to F D, meeting 1 O in the point 5. From 5 draw the line 5 6 parallel to G E, meeting the line 2 O in the point 6. From 6 draw the line 6 7 parallel to F D, meeting the line *c* 3 in the point 7. Proceed in this manner, obtaining the remaining points, 8, 9, 10, 11 and 12. These points form the centers by which the outer line of the volute proper is drawn. From 1 as center, and with radius 1 F', describe the quarter circle F' G'. Then form 2 as center, and with radius 2 G' describe the quarter circle G' D', and so continue striking a quarter circle from each of the centers above described, until the last arc meets the circle first drawn. To obtain the centers by which the inner line of of the volute is struck, and which gradually approaches the outer line throughout its course, proceed as follows: Produce the line 3 *c* until it intersects 1 2 in the point 1', which mark. This operation gives also the points 9' and 5' of intersection with the lines parallel to 1 2, which also mark. In like manner produce 4 *c*, 1 *c* and 2 *c*, as shown by the dotted lines, and mark the several points of intersection formed with the cross lines. Then the points 1', 2', 3', 4', etc., thus obtained are the centers for the inner line of the volute, which use in the same manner as described for producing the outer line. Although this rule is in quite general use for describing the scroll sides of brackets and modillions, that given in Section 310 will be found more satisfactory.

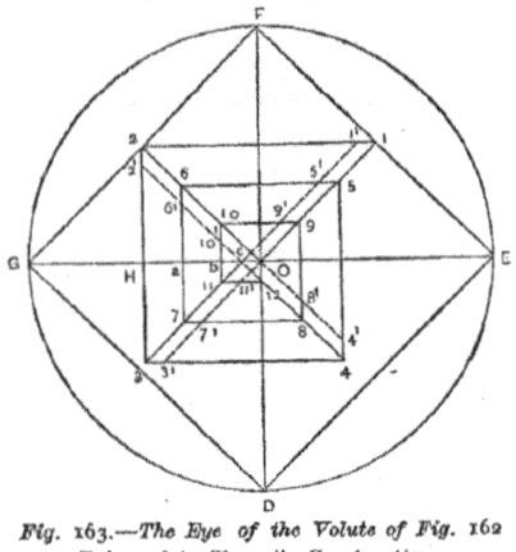


*Fig. 163.—The Eye of the Volute of Fig. 162 Enlarged to Show its Construction.*

308. *To Draw a Spiral from Centers with Compasses.*—Divide the circumference of the primary—sometimes called the eye of the spiral—into any number of equal parts; the larger the number of parts the more regular will be the spiral. Fig. 164 shows the primary divided into six equal parts. Fig. 165 is an enlarged view of a portion of the preceding figure. Complete the polygon by drawing the lines 1 2, 2 3, 3 4, etc., producing them outside of the primary, as shown by A, B, D, F, C and E. From 2 as center, with 2 1 as radius, describe the arc A B. From 3 as center, and 3 B as radius, describe the arc B D; and with 4 as center, with radius 4 D, describe the arc D F. In this manner the spiral may be drawn any number of revolutions. Use 1, 2, 3, 4, 5 and 6 as centers, describing from each in turn an arc contained between two sides.

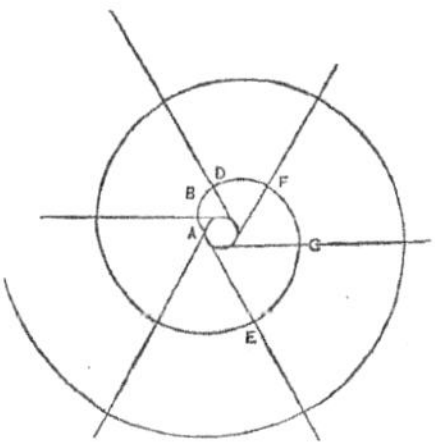


*Fig. 164.—To Draw a Spiral from Centers with Compasses.*

309. *To Draw a Spiral, by Means of a Spool and Thread.*—Set the spool, as shown by A D B in Fig. 166, and wind a thread around it. Make a loop, E, in the end of the thread, in which place a pencil, as shown. Hold the spool firmly and move the pencil around it, unwinding the thread. A curve will be described, as shown in the dotted lines of the engraving. It is evident that the proportions of the figure are determined by the size of the spool. Hence a larger or smaller spool is to be used, as circumstances require.

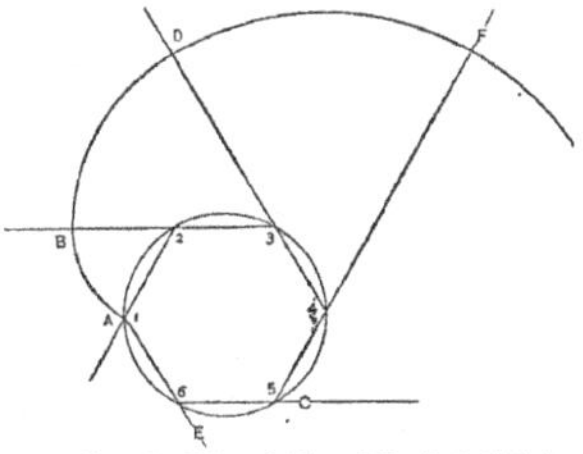


*Fig. 165.—An Enlarged View of the Central Part of Fig. 164.*

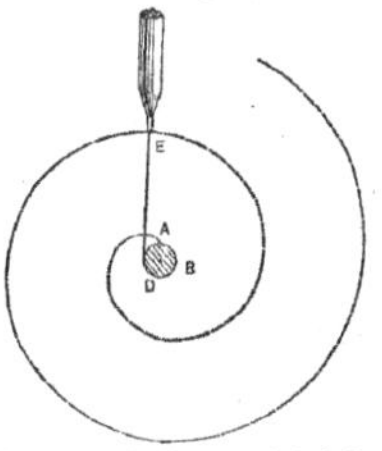


*Fig. 166.—To Draw a Spiral by Means of a Spool and Thread.*

310. *To Draw a Scroll to a Specified Width, as for a Bracket or Modillion.*—In Fig. 167, let it be required to construct a scroll which shall touch the line D B at the top, E A at the bottom and A B at the side, the length of A B, which determines the length of the top and bottom line, being given. Bisect A B, obtaining the point C. Let the distance between the beginning and ending of the first revolution of the scroll, shown by *a e*, be established at pleasure. Having determined this distance, take one-eighth of it and set it off upward from C on the line A B, thus obtaining the point *b*. From *b* draw a horizontal line of any convenient length, as shown by *b h*. With the point of the compasses set at *b*, and with *b* A as radius, describe an arc cutting the line *b h* in the point 1. In like manner, from the same center, with radius *b* B, describe an arc cutting the line *b h* in the point 2. Upon 1 2 as a base erect a square, as shown by 1 2 3 4. Then from 1 as center, with 1 *a* as radius, describe an arc *a b*; and from 2 as center, with 2 *b* as radius, describe the arc *b c*. From 3 as center, with radius 3 *c*, describe the arc *c d*. From 4 as center, with radius 4 *d*, describe the arc *d e*. If the curve were continued from E, being struck from the same centers, it would run parallel to itself; but as one line of the scroll runs parallel to the outer line, its width may be set off at pleasure, as shown by *a a*¹, and the inner line may be drawn by the same centers as already used for the outer, and continued until it is intersected by the outer curve. To find the centers from which to complete the outer curve, construct upon the line of the last radius above used (4 *e*) a smaller square within the larger one, as shown by 5 6 7 8. This is better illustrated by the larger diagram, Fig. 168, in which like figures represent the same points. Make the distance from 5 to 8 equal to one-half of the space from 4 to 1; and make 4 to 8 equal the

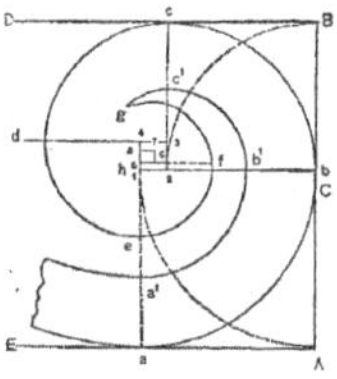


*Fig. 167.—To Draw a Scroll to a Specified Width, as for a Bracket or Modillion.*

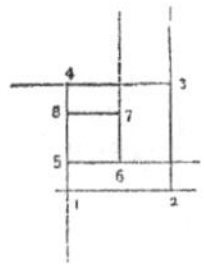


*Fig. 168.—The Center of Fig. 167, Enlarged to better Illustrate its Construction.*

distance of 5 to 1. Make 5 to 6 equal the distance from 8 to 5. After obtaining the points 5, 6, 7, etc., in this manner, so many of them are to be used as are necessary to make the outer curve intersect the inner one, as shown at *g*. Thus 5 is used as a center for the arc *e f*, and 6 as a center for the arc *f g*. If the distance *a a'* were taken less than here given, it is easy to see that more of the centers upon the small square would require to be used to arrive at the intersection.

## THE CONSTRUCTION OF REGULAR POLYGONS.

### I.—BY THE USE OF COMPASSES AND STRAIGHT-EDGE.

311. The most common rules in use for the construction of polygons, whether drawn within circles or erected upon given sides, are those which employ the straight-edge and compasses only. In some instances these rules are the best for the pattern cutter to employ. In other cases his ends are better served by rules making use of other instruments. Accordingly, we divide our remarks upon the construction of polygons into four parts, arranging them according to the tools employed. By this presentment the student will have no difficulty in seeing the relative advantages of the different methods, and by becoming expert in the use of different instruments, will be able to select the best rules for his purpose as circumstances arise.

312. *To Draw an Equilateral Triangle within a Given Circle.*—In Fig. 169, let A B D be any given circle, within which an equilateral triangle is to be drawn. From any point in the circumference, as E, with a radius equal to the radius of the circle, describe the arc D C B, cutting the given circle in the points D and B. Draw the line D B, which will be one side of the required triangle. From D or B as center, and with D B as radius, cut the circumference of the given circle, as shown at A. Draw A B and A D, which will complete the figure.

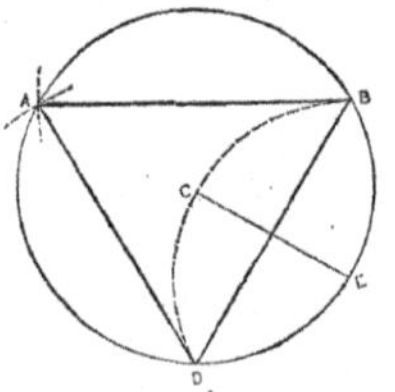


*Fig. 169.—To Draw an Equilateral Triangle within a Given Circle.*

313. *To Draw a Square within a Given Circle.*—In Fig. 170, let A C B D be any given circle within which it is required to draw a square. Draw any two diameters at right angles with each other, as C D and A B. Join the points C B, B D, D A and A C, which will complete the required figure.

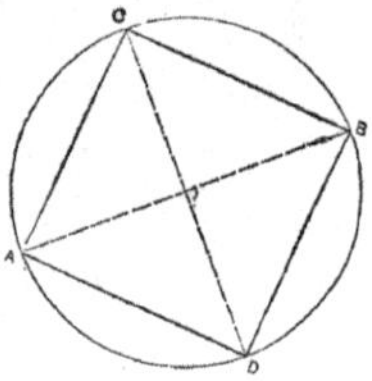


*Fig. 170.—To Draw a Square within a Given Circle.*

314. *To Draw a Regular Pentagon within a Given Circle.*—In Fig. 171, A D G B C represents a circle in which it is required to draw a regular pentagon. Draw any two diameters at right angles to each other, as A B and D C. Bisect the radius A H, as shown at E. With E D as radius strike the arc D F, and with the chord D F as radius strike the arc F G, cutting the circumference of the given circle at the point G. Draw D G, which will equal one side of the required figure. With the dividers set equal to D G, step off the spaces in the circumference of the circle, as shown by the points I K L. Draw D I, I K, K L and L G, thus completing the figure.

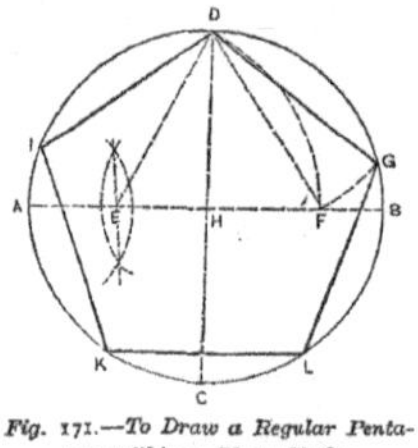


*Fig. 171.—To Draw a Regular Pentagon within a Given Circle.*

315. *To Draw a Regular Hexagon within a Given Circle.*—In Fig. 172, let A B D E F G be any given circle within which a hexagon is to be drawn. From any point in the circumference of the circle, as at A, with a radius equal to the radius of the circle, describe the arc C B, cutting the circumference of the circle in the point B. Connect the points A and B. Then A B will be one side of the hexagon. With the dividers set to the distance A B, step off in the circumference of the circle the points G, F, E and D. Draw the connecting lines A G, G F, F E, E D and D B, thus completing

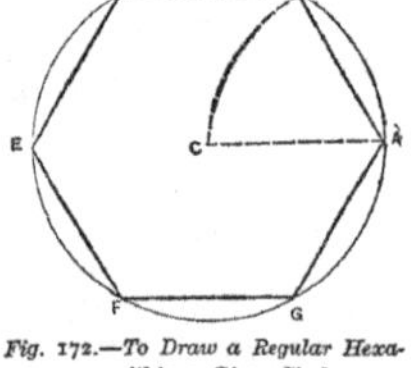


*Fig. 172.—To Draw a Regular Hexagon within a Given Circle.*

the figure. By inspection of this figure it will be noticed that the radius of a circle is equal to one side of the regular hexagon which may be inscribed within it. Hence it follows that drawing the arc C B may be dispensed with. Set the dividers to the radius of a circle and step around the circumference, connecting the points thus obtained.

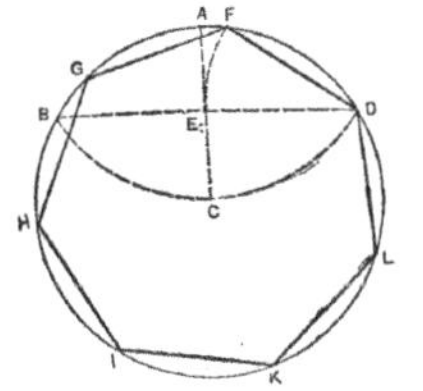

*Fig. 173.—To Draw a Regular Heptagon within a Given Circle.*

316. *To Draw a Regular Heptagon within a Given Circle.*—In Fig. 173, let F A G B H I K L D be the given circle. From any point, A, in the circumference, with a radius equal to the radius of the circle, describe the arc B C D, cutting the circumference of the circle in the points B and D. Draw the chord B D. Bisect the chord B D, as shown at E. With D as center, and with D E as radius, strike the arc E F, cutting the circumference in the point F. Draw D F, which will be one side of the heptagon. With the dividers set to the distance D F, set off in the circumference of the circle the points G H I K L, and draw the connecting lines F G, G H, H I, I K, K L and L D, thus completing the figure.

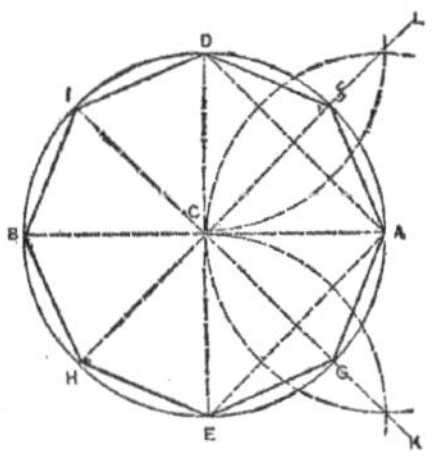

*Fig. 174.—To Draw a Regular Octagon within a Given Circle.*

317. *To Draw a Regular Octagon within a Given Circle.*—In Fig. 174, let B I D F A G E H be the given circle within which an octagon is to be drawn. Draw any two diameters at right angles to each other, as B A and D E. Draw the chords D A and A E. Bisect D A, as shown, and draw L H. Bisect A E and draw K I. Then connect the several points in the circumference thus obtained by drawing the lines D I, I B, B H, H E, E G, G A, A F and F D, which will complete the figure.

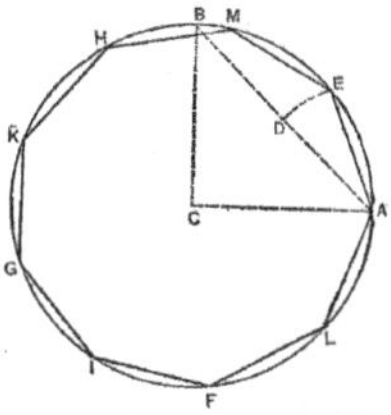

*Fig. 175.—To Draw a Regular Nonagon within a Given Circle.*

318. *To Draw a Regular Nonagon within a Given Circle.*—In Fig. 175, let M G E be the given circle. Draw any two radii at right angles to each other, as B C and A C, and draw the chord B A. From A as center, and with a radius equal to one-half the chord A B, as shown by A D, strike the arc D E, cutting the circumference of the circle at the point E. Draw A E, which will be one side of the nonagon. Set the dividers to the distance A E and step off the points M, H, K, G, I, F and L, and draw the connecting lines, as shown, thus completing the figure.

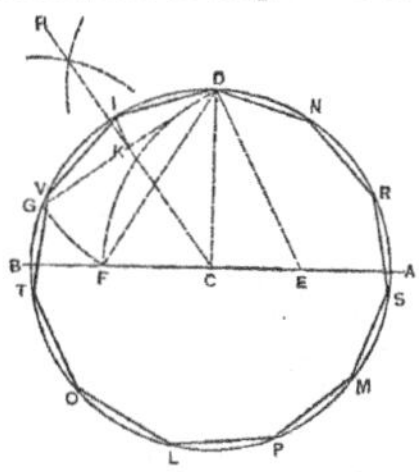

*Fig. 176.—To Draw a Regular Decagon within a Given Circle.*

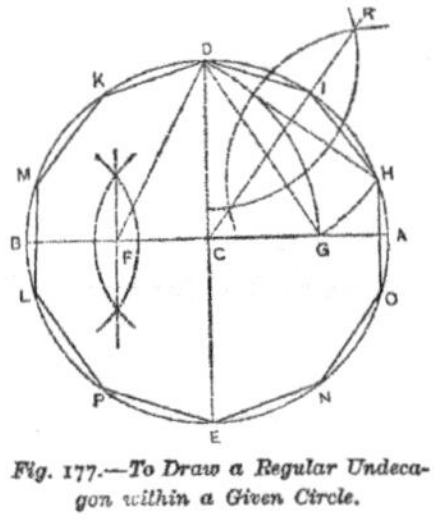

*Fig. 177.—To Draw a Regular Undecagon within a Given Circle.*

319. *To Draw a Regular Decagon within a Given Circle.*—In Fig. 176, let D B E A be any given circle in which a decagon is to be drawn. Draw any two diameters through the circle at right angles to each other, as shown by B A and D E. Bisect B C, as shown at F, and draw F D. With F as center, and F D as radius, describe the arc D G, cutting B A in the point G. Draw the chord D G. With D as center, and D G as radius, strike the arc G H, cutting the circumference in the point H. Connect D and H, as shown. Bisect D H and draw the line C R, cutting the circumference in the point I. Draw the lines H I and I D, which will then be two sides of the required figure. Set the dividers to the distance H I and space off the circumference of the circle, as shown, and draw the connecting lines D K, K M, M L, L P, P E, E N, N O and O H, thus completing the figure.

320. *To Draw a Regular Undecagon within a Given Circle.*—In Fig. 177, let B D A L be any given circle

in which a regular figure of eleven sides is to be drawn. Draw any diameter, as B A, and draw a radius, as D C, at right angles to B A. Bisect C A, thus obtaining the point E. From E as center, and with E D as radius, describe the arc D F, cutting B A in the point F. With D as center, and D F as radius, describe the arc F G, cutting the circumference in the point G. Draw the chord G D and bisect it, as shown by H C, thus obtaining the point K. From D as center, and with D K as radius, cut the circumference in the point I. Draw I D. Then I D will be equal to one side of the required figure. Set the dividers to this space and step off the points in the circumference, as shown by N, R, S, M, P, L, O, T, J and G, and draw the connecting arcs, as shown, thus completing the figure.

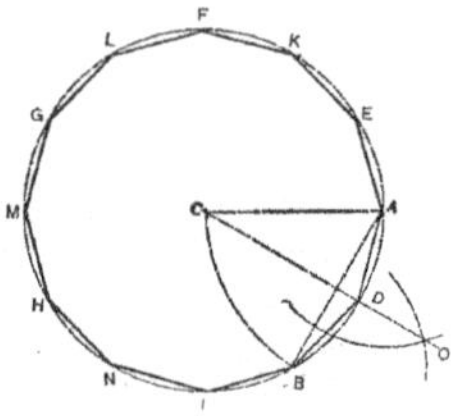


*Fig. 178.—To Draw a Regular Dodecagon within a Given Circle.*

321. *To Draw a Regular Dodecagon within a Given Circle.*—In Fig. 178, let M F A I be any given circle in which a dodecagon is to be drawn. From any point in the circumference, as A, with a radius equal to the radius of the circle, describe the arc C B, cutting the circumference in the point B. Draw the chord A B, which bisect as shown, and draw the line O C, cutting the circumference in the point D. Draw A D, which will then be one side of the given figure. With the dividers set to this space step off in the circumference the points B, I, N, H, M, G, L, F, K and E, and draw the several chords, as shown, thus completing the figure.

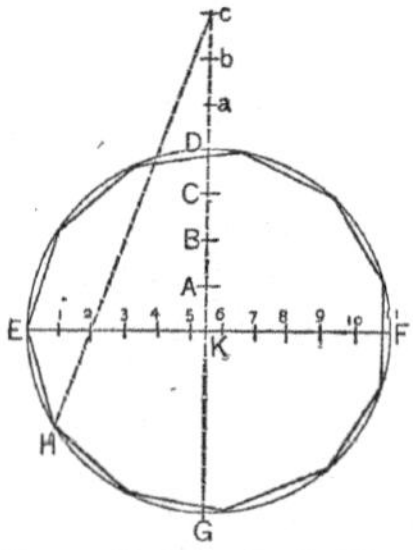


*Fig. 179.—To Draw a Regular Polygon of Eleven Sides within a Given Circle by General Rule Given in Section 322.*

322. *General Rule for Drawing any Regular Polygon in a Circle.—Rule.*—Through the given circle draw any diameter. At right angles to this diameter draw a radius. Divide that radius into four equal parts, and prolong it outside the circle to a distance equal to three of those parts. Divide the diameter of the circle into the same number of equal parts as the polygon is to have sides. Then from the end of the radius prolonged, as above described, through the second division in the diameter, draw a line cutting the circumference. Connect this point in the circumference and the nearest end of the diameter. The line thus drawn will be one side of the required figure. Set the dividers to this space and step off on the circumference of the circle the remaining number of sides and draw connecting lines, which will complete the figure.

323. *To Draw a Regular Polygon of Eleven Sides within a Given Circle by the General Rule just given.*—Through the given circle, E D F G in Fig. 179, draw any diameter, as E F, which divide into the same number of equal parts as the figure is to have sides, as shown by the small figures. At right angles to the diameter just drawn draw the radius D K, which divide into four equal parts. Prolong the radius D K outside the circle to the extent of three of those parts, as shown by *a b c*, thus obtaining the point *c*. From *c*, through the second division in the diameter, draw the line *c* H, cutting the circumference in the point H. Connect H and E. Then H E will be one side of the required figure. Set the dividers to the distance H E and step off the circumference, as shown, thus obtaining the points for the other sides, and draw the connecting arcs, all as illustrated in the figure.

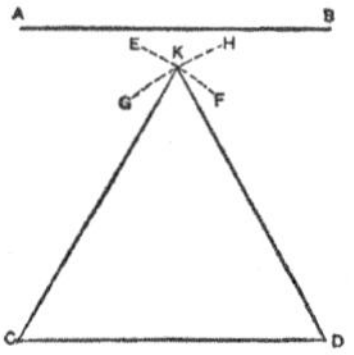


*Fig. 180.—Upon a Given Side to Construct an Equilateral Triangle.*

324. *Upon a Given Side to Construct an Equilateral Triangle.*—In Fig. 180, let A B represent the length of the given side. Draw any line, as C D, making it equal to A B. Take the length A B in the dividers, and placing one foot upon the point C, describe the arc E F. Then from D as center, with the same radius, describe the arc G H, intersecting the first arc in the point K. Draw K C and K D. Then C D K will be the required triangle.

325. *To Construct a Triangle, the Length of the Three Sides being Given.*—In Fig. 181, let A B, C D and E F be the given sides from which it is required to construct a triangle. Draw any straight line, G H, making

it in length equal to one of the sides, E F. Take the length of one of the other sides, as A B, in the compasses, and from one end of the line just drawn, as G, for center describe an arc, as indicated by L M. Then, setting the compasses to the third side, C D, from the opposite end of the line first drawn, as H, describe a second arc, as I K, intersecting the first in the point O. Connect O G and O H. Then O G H will be the required triangle.

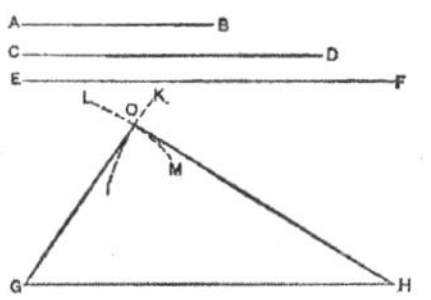


*Fig. 181.—To Construct a Triangle, the Length of the Three Sides being Given.*

326. *Upon a Given Side to Draw a Regular Pentagon.*—In Fig. 182, let A B represent the given side upon which a regular pentagon is to be constructed. With B as center and B A as radius, draw the semicircle A D E. Produce A B to E. Bisect the given side A B, as shown at the point F, and erect a perpendicular, as shown by F C. Also erect a perpendicular at the point B, as shown by G H. With B as center, and F B as radius, strike the arc F G, cutting the perpendicular H G in the point G. Draw G E. With G as center, and G E as radius, strike the arc E H, cutting the perpendicular in the point H. With E as center, and E H as radius, strike the arc H D, cutting the semicircle A D E in the point D. Draw D B, which will be the second side of the pentagon. Bisect D B, as shown, at the point K, and erect a perpendicular, which produce until it intersects the perpendicular F C, erected upon the center of the given side in the point C. Then C is the center of the circle which circumscribes the required pentagon. From C as center, and with C B as radius, strike the circle, as shown. Set the dividers to the distance A B and step off the circumference of the circle, obtaining the points M and L. Draw A M, M L and L D, which will complete the figure.

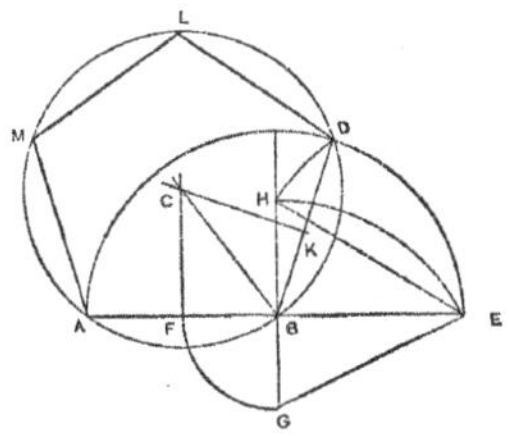


*Fig. 182.—Upon a Given Side to Draw a Regular Pentagon.*

327. *Upon a Given Side to Draw a Regular Hexagon.*—In Fig. 183, let A B be the given side upon which a regular hexagon is to be erected. From A as center, and with A B as radius, describe the arc B C. From B as center, and with the same radius, describe the arc A C, intersecting the first arc in the point C. C will then be the center of the circle which will circumscribe the required hexagon. With C as center, and C B as radius, strike the circle, as shown. Set the dividers to the space A B and step off the circumference, as shown, obtaining the points E, G, F and D. Draw the chords A E, E G, G F, F D and D B, thus completing the required figure.

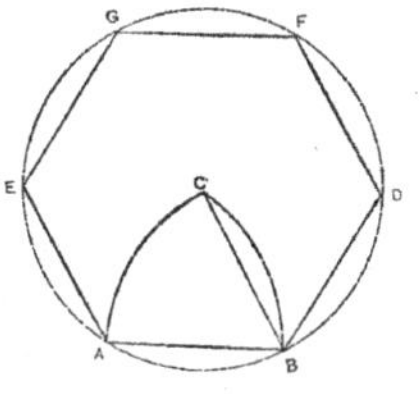


*Fig. 183.—Upon a Given Side to Draw a Regular Hexagon.*

328. *Upon a Given Side to Draw a Regular Heptagon.*—In Fig. 184, A B represents the given side upon which a regular heptagon is to be drawn. From B as center, and with B A as radius, strike the semicircle A E D. Produce A B to D. From A as center, and with A B as radius, strike the arc B F, cutting the semicircle in the point F. Bisect the given side A B, obtaining the point G. Draw G F, producing it indefinitely in the direction of C. From D as center, and with radius G F, cut the semicircle in the point E. Draw the line E B, which is another side of the required heptagon. Bisect E B, and upon its middle point erect a perpendicular, which produce until it meets the perpendicular erected upon the center of the given side A B in the point C. Then C is the center of a circle which will circumscribe the required heptagon. From C as center, and with C B as radius, strike the circle. Set the dividers to the distance A B and step off the circumference, as shown, obtaining the points K, N, M and L. Draw the connecting arcs A K, K N, N M, M L and L E, thus completing the figure.

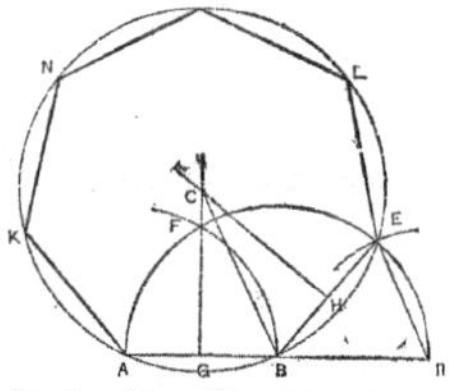


*Fig. 184.—Upon a Given Side to Draw a Regular Heptagon.*

329. *Upon a Given Side to Draw a Regular Octagon.*—In Fig. 185, let A B represent the given side upon which a regular octagon is to be constructed. Produce A B indefinitely in the direction of D. From B as center, and with A B as radius, describe the semicircle A E D. At the point B erect a perpendicular to A B, as shown, cutting the circumference of the semicircle in the point E. Bisect the arc E D, obtaining the point F. Draw F B, which is another side of the required octagon. Bisect the two sides now obtained and erect perpendiculars to their middle points, G and H, which produce until they intersect at the point C. C then is the center of the circle that will circumscribe the octagon. From C as center, and with C B as radius, strike the circle, as shown. Set the dividers to the space A B and step off the circumference, obtaining the points L, K, M, O and N. Draw the connecting arcs A L, L K, K M, M O, O N and N F, thus completing the required figure.

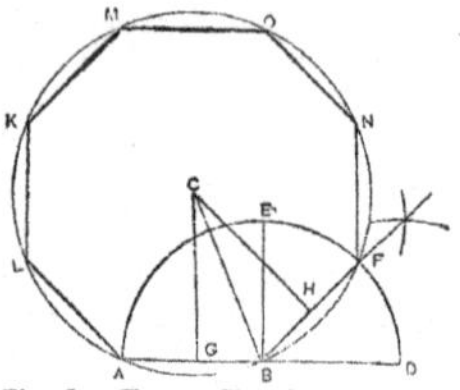

*Fig. 185.—Upon a Given Side to Draw a Regular Octagon.*

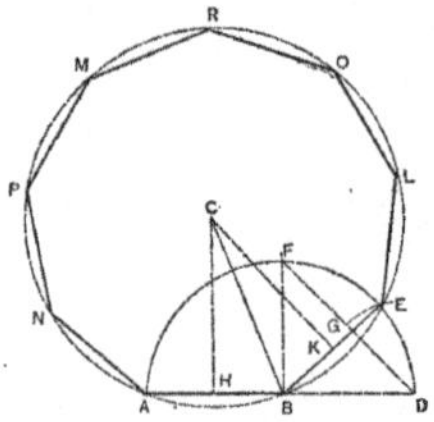

*Fig. 186.—Upon a Given Side to Draw a Regular Nonagon.*

330. *Upon a Given Side to Draw a Regular Nonagon.*—In Fig. 186, A B is any given side upon which it is required to draw a regular nonagon. Produce A B indefinitely in the direction of D. From B as center, and with B A as radius, strike the semicircle A F D. At the point B erect a perpendicular to A B, cutting the semicircle in the point F. Draw the arc F D, which bisect, obtaining the point G. From D as center, and with D G as radius, cut the semicircle in the point E. Draw E B, which will be another side of the required figure. From the middle points of the two sides now obtained, as H and K, erect perpendiculars, which produce until they intersect at the point C. Then C is the center of the circle which will circumscribe the required nonagon. From C as center, and with C B as radius, strike the circle B O P A. Set the dividers to the space A B and step off the circle, as shown, obtaining the points N, P, M, R, O and L. Draw the connecting chords, A N, N P, P M, M R, R O, O L and L E, thus completing the figure.

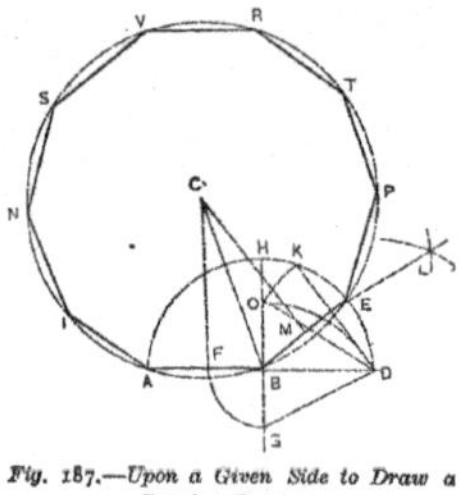

*Fig. 187.—Upon a Given Side to Draw a Regular Decagon.*

331. *Upon a Given Side to Draw a Regular Decagon.*—In Fig. 187, A B is the given side upon which a regular decagon is to be drawn. Produce A B indefinitely in the direction of D. From B as center, and with B A as radius, strike the semicircle A H D. Bisect the given side A B, obtaining the point F. Through the point B draw the line H B G, perpendicular to A B. From B as center, and with B F as radius, strike the arc F G, cutting the perpendicular H G in the point G. From G as center, and with G D as radius, strike the arc D O, cutting the perpendicular H G in the point O. From D as center, and with D O as radius, strike the arc O K, cutting the semicircle in the point K. Draw the line K D, which bisect with the line B L, cutting the semicircle in the point E. Then E B will be another side of the decagon. Upon the middle points, F and M, of the two sides now obtained erect perpendiculars, which produce until they intersect at the point C. Then C is the center of the circle which will circumscribe the required decagon. From C as center, and with C B as radius, strike the circle, as shown. Set the dividers to the space A B and step off the circle, obtaining the several points, I, N, S, V, R, T and P. Draw the connecting lines, A I, I N, N S, S V, V R R T, T P and P E, thus completing the figure.

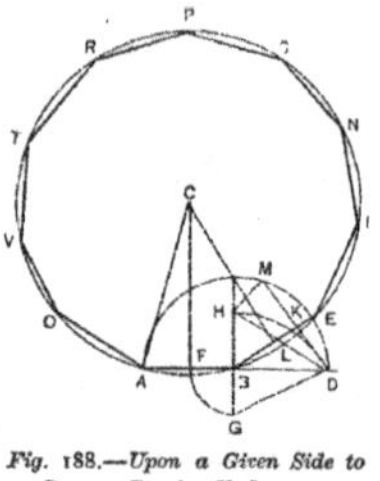

*Fig. 188.—Upon a Given Side to Draw a Regular Undecagon.*

332. *Upon a Given Side to Draw a Regular Undecagon.*—In Fig. 188, A B represents the given side upon which a regular undecagon is to be drawn. Produce A B indefinitely in the direction of D. From B as center, and with B A as radius, draw the semicircle A M D. Through the point

B, perpendicular to A B, draw the line H G indefinitely. From B as center, and with B F as radius, strike the arc F G, cutting the perpendicular H G in the point G. From G as center, and G D as radius, strike the arc D H, cutting the perpendicular H G in the point H. With D as center, and D H as radius, strike the arc H M, cutting the semicircle in the point M. Draw M D, which bisect, obtaining the point K, through which, from B, draw the line B K, and produce it until it cuts the semicircle in the point E. Then B E will be another side of the required figure. Bisect the two sides now obtained and erect perpendicular lines, producing them until they intersect, as shown by F C and L C. Then C, the point of intersection, is the center of the circle which circumscribes the undecagon. From C as center, and with C A as radius, strike the circle, as shown. Set the dividers to the space A B and step off the circumference, obtaining the points O, V, T, R, P, S, N and I. Draw the chords A O, O V, V T, T R, R P, P S, S N, N I and I E, thus completing the figure.

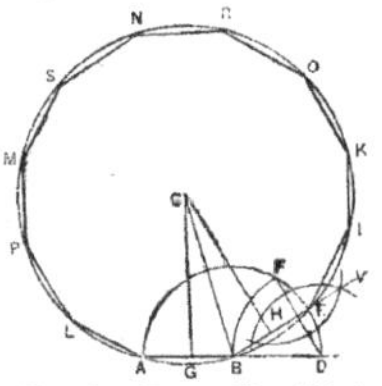

*Fig. 189.—Upon a Given Side to Draw a Regular Dodecagon.*

333. *Upon a Given Side to Draw a Regular Dodecagon.*—In Fig. 189, let A B represent the given side upon which a regular dodecagon is to be drawn. Produce A B indefinitely in the direction of D. From B as center, and with B A as radius, describe the semicircle A E D. From D as center, and with D B as radius, describe the arc B F, cutting the semicircle in the point F. Draw F D, which bisect by the line V B, cutting the semicircle in the point E. Then E B is another side of the dodecagon. From the middle points of the two sides now obtained, as G and H, erect perpendiculars, as shown, cutting each other at the point C. This point of intersection, C, then is the center of the circle which will circumscribe the required dodecagon. From C as center, and with C B as radius, strike the circle, as shown. Set the dividers to the distance A B and space off the circumference, thus obtaining the points L, P, M, S, N, R, O, K and I. Draw the connecting lines L P, P M, M S, S N, N R, R O, O K, K I and I E, thus completing the figure.

334. *General Rule by which to Draw any Regular Polygon, the Length of a Side being Given.*—With a radius equal to the given side describe a semicircle, the circumference of which divide into as many equal parts as the figure is to have sides. From the center by which the semicircle was struck draw a line to the second division in the circumference. This line will be one side of the required figure, and one-half of the diameter of the semicircle will be another, and the two will be in proper relationship to each other. Therefore, bisect each, and through their centers erect perpendiculars, which produce until they intersect. The point of intersection will be the center of the circle which will circumscribe the polygon. Draw the circle, and setting the dividers to the length of one of the sides already found, step off the circumference, thus obtaining points by which to draw the remaining sides of the figure.

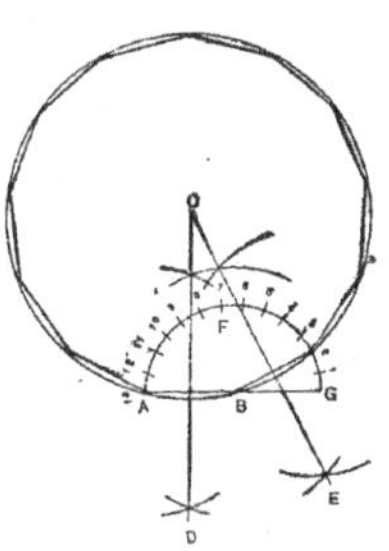

*Fig. 190.—To Construct a Regular Polygon of Thirteen Sides, the Length of a Side being Given, by the General Rule in Section 334.*

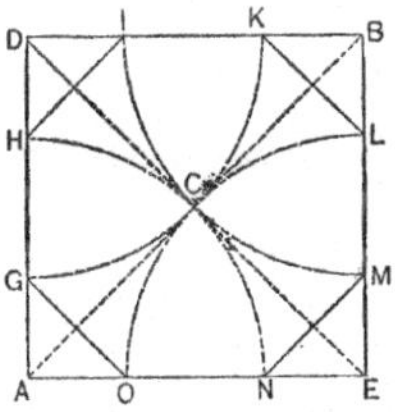


*Fig. 191.—Within a Given Square to Draw a Regular Octagon.*

335. *To Construct a Regular Polygon of Thirteen Sides, the Length of a Side being Given, by the General Rule in Section 334.*—In Fig. 190, let A B be the given side. With B as center, and with B A as radius, describe the semicircle A F G. Divide the circumference of the semicircle into thirteen equal parts, as shown by the small figures, 1, 2, 3, 4, etc. From B draw a line to the second division in the circumference, as shown by B 2. Then A B and B 2 are two of the sides of the required figure, and are in correct relationship to each other. Bisect A B and B 2, as shown, and draw D C and E C through their central points, prolonging them until they intersect at the point C. Then C is the center of the circle which will circumscribe the required polygon. Strike the circle, as shown. Set the dividers to the space A B, and step off corresponding spaces in the circumference of the circle, as shown, and connect the several points so obtained by lines, thus completing the figure.

336. *Within a Given Square to Draw a Regular Octagon.*—In Fig. 191, let A D B E be any given square, within which it is required to draw an octagon. Draw the diagonals D E and A B, intersecting at the point C. From A, D, B and E as centers, and with radius equal to one-half of one of the diagonals, as A C, strike the several arcs H N, G K, I M and L O, cutting the sides of the square, as shown. Connect the points thus obtained in the sides of the square by drawing the lines G O, H I, K L, and M N, thus completing the figure.

II.—BY THE USE OF THE T-SQUARE AND TRIANGLES, OR SET-SQUARES.

337. In another part of the book (Section 72) we described the division of the circle, for the measurement of angles, into spaces called degrees, and, in connection with our description of drawing tools, we described certain triangles or set-squares (Section 233) which are in common use, naming them by the degrees which their angles contain. These set-squares, in connection with the T-square or a straight-edge, can be used advantageously for constructing various polygons, whether inscribed or circumscribed. They are derived directly from the circle; that is, they represent certain fixed portions of the circle, and therefore may be employed in dividing a circumference for the purpose of constructing polygons. To make their use for this purpose entirely clear, we will first describe their origin and afterward give illustrations of their employment.

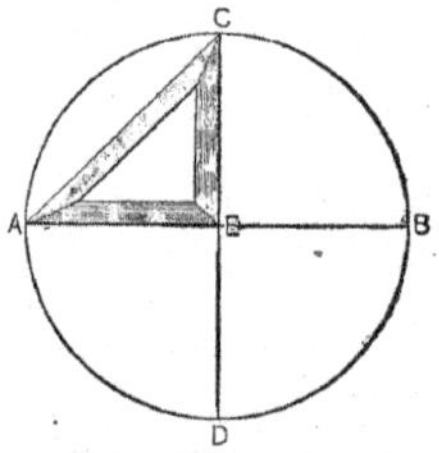

*Fig. 192.—A Circle Divided into Four Equal Parts by the Use of a 90-Degree Set-Square.*

338. Since the circle consists of 360 degrees, a quarter of it is represented by 90 degrees. In Fig. 192, the circle A C B D is divided into quarters by the diameters A B and C D, drawn at right angles to each other. It will be seen that the same result might be accomplished by using the set-square A E C, by bringing its right angle to the center of the circle E, and scribing along its sides for E C and E A, and then shifting it for the other parts. The instrument A E C corresponds to a quarter circle, and is therefore called a 90-degree set-square. If we divide the circle into eight equal parts by diameters, as shown in Fig. 193, each angle will represent one-eighth of 360 degrees, or 45 degrees. Hence, the instrument which corresponds to one of these angles, as A E F in Fig. 193, is called a 45-degree set-square. If we divide the circle into twelve equal parts by diameters, as shown in Fig. 194, each angle will represent one-twelfth of 360 degrees, or 30 degrees, which gives the name to the angle of the set-square corresponding to it, as shown. In like manner, if the circle were divided into six equal parts, each of the angles would measure 60 degrees, which gives name to another angle of the set-square, which is shown by A B E in Fig. 194. Still other set-squares might be employed, but the two which contain the four angles we have described are found entirely adequate for all ordinary requirements.

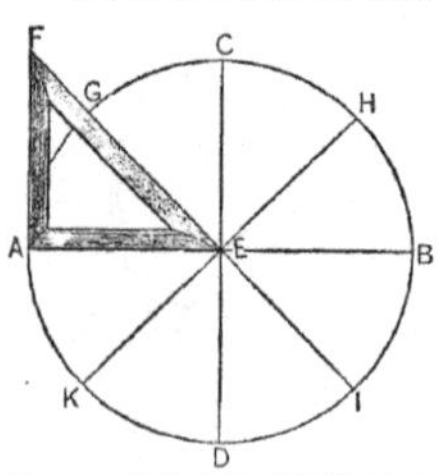

*Fig. 193.—A Circle Divided into Eight Equal Parts by the Use of a 45-Degree Set-Square.*

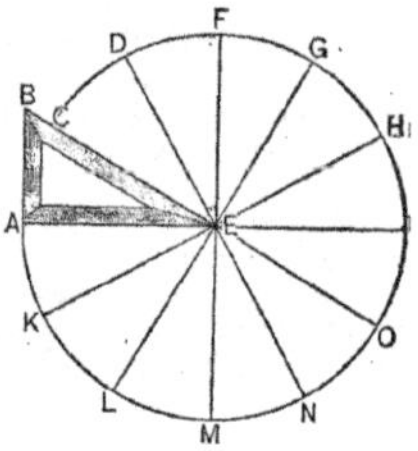

*Fig. 194.—A Circle Divided into Twelve Equal Parts by the Use of a 30-Degree Set-Square.*

339. A governing principle upon which this use of the set-square depends, may be briefly referred to in this connection with advantage. We have described the set-squares as 45, 45 and 90, and 30, 60 and 90 degrees respectively. It will be observed that the sums of these sets of figures are the same; that is, $45 + 45 + 90 = 180$, and $30 + 60 + 90 = 180$. Further, it will be discovered, upon investigation, that the sum of the angles of any triangle whatsoever also equals 180 degrees. Each of the set-squares contains a right angle. Hence, in working from a T-square or other straight line, by means of it lines may be drawn at right angles, and also at the several intermediate angles represented by their other sides. The sum of the angles of the set-squares always being 180 degrees, addition, subtraction and division in the calculation of angles become a very simple matter; but for the most part these operations are performed graphically, as will appear further on.

340. Inasmuch as each of the set-squares contains an angle of 90 degrees, instead of describing them as 45, 45 and 90-degree, and 30, 60 and 90-degree set-squares, the form is abbreviated in the first instance to a "45-degree set-square," and in the second to a "30-degree set-square," or a "60-degree set-square," as the case may be, the latter terms for the second instrument being used interchangeably. With a right angle (90 degrees) in the set-square and an angle of 45 degrees, the third angle must be 45 degrees also, in order to complete the sum, 180 degrees. In like manner, given a set-square with an angle of 90 degrees (a right angle) and another of 60 degrees, the remaining angle must be 30 degrees, and *vice versa.* Therefore, no confusion can possibly arise in calling these tools set-squares of 45 degrees and of 30 degrees, or 60 degrees, as the case may be.

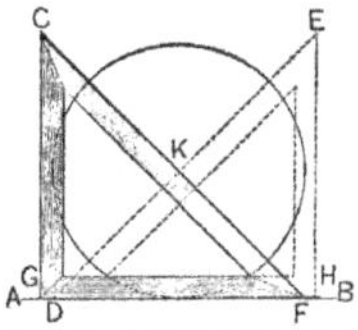

*Fig. 195.—A Circle Divided into Four Equal Parts by a 45-Degree Triangle.*

341. In describing the angle of 90 degrees in the set-square, we compared the division of the circle into four equal parts by two diameters drawn at right angles, with the same result accomplished by the use of this tool placed as shown in Fig. 192. Such a plan of dividing the circle by the use of the set-square, that is, by bringing the right angle of the set-square against its center, is quite inconvenient. A better method, and one which makes use of the same principles in the set-square, is shown in Fig. 195. A straight-edge, as, for instance, a T-square, is placed tangent to or near the circle, as shown by A B. One side of a 45-degree set-square is placed against it, as shown, its side C F being brought against the center. The line C F is then drawn. By reversing the set-square, as shown by the dotted lines, the line E D is drawn at right angles to C F, thus dividing the circle into quarters.

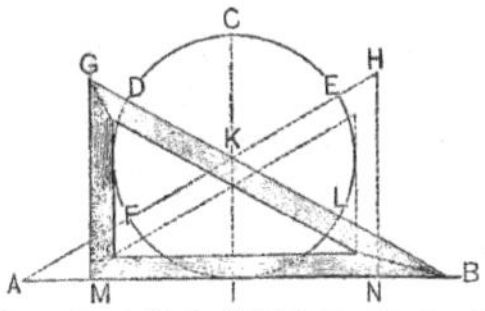

*Fig. 196.—A Circle Divided into Six Equal Parts by the Use of a 30-Degree Set Square.*

342. A similar use of the second set-square above described is shown in Fig. 196, by which a circle is divided into six equal parts. Place a straight-edge tangent to or near the circle, as shown by A B. Then place the set-square as shown by G B M, bringing the side G B against the center of the circle, drawing the line D L. Then place it as shown by the dotted lines, bringing the side A H against the center, scribing the line F E. Then, by reversing the set-square, placing the side G M against the straight-edge, erect the perpendicular C I, completing the division. A few of the problems to which these principles may be advantageously applied will now be demonstrated.

343. *To Draw an Equilateral Triangle within a Given Circle.*—In Fig. 197, let D be the center of the given circle. Set the side C F of a 30-degree set-square against the T-square, as shown, and move it along until the side E G touches D. Mark the point B. Reverse the set-square so that the point E will come to the right of the side F G. Move the set-square along in the reversed position until the side E G again meets the point D, and mark the point C. Move the T-square upward until it touches the point D, and mark the point A. Then A B and C are points which divide the circle into three equal parts. The triangle may be easily completed from this stage by drawing lines connecting A B, B C and C A, with any straight-edge or rule, but greater accuracy is obtained by the further use of the set-square, as follows: Place the side F G of the set-square against the T-square, as shown in Fig. 198, and move it along until the side E G touches the points A and C, as shown. Draw A C, which will be one side of the required triangle. Set the side E F of the set-square against the T-square, and move it along until the side F G coincides with the points C and B. Then draw C B, which will be the second side of the triangle. Place the side F G of the set-square against the T-square, with the side

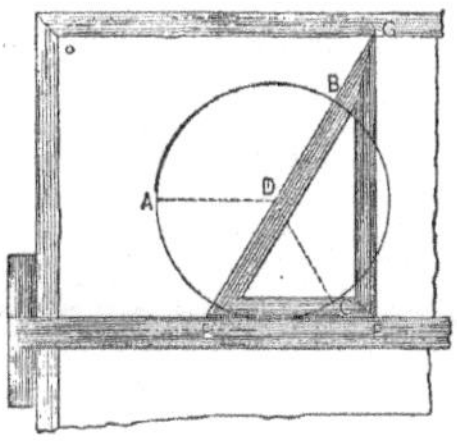

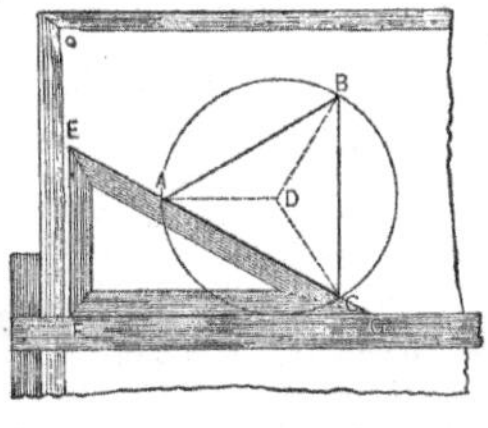

*Figs. 197 and 198.—To Draw an Equilateral Triangle within a Given Circle.*

E F to the right, and move it along until the side E G coincides with the points A and B. Then draw A B, thus completing the figure. The same results may be accomplished by first establishing the point A, by bringing the T-square against the center and using the set-square, as shown in Fig. 198. We present the different methods here given, in order to more clearly illustrate the use of the tools employed.

344. *To Draw a Square within a Given Circle.*—Let D, in Fig. 199, be the center of the given circle. Place the side E F of a 45-degree set-square against the T-square, as shown, and move it along until the side E G meets the point D. Mark the points A and B. Reverse the set-square, and in a similar manner mark the points C and H. The points A, H, B and C are corners of the required square. Move the T-square upward until it coincides with the points A and H and draw A H, as shown in Fig. 200. In like manner draw C B. With the side E F of the set-square against the T-square, move it along until the side G F coincides with the points B and H, and draw B H. In a similar manner draw C A, thus completing the figure.

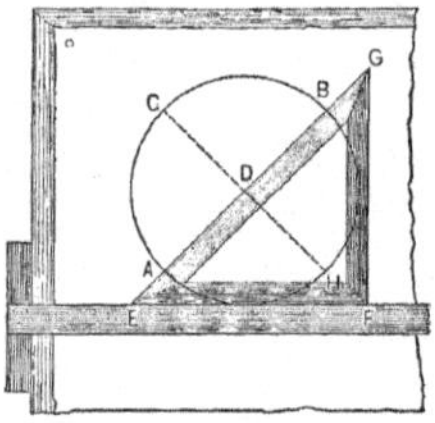


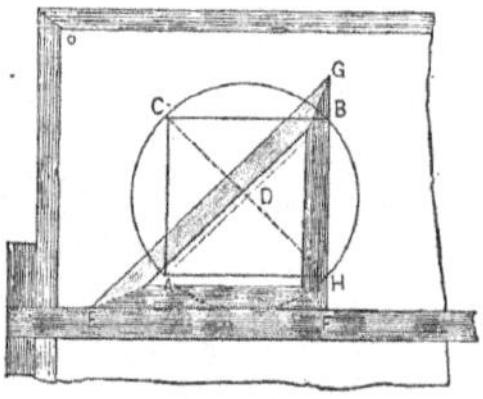


*Figs.* 199 *and* 200.—*To Draw a Square within a Given Circle.*

345. *To Draw a Hexagon within a Given Circle.*—In Fig. 201, let O be the center of the given circle. Place the side E F of a 30-degree set-square against the T-square, as shown. Move the set-square along until the side E G meets the point O. Mark the points A and B. Reverse the set-square, and in like manner mark the points C and D. With the side F G of the set-square against the T-square, move it along until the side E F meets the point O, and mark I and H.

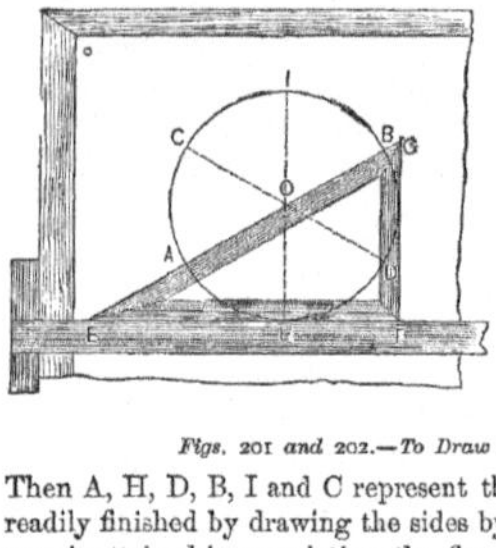


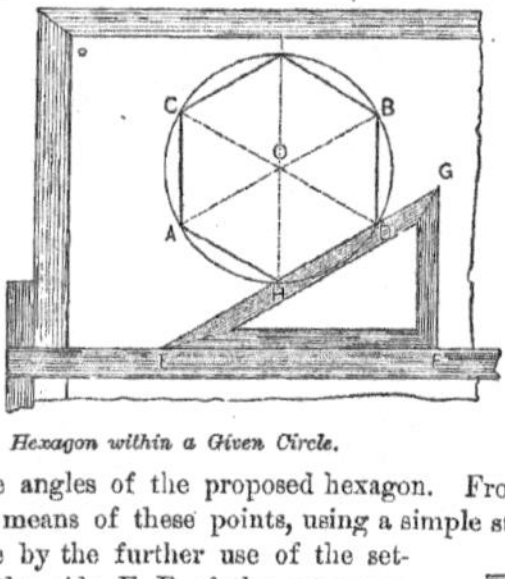


*Figs.* 201 *and* 202.—*To Draw a Hexagon within a Given Circle.*

Then A, H, D, B, I and C represent the angles of the proposed hexagon. From this stage the figure may be readily finished by drawing the sides by means of these points, using a simple straight-edge; but greater accuracy is attained in completing the figure by the further use of the set-square, as shown in Fig. 202. With the side E F of the set-square against the T-square, as shown, draw the line H D, and, by moving the T-square upward, draw the side C I. Reversing the set-square so that the point G is to the left of the point E, draw the side A H, and also, by shifting the T-square, the side I B. With the edge E F of the set-square against the T-square, move it up until the side G F coincides with the points B and D, and draw the side B D. In like manner draw A C, thus completing the figure. In this figure, as with the triangle, the same results may be reached by establishing some point, as H, by means of a diameter drawn at right angles to the T-square, as shown in the engravings, and using it as a base, employing the set-square, as shown in Fig. 202. The combination method we have shown is, however, to be preferred in many instances, on account of its greater accuracy.

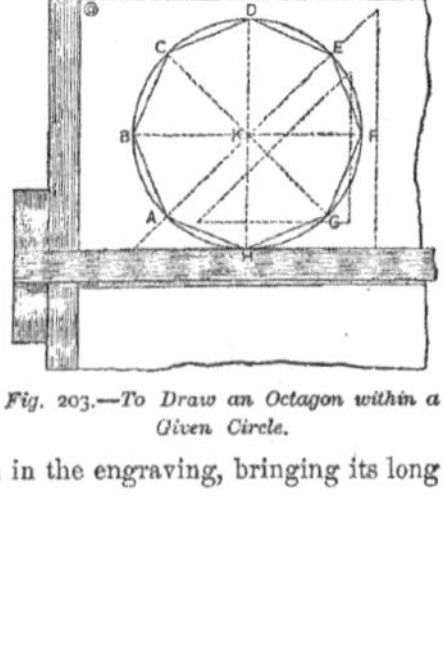


*Fig.* 203.—*To Draw an Octagon within a Given Circle.*

346. *To Draw an Octagon within a Given Circle.*—In Fig. 203, let K be the center of the given circle. Place a 45-degree set-square as shown in the engraving, bringing its long

side in contact with the center, and mark the points E and A. Keeping it in the same position, move it along until its vertical side is in contact with K, and mark the points D and H. Reverse the set-square from the position shown in the engraving, and mark the points C and G. Move the T-square upward until it touches the point K, and mark the points B and F. Then A, H, G, F, E, D, C and B are corners of the octagon. The figure may now be readily completed by drawing the sides, by means of these points, using any rule or straight-edge for the purpose, all as shown by A H, H G, G F, F E, E D, D C, C B and B A.

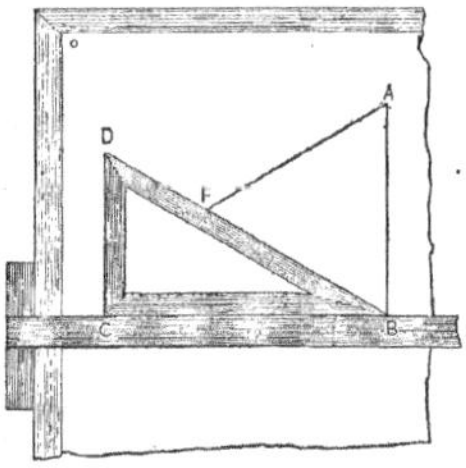


*Fig. 204.—To Draw an Equilateral Triangle upon a Given Side.*

347. *To Draw an Equilateral Triangle upon a Given Side.*—In Fig. 204, let A B be the given side. Set the edge C B of a 30-degree set-square against the T-square, and move it along until the edge B D meets the point B, and draw the line B F. Reverse the set-square, still keeping the side C B against the T-square, and move it along until the side B D meets the point A, and draw the line A F, thus completing the figure.

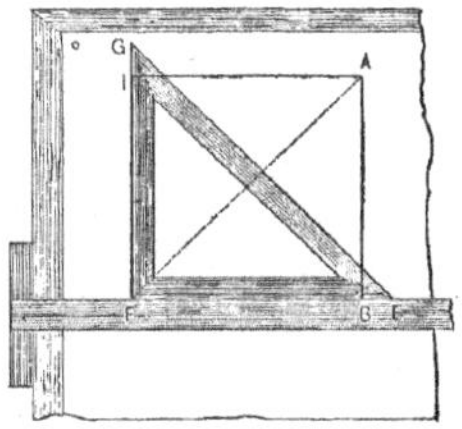


*Fig. 205.—To Draw a Square upon a Given Side.*

348. *To Draw a Square upon a Given Side.*—In Fig. 205, let A B be the given side. Set the edge E F of a 45-degree set-square against the T-square, as shown, and move it along until the side E G meets the point B, and draw B I indefinitely. Reverse the set-square, and bringing the side E G against the point A, draw A F indefinitely. Bring the T-square against the point B and draw B F, producing it until it meets the line A F in the point F. In like manner draw A I, meeting the line B G in the point I. Then with the set-square, placed as shown in the engraving, connect I and F, thus completing the required figure.

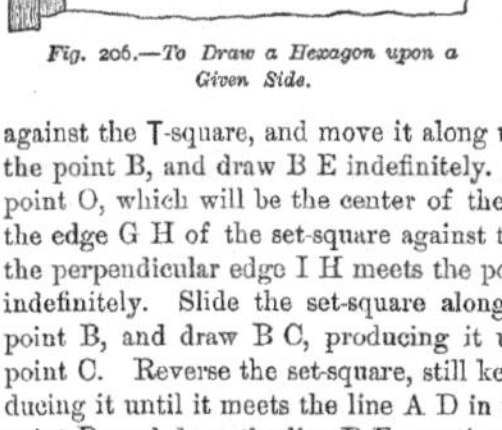


*Fig. 206.—To Draw a Hexagon upon a Given Side.*

349. *To Draw a Hexagon upon a Given Side.*—In Fig. 206, let A B be the given side. Set the edge G H of a 30-degree set-square against the T-square, as shown, and move it along until the edge I G coincides with the point A, and draw the line A D indefinitely. Reverse the set-square, still keeping the edge G H against the T-square, and move it along until the side I G coincides with the point B, and draw B E indefinitely. These lines will intersect in the point O, which will be the center of the required figure. Still keeping the edge G H of the set-square against the T-square, move it along until the perpendicular edge I H meets the point O, and through O draw F C indefinitely. Slide the set-square along until the edge I G meets the point B, and draw B C, producing it until it meets the line F C in the point C. Reverse the set-square, still keeping the edge G H against the T-square, and draw the line C D, producing it until it meets the line A D in the point D. Slide the set-square along until the side I H meets the point D, and draw the line D E, meeting the line B E in the point E. Move the set-square along until the edge I G meets the point E, and draw the line E F, meeting the line C F in the point F. Reverse the set-square and slide it along until the edge F G meets the point F, and draw F A, meeting the given side in the point A, thus completing the required figure.

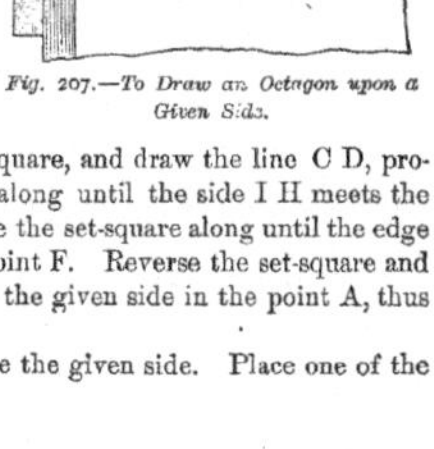


*Fig. 207.—To Draw an Octagon upon a Given Side.*

350. *To Draw an Octagon upon a Given Side.*—In Fig. 207, let C D be the given side. Place one of the

short sides of a 45-degree set-square against the T-square, as shown in the engraving. Move the set-square along until its long side coincides with the point C. Draw the line C B, and make it in length equal to C D. With the T-square draw the line A B, also in length equal to C D. Reverse the set-square, and bring the edge against the point A. Draw A H in length the same as C D. Still keeping a short side of the set-square against the T-square, slide it along until the other short side meets the point H, and draw H G, also of the same length. Then, using the long side of the set-square, draw G F of corresponding length. By means of the T-square draw F E, and by reversing the set-square draw E D, both in length equal to the original side, C D, joining it in the point D, thus completing the required octagon.

351. *To Draw an Equilateral Triangle about a Given Circle.*—In Fig. 208, let O be the center of the given circle. Place the edge E F of a 30-degree set-square against the T-square, as shown, and move it along until the edge F G meets the center O, and mark the point A. Reverse the set-square, still keeping the edge E F against the T-square, and in like manner mark the point B. Move the T-square upward until it meets the point O, and mark the point C. The required figure will be described by drawing lines tangent to the circle at the points A, B and C, which may be done in the manner following, as indicated in Fig. 209. Place the edge E G of the set-square against the T-square, and slide it along until the edge F G touches the circle in the point B. Draw I K indefinitely. Reverse the set-square, keeping the same edge against the T-square, and move it along until its edge F G touches the circle in the point A, and draw I L, intersecting I K in the point I, the other end being indefinite. Then, placing the edge F E of the set-square against the T-square, bring its edge E G against the circle in the point C, and draw L K, intersecting I L in the point L and I K in the point K, thus completing the figure.

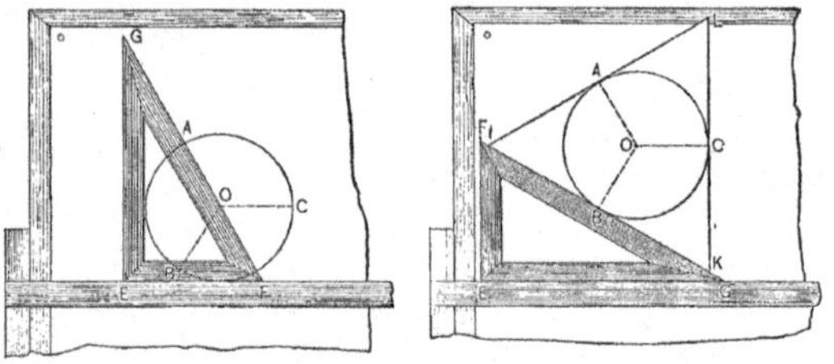


*Figs. 208 and 209.—To Draw an Equilateral Triangle about a Given Circle.*

352. *To Draw a Hexagon about a Given Circle.*—In Fig. 210, let O be the center of the given circle. Place the edge E F of a 30-degree set-square against the T-square, and slide it along until the edge F G meets the point O, and mark the points B and A. Reverse the set-square, still keeping the edge E F against the T square, and in like manner mark the points C and D. Bring the edge of the T-square against O, and mark the points I and K. Then C, A, K, D, B and I are six points in the circumference of the circle, corresponding to the six sides of the required figure. The hexagon is completed by drawing a side tangent to the circle at each of these several points, which may be done by using the set-square as follows, and as shown in Fig. 211: With the edge E G of the set-square against the T-square, bring the edge F G against the circle at the point C, as shown, and draw L M indefinitely. Reverse the set-square, and in like manner bring it against the circle at the point A, and draw M N, cutting L M in the point M, and extending indefinitely in the direction of N. Slide the set-square along until the edge E F meets the

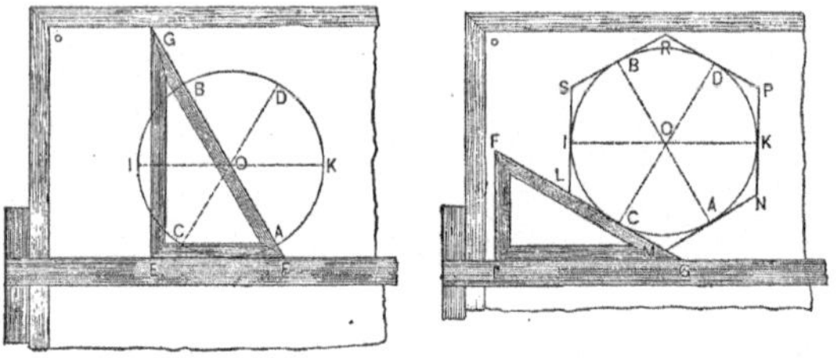


*Figs. 210 and 211.—To Draw a Hexagon about a Given Circle.*

circle in the point K, and draw N P, intersecting M N in the point N, and extending in the direction of P indefinitely. Still keeping the edge E G of the set-square against the T-square, slide it along until the edge F G meets the circle in the point D, and draw R P, cutting N P in the point P, but being indefinite in the direction of R. Reverse the set-square, and in like manner draw R S tangent to the circle in the point B, cutting P R in the point R, and extending in the direction of S indefinitely. Slide the set-square along until its edge E F meets the circle in the point I, and draw S L, cutting R S in the point S and L M in the point L, thus completing the required figure.

353. *To Draw an Octagon about a Given Circle.*—In Fig. 212, let O be the center of the given circle. With the edge E F of a 45-degree set-square against the T-square, as shown, move it along until the side E G meets the point O, and mark the points A and B. Reverse the set-square, and in like manner mark the points C and D. Slide the set-square along until the vertical side G F meets the point O, and mark the points H and I. Move the T-square up until it meets the point O, and mark the points K and L. Then A, I, D, L, B, H, C and K are points in the circumference of the given circle corresponding to the sides of the required figure. The octagon is then to be completed by drawing lines tangent to the circle at these several points, as shown in Fig. 213, which may be done by the use of the set-square, as follows: With the edge E F of the set-square against the T-square, as shown, bring the edge E G against the circle in the point D, and draw M N indefinitely. Sliding the set-square along until the vertical edge F G meets the circle in the point L, draw N P, cutting M N in the point N, and extending in the opposite direction indefinitely. Reverse the set-square, and bringing the edge E G against the circle in the point B, draw P R, cutting N P in the point P, and extending indefinitely in the direction of R. Move the T-square upward until it meets the circle in the point H, and draw the line S R, meeting P R in the point R, and extending indefinitely in the opposite direction. Then, with the set-square placed as shown in the engraving, move it until its edge E G meets the circle in the point C, and draw S T, meeting S R in the point S, and continuing indefinitely in the direction of T. With the set-square in the same position, move it along until its edge G F meets the circle in the point K, and draw T U, cutting S T in the point T, and extending in the opposite direction indefinitely. Reverse the set-square, and bringing its long side against the circle in the point A, draw U V, cutting T U in the point U, and continuing indefinitely in the opposite direction. Bring the T-square against the circle in the point I, and draw V M, connecting U V and M N in the points V and M respectively, thus completing the figure. The above rule will be found very convenient for use, although, as the student may discover, some points are obtained in the first operation not absolutely necessary.

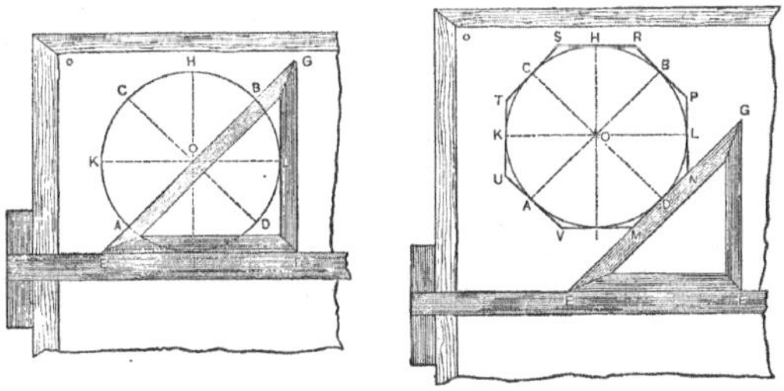


*Figs. 212 and 213.—To Draw an Octagon about a Given Circle.*

354. *To Draw a Square about a Given Circle.*—In Fig. 214, let O be the center of the given circle. Place the blade of the T-square against the point O, and draw the line A O B. With one of the shorter sides, E F, of a 45-degree set-square against the T-square, and with the other short side against the point O, draw the line D O C. Move the T-square upward until it strikes the point C, and draw the line H C I. Move it down until it strikes the point D, and draw the line E D K. With the side E F of the set-square against the T-square, as shown in the engraving, bring the side E G againt the point A, and draw E A H. In like manner bring it against the point B, and draw K B I, thus completing the figure. It is

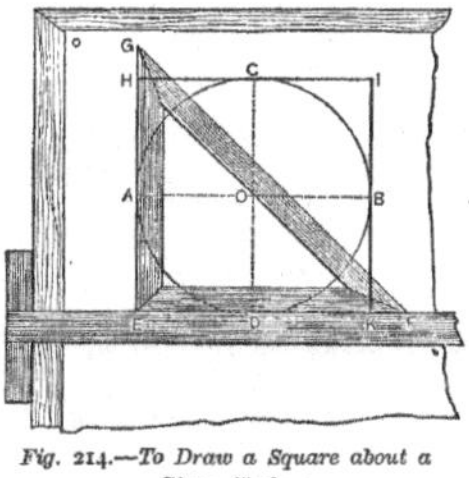


*Fig. 214.—To Draw a Square about a Given Circle.*

to be observed that the several lines composing the sides of the square are tangent to the circle in the points A C B D respectively. The only object served by drawing the diameters A B and C D is that of obtaining greater accuracy, in locating the points just named, than it is possible to secure in drawing the figure around the circle without them.

355. *To Draw a Square upon a Given Side.*—Let A B of Fig. 215 be the given side. Place one of the shorter edges of a 45-degree set-square against the T-square, as placed for drawing the given side, and slide it along until the long edge touches the point A, and draw the diagonal line A C indefinitely. Place the T-square so that its stock comes against the left side of the board, as shown by the dotted lines in the engraving, and, bringing the blade against the point A, draw A D indefinitely. Then bringing the blade against the point B, draw B C, stopping this line at the point of intersection with the line A C, as shown at C. Bring the T-square back to the original position and draw the line C D, thus completing the figure. In the case of a large drawing board, unless the figure is to be located very near one corner of it, and in the case of a drawing board of which the adjacent sides are not at right angles, it will be desirable to use the right angle of the set-square, instead of changing the T-square from one side to the other, as above described. The object of drawing the diagonal line A C is to determine the length of the side C B. This also may be done by the use of the compasses instead of the set-square, as shown by the dotted arc A O C. From B as center, with B A as radius, describe the arc A O C. Place the T-square as shown by the dotted lines, and, bringing it against the point B, draw B C, producing it until it intercepts the arc A O C in the point C. The remaining steps are then to be taken in the manner above described.

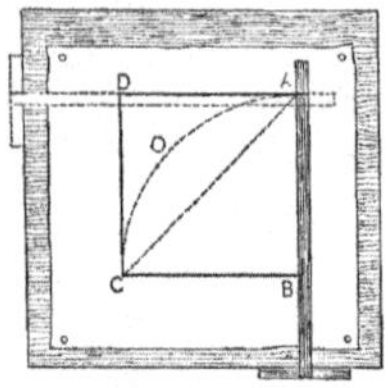

*Fig. 215.—To Draw a Square upon a Given Side.*

III.—BY MEANS OF THE PROTRACTOR.

356. The protractor, which has been already described (Section 244), is an instrument for measuring angles. The most usual form in which this instrument is constructed is that of a semicircle with a graduated edge, the divisions being more or less numerous, according to the size of the article. In instruments of ordinary size the divisions are single degrees, numbered by 5s or by 10s, while in larger sizes the divisions are made to fractions of degrees.

357. Since the protractor by its divisions represents the divisions of the circle, it may be conveniently employed in the construction of polygons. It is especially useful in drawing polygons within given circles, but it may also be employed in drawing polygons about given circles, as well as for constructing them upon given sides. The latter two cases we shall not attempt to illustrate, as they are rules less advantageous for the pattern cutter's use than other methods of doing the same thing elsewhere described in this work. Of the first, namely, constructing polygons within given circles, we shall give a few instances, enough to illustrate the use of the instrument in a manner which will enable the reader to make application in other cases as they may arise.

358. The general plan of using the protractor may be described as measuring from a given point, which represents one angle of the required figure, by means of the degrees marked upon it, to another point, and so on until the circuit of the circle is completed. Thus, in an equilateral triangle, three spaces of 120 degrees are required ($3 \times 120 = 360$), and in a square, four spaces of 90 degrees are required ($4 \times 90 = 360$), while in an octagon, eight spaces of 45 degrees are required ($8 \times 45 = 360$), and so on for other polygons.

359. Since for the purposes of pattern cutting, and perhaps also in some other instances, it is desirable to have one side of the polygon fall either to the right or to the left of the figure and parallel to a vertical line drawn through the center of it, there are some points to be observed in the manner of making application of the simple principles just described, which we will attempt to make plain in the few demonstrations following.

360. *To Draw an Equilateral Triangle within a Given Circle.*—In Fig. 216, let O be the center of the given circle. Through O draw a diameter, as shown by C O D. Place the protractor so that its center point shall coincide with O, and turn it until the point marking 60 degrees falls upon the line C O D. Then mark points in the circumference of the circle corresponding to O and 120 degrees of the protractor, as shown by B and E respectively. Draw the lines C E, E B and B C, thus completing the required figure. The reasons

for these several steps are quite evident. The circle consists of 360 degrees. Then each side of an equilateral triangle must represent one-third of 360 degrees, or 120 degrees. We assume the point C for one of the angles, and draw the line C O D. Then, by the nature of the figure to be drawn, D must fall opposite the center of one side. Therefore, since 60 is the half of 120 (the length of one side in degrees) we place 60 opposite the point D, and mark 0 and 120 for the other angles. We then complete the figure by drawing the lines as shown. Since in many cases the protractor is much smaller than the circle in which the figure is to be constructed, it becomes necessary to mark the points at the edge of the instrument, and carry them to the circumference by drawing lines from the center of the circle through the points, producing them until the circle is reached.

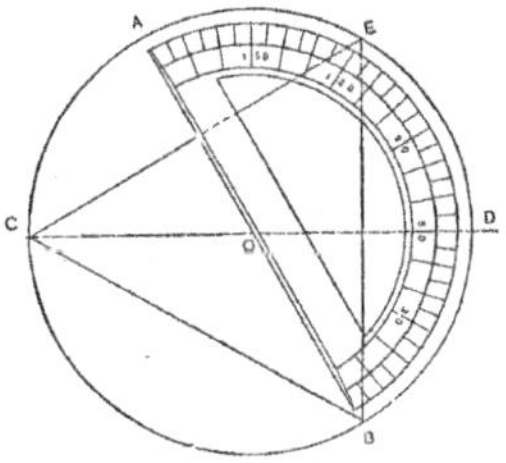

*Fig. 216.—To Draw an Equilateral Triangle within a Given Circle.*

361. *To Draw a Square within a Given Circle.*—In Fig. 217, let O be the center of the given circle. Through O draw a diameter, as shown by C O D. Place the protractor so that its center point coincides with O, and turn it until the point marking 45 degrees falls upon the line C O D. Mark points in the circumference of the circle corresponding to 0, 90 and 180 degrees of the protractor, as shown by F, G and E respectively. From G, through the center O, draw G O H, cutting the circumference of the circle in the point H. Then E, G, F and H are the angles of the required figure, which is to be completed by drawing the sides E G, G F, F H and H E. Since the circle is composed of 360 degrees, one side of an inscribed square must represent one-fourth part of 360 degrees, or 90 degrees. The half of 90 degrees is 45 degrees. Hence, in setting the protractor we placed the point representing 45 degrees opposite the point in which we desired the center of one of the sides to fall, or, in other words, upon the line C O D. Then, having marked points 90 degrees removed from each other, or, as explained above, opposite the points 0, 90 and 180 of the protractor, as shown by F, G and E, the fourth point was obtained by the diagonal line. It is evident that H must fall opposite G, upon a line drawn through the center. Or we might have accomplished the same by moving the protractor around, and by means of it measured a space of 90 degrees from either F or E, which, as will be clearly seen, would have given the same point, H.

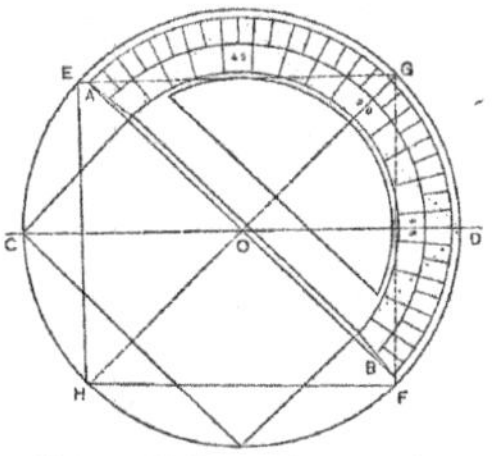

*Fig. 217.—To Draw a Square within a Given Circle.*

362. *To Draw an Octagon within a Given Circle.*—Through the center O of the given circle, Fig. 218, draw a diameter, A O B, upon which the center of one side is required to fall. Place the protractor so that its center point shall coincide with the center O, and turn it so that the point representing 22½ degrees shall fall on the line A O B. Then mark points in the circumference of the circle corresponding to 0, 45, 90, 135 and 180 degrees of the protractor, as shown by E, G, H, I and F. Reverse the protractor, and in like manner mark the points M, L and K; or these points may be obtained by drawing lines from I, H and G respectively through the center O, cutting the circumference in M, L and K. The figure is to be completed by drawing the sides F I, I H, H G, G E, E M, M L, L K and K F. Since the circle consists of 360 degrees, an octagon must represent 45 degrees, or one-eighth of 360, in each of its sides. The half of 45 is 22½. Hence, we placed the point of the protractor representing 22½ degrees upon the line A O B, which represents the center of one side of the required figure. Having thus established the position of one side, the other sides of the figure are located by marking points in the circumference of the circle opposite points in the protractor at regular intervals of 45 degrees.

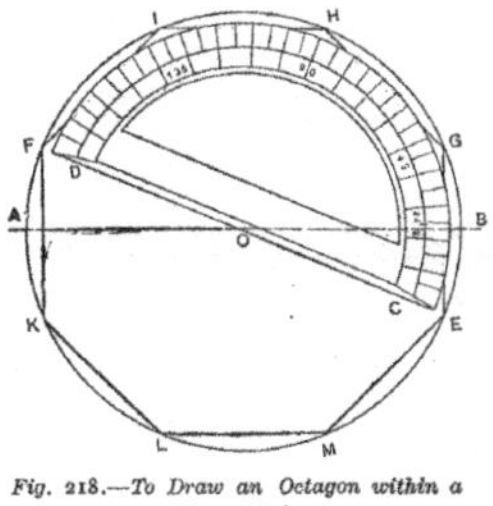

*Fig. 218.—To Draw an Octagon within a Given Circle.*

363. *To Draw a Dodecagon within a Given Circle.*—In Fig. 219, let O be the center of the given circle. Through O draw the Diameter A O B, at right angles to which one of the sides of the polygon is required to be. Set the protractor so that the center point of it coincides with the center O, and revolve it until the point marking 15 degrees falls upon the line A O B. With the protractor in this position, mark points in the circumference of the circle opposite the points in the protractor representing 0, 30, 60, 90, 120, 150 and 180 degrees, as shown by E, F, G, H, I, K and L. Then these points will represent angles of the required polygon. The remaining angles may be obtained by placing the protractor in like position in the opposite half of the semicircle, or they may be determined by drawing lines from the points F, G, H, I and K through the center O, producing them until they cut the circumference in the points M, N, P, R and S, which are the remaining angles. The figure is now to be completed by drawing the sides, as shown. In a dodecagon, or twelve-sided figure, each side must occupy a space represented by one-twelfth of 360 degrees, or 30 degrees of the protractor. As the side F E was required to be located in equal parts upon opposite sides of A O B, we placed the middle of one division of the protractor representing a side (that is, 15 degrees, or one-half of 30 degrees) upon the line A O B. Having thus established the position of one side, the others are measured off in the manner above described.

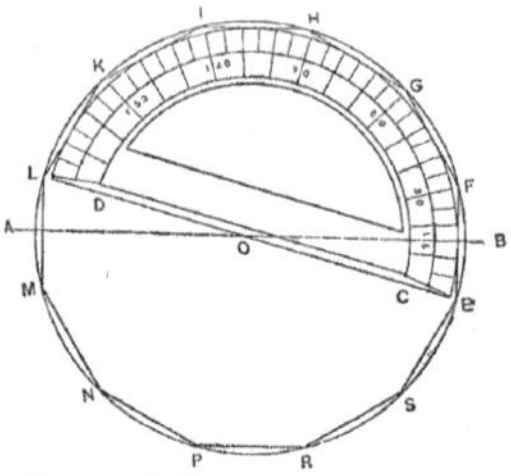

*Fig. 219.—To Draw a Dodecagon within a Given Circle.*

IV.—BY THE USE OF THE CARPENTER'S SQUARE.

364. All of the regular polygons may be constructed by the use of a carpenter's square, and the employment of this tool for the purpose is frequently of great advantage to the pattern cutter. We shall not attempt to give rules for all of the polygons which occur in regular work, but shall limit our remarks, presenting only so much as is necessary to illustrate the principles upon which the use of this tool depends. We append a table showing the figures upon the square to be used for some of the other polygons than those we describe in full, thus enabling any one who is so disposed to experiment further than here illustrated.

365. *To Construct an Equilateral Triangle, the Length of a Side being Given.*—In Fig. 220, let A B in the straight line D C be the length of the given side. With 12 of the blade placed against the line D C, and with 7 of the tongue brought against the point B, draw the line B E indefinitely. Reverse the square, as shown by the dotted lines, maintaining the same points, but bringing 7 of the tongue against the point A, and draw A E, which produce until it cuts the line B E, previously drawn in the point E. Then A E B will be the required equilateral triangle.

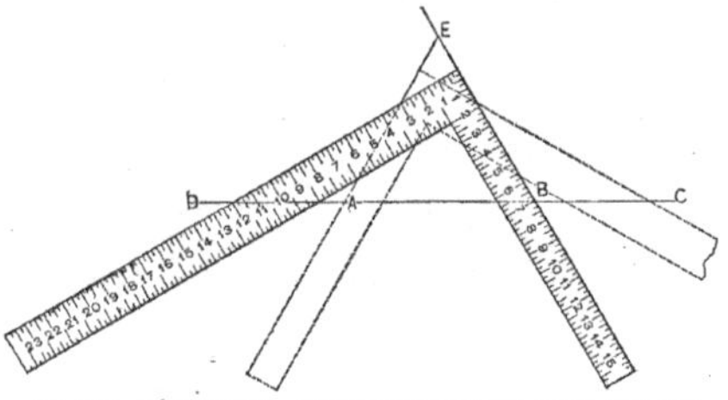

*Fig. 220.—To Construct an Equilateral Triangle, the Length of a Side being Given.*

366. *To Construct a Hexagon, the Length of a Side being Given.*—In Fig. 221, let B E in the line G H be the length of the given side. Take 12 on the blade of the square and 7 of the tongue, and placing the latter against the point D, bring the former to the line G H, as shown in the engraving. Then draw the line D C, making it in length equal to the given side. Next place the square, as shown by the dotted lines, with 12 of the blade against the line G H and 7 of the tongue against the point E, and draw E F, which also make equal to the given side. Continue in this way until the several sides of the figure are drawn. In pattern cutting the mechanic more frequently requires the joint line than the outline of the figure itself. The use of the square affords him a ready means of obtaining this, without the tedious process of first laying off the polygon. In the case

of the figure we have just described, since a hexagon is composed of six equilateral triangles, it follows that what we have shown in Fig. 220 is all that is necessary when the miter joints in this shape are required. We will, however, for the sake of better illustrating this principle, introduce an additional engraving, showing a different mode of constructing a hexagon from that just described.

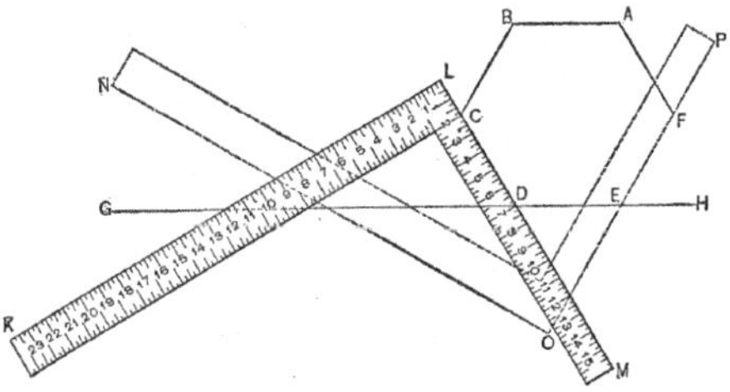


*Fig. 221.—To Construct a Hexagon, the Length of a Side being Given.*

367. *To Construct a Hexagon by Means of Six Equilateral Triangles, the Length of a Side being Given.*—In Fig. 222, let C D in the line A B be the length of the given side. Place 7 of the tongue and 12 of the blade against the line A B, as shown, making the latter point fall upon C, and draw C N indefinitely. Next place the square, as shown by the dotted lines, with 12 of the tongue and 7 of the blade against the line A B, the latter point falling upon D. Draw D R indefinitely, cutting the line C N, previously drawn, in the point E. Then E is the center of the circle which, if drawn, will circumscribe the hexagon. From E as center, with E C as radius, draw the circle C F H K G D. Take the length C D in the dividers, and step off the circle for the other points. It is evident, upon inspection, that by producing the line C N and D R until they cut opposite sides of the circle, the points H and K will be obtained, thus making it necessary to determine only the points F and G by means of the dividers. From what has preceded, it is also evident that it makes no difference upon which arm of the square the longer dimension is taken. The principle involved is simply that of a right-angled triangle and its hypothenuse. Other lengths than those we have described may be employed for the purposes indicated, it being necessary simply to maintain like proportions. In the above problems we have used 12 on one arm of the square, suiting the length on the other to it. In the table given below we have also pursued the same plan. We advise the use of 12 as one of the dimensions, because it is easily kept in mind, and therefore somewhat simplifies the rules.

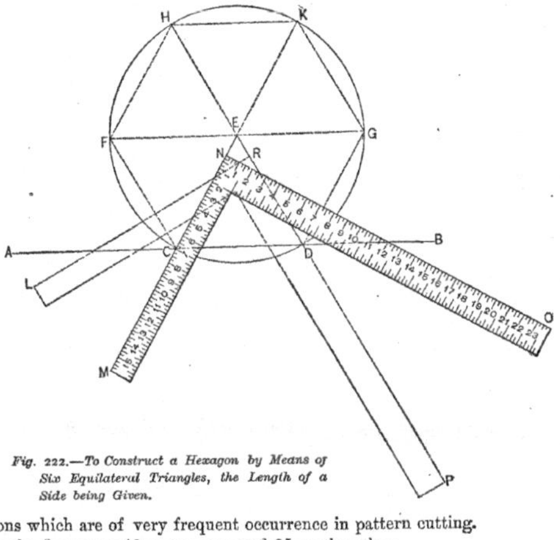


*Fig. 222.—To Construct a Hexagon by Means of Six Equilateral Triangles, the Length of a Side being Given.*

368. The following table shows the divisions upon the square to be used for constructing some of the polygons which are of very frequent occurrence in pattern cutting.

Five sides, pentagon, for the figure use 12 on one arm and $3\frac{7}{8}$ on the other.
Seven " heptagon, " " " 12 " " $9\frac{9}{16}$ " "
Eight " octagon, " " " 12 " " 12 " "
Five " pentagon, for the joint line use 12 on one arm and $8\frac{3}{4}$ on the other.
Seven " heptagon, " " " 12 " " $5\frac{3}{4}$ " "
Eight " octagon, " " " 12 " " 5 " ".

Mathematical accuracy is not claimed for these rules, although they approach the correct result so closely that with ordinary measuring appliances the difference can scarcely be detected. They are sufficiently accurate for all the purposes in connection with sheet-metal pattern cutting.

369. *Adjusting the Drawing of a Polygon to Suit the Requirements of Miter Cutting.*—By rules commonly employed for drawing polygons, the figures are frequently so turned as to prevent the use of a T-square from the sides of the board for dropping points, and drawing the stretchout and measuring lines. The plans are produced as shown in Fig. 223, while for convenience in pattern cutting they should be as shown in Figs. 224 and 225.

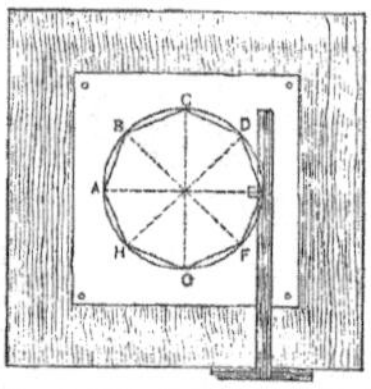

*Fig. 223.—The Polygon in Position, as Drawn by Some of the Rules.*

370. There are two ways of overcoming this difficulty. One is by redrawing the figure, and the other by shifting the paper. The former, while it involves considerably more work than the latter, is more frequently employed than it, because of other drawings or lines upon the same sheet, as, for instance, the elevation or profile, which would not be in correct position for use after the paper was shifted.

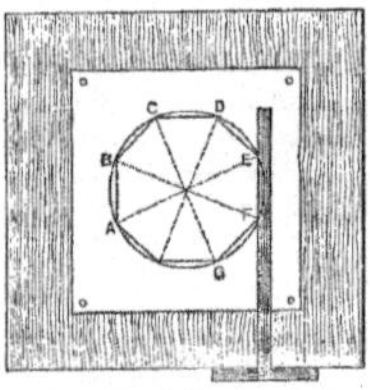

*Fig. 224.—Putting the Plan in Correct Position by Redrawing.*

371. To redraw the figure, proceed as follows: Take the length of one side in the dividers, as E F, Fig. 223, and bringing the T-square across the edge of the circle, as shown in Fig. 224, move it until it cuts the circumference in a chord, the length of which is equal to E F. Draw the line E F, which will be one side of the required figure in correct position. Then step off the circle in the usual manner for the other sides. For miter cutting, the side E F and the two radii are alone sufficient, as will be explained further on.

372. It may be observed in this connection that when a polygonal plan is being drawn for the purpose of miter cutting, and which it is known will not be in the proper position when finished, it is not necessary to proceed with the figure further than to obtain the length of one side and the radius of the circumscribing circle, before making the adjustment by means of the T-square, as illustrated in Fig. 224.

373. To bring the figure into proper position by shifting the paper, which is illustrated in Fig. 225, proceed as follows: Place the T-square in position against one side of the board. Then bring one side of the polygon against the edge of the blade of the T-square, as shown by D E in the engraving. Carefully hold the paper in this position while fastening it at the corners with thumb tacks.

*Fig. 225.—Shifting the Position by Moving the Sheet of Paper.*

## THE ELLIPSE.

374. Perhaps we cannot do better, in explaining this figure, which in one form or another is so common in the pattern cutter's work, than to put our remarks in the shape of a familiar talk about it, giving illustrations first of the definitions of an ellipse, and following with several of the methods in common use for drawing it.

375. A definition of the ellipse very frequently encountered is, "a figure bounded by a regular curve, generated from two points, called foci." The idea presented to the mind by this definition immediately contrasts the ellipse with the circle. Both are figures bounded by regular curves, but while the ellipse is generated from two points, the circle is generated from only one. To carry this comparison a step further, in order to make the properties of the ellipse more apparent than perhaps we can do in any other way, let us consider for a moment how a circle is drawn by the use of a string and pencil, and then we will see how an ellipse may be drawn by the same means.

376. To draw a circle by a string and pencil, we first determine where we want the center of the figure, and then, fastening one end of the string at that place, we attach the pencil to the string at a point just as far

removed from the center as one-half of the diameter of the circle we propose to draw. After the pencil and string are thus arranged, we move the pencil around the center, keeping the string straight all the time. Or, to state it in a little different form, if we desire to draw a circle twelve inches in diameter, we tie the pencil to the string six inches removed from the center, and then, while keeping the string taut, move the pencil. The resulting line will be a circle.

377. In Fig. 226 is shown the method of drawing an ellipse with string and pencil. By examination of the engraving it will be seen that the string (represented by the dotted lines) is controlled by the two points F and G, which, as already stated, are called foci. To arrange these points, and to adjust the string so as to produce a figure of specified dimensions, constitutes the art of drawing an ellipse with string and pencil. In drawing a circle by the plan described, there being but one point and but one dimension, the calculations required in getting the position of the pencil with relation to the center are very simple. In drawing an ellipse by the same general method, there being two points which, by means of the string, control the pencil, and two dimensions to the figure to be produced, the calculations are a little more complex. Having thus indicated some of the points of similarity and contrast between the circle and the ellipse, we think the following rule for drawing an ellipse with string and pencil will be readily comprehended.

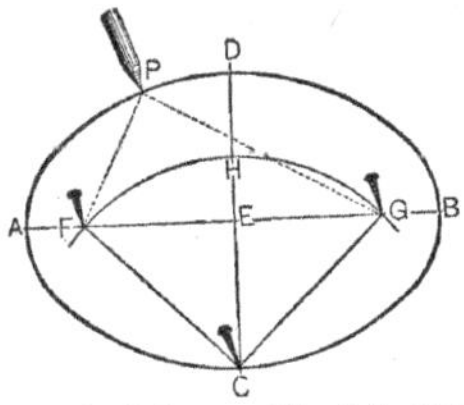

*Fig. 226.—To Draw an Ellipse to Specified Dimensions with a String and Pencil.*

378. *To Draw an Ellipse to Specified Dimensions with a String and Pencil.*—In Fig. 226, let it be required to draw an ellipse, the length of which shall be equal to the line A B, and the width of which shall be equal to the line D C. Lay off A B and D C at right angles to each other, their middle points intersecting, as shown at E. With the compasses set to one-half the length of the required figure, as A E, and from either D or C as center, strike an arc, cutting A B in the points F and G. These points, F and G, then are the two foci, into which drive pins, as shown. Drive a third pin at C. Then pass the string around the three points F, G and C and tie it. Remove the pin C and substitute the pencil, as shown by P.

379. Another definition of the ellipse, which Fig. 226 also illustrates, and which we call attention to at this time because it explains the reason for some of the steps we have just described, is, "a figure bounded by a regular curve, from any point in which, if straight lines be drawn to two fixed points, their sum will always be the same." A moment's examination of the engraving will demonstrate this. If we take the dividers and set off on a straight line the lengths from F and G to the several points in the boundary of the figure which can be conveniently measured, we shall find their sums equal. For example, the sums of P F and P G, A F and A G, C F and C G, B F and B G, are all the same.

380. Now, without stopping to demonstrate it, we will simply call attention to a fact, which is quite evident upon inspection of the engraving, and which can be readily proven by the use of the dividers. The sum of the distance from any point in the boundary to the two foci is equal to the length of the figure. In other words, the sum of P F and P G, or C F and C G, is equal to the length A B. By inspection of the figure it is evident that each of the two foci must be equally distant from the extreme point in the side of the figure, as, for instance, C. Therefore the distance from C to F and from C to G must each be equal to half of the length of the figure. Hence, in order to obtain the position of F and G, we take one-half of A B in the dividers, and, with C as center, cut A B by the arc in these points.

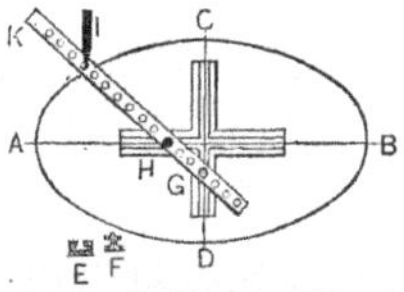

*Fig. 227.—To Draw an Ellipse to Given Dimensions by Means of a Trammel.*

381. An ellipse is sometimes described as "a figure bounded by a regular curve, generated from a moving center." This definition necessarily implies that the movement of the center and of the point or pencil which describes the curve must be entirely in harmony with each other. The most convenient illustration of this definition which can be given, is a description of the use of a trammel for drawing an ellipse to given dimensions.

382. In Fig. 227 we show a trammel as commonly constructed. E is a section through the arms, showing the groove in which the head of the bolt F moves. H and G are bolts and pins by which the movement is controlled and regulated. In the engraving the bar K is shown with holes at fixed distances, through which

the governing pins are passed. An improvement upon this plan of construction consists of such a device in connection with the pins as will clamp them firmly to the bar at any point, thus providing for an adjustment of the most minute variations.

383. Referring now to the definition of the ellipse before given, H may be regarded as the moving center from which the curve bounding the figure is generated. Its motion is lengthwise of the figure, or, in other words, from A toward B while describing the upper part of the curve, and the reverse while describing the lower part. G is simply the regulator or governor by which harmony of movement is maintained between the center H and the pencil I. We will now give the rule for drawing an ellipse with a trammel.

384. *To Draw an Ellipse to Given Dimensions by Means of a Trammel.*—In Fig. 227, let it be required to describe an ellipse, the length of which shall be equal to A B and the breadth of which shall be C D. Draw A B and C D at right angles, intersecting at their middle points. Place the trammel as shown in the engraving, so that the center of the arms shall come directly over the lines. First place the rod along the line A B, so that the pencil or point I shall coincide with either A or B. Then place the pin G directly over the intersection of A B and C D. Next place the rod along the line C D, bringing the pencil or point I to either C or D, and put the pin H over the intersection of A B and C D. The instrument is then ready for use, and the curve is described by the pencil I moved by the hand, but controlled by the pins working in the grooves.

385. *An Improvised Trammel.*—It frequently happens that an ellipse is wanted of specified dimensions, under conditions which make the use of a trammel desirable. When a trammel is not convenient, a very fair substitute is afforded by the use of a common steel square and a thin strip of wood, like a lath. This method of drawing an ellipse is also quite useful under ordinary circumstances when only a part of the figure is required for use, as in the shape of the top of a window frame to which a cap is to be fitted, in which half of the figure would be employed, or in the shaping of a member of a molding in which a quarter, or less than quarter, of the figure would be used. In presenting the rule, we show how to produce the complete figure, but the application of it to the other purposes cited is so self-evident that no difficulty can arise which would require special explanation.

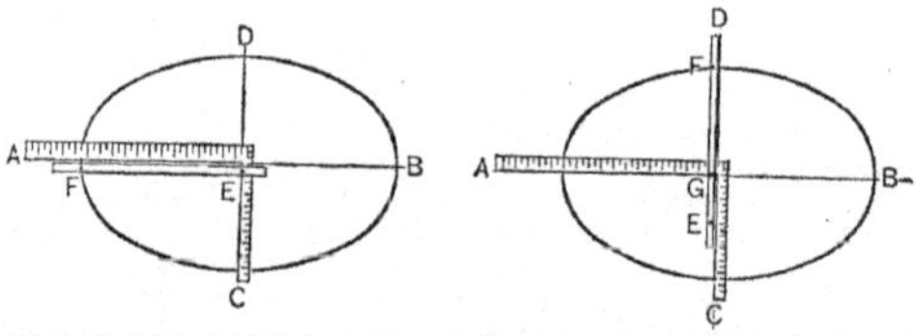


*Figs. 228 and 229.—To Draw an Ellipse of Given Dimensions by Means of a Square and a Strip of Wood.*

386. *To Draw an Ellipse of Given Dimensions by Means of a Square and a Strip of Wood.*—In Fig. 228, set off the length of the figure, and at right angles to it, through its middle point, draw a line representing the width of the figure. Place a square as shown by A E C, its inner edge corresponding to the lines. Lay the strip of wood as shown by F E, putting a pencil at the point F, corresponding to one end of the figure, and a pin at E, corresponding to the inner angle of the square. Then place the stick across the figure, as shown in Fig. 229, making the pencil, F, correspond with one side of the figure, and put a pin at G, corresponding with the inner angle of the square. In drawing the figure the square must be changed in position for each quarter of the curve. As shown in the engravings, it is correct for the quarter of the curve represented by F D. It must be changed for each of the other sections, its inner edge being brought against the lines each time, as shown.

387. Still another definition of an ellipse is that "it is a figure bounded by a regular curve, which corresponds to an oblique projection of a circle."

388. An oblique projection of a circle, perhaps, will be most readily understood if explained by referring to a cylinder, as a piece of stove pipe, for example. If the piece of pipe is cut square across and the end placed upon a board, and we scribe a line around it, the resulting figure will be a circle. If we now cut the pipe obliquely, as, for example, to make a square elbow, or any elbow for that matter—for the angle of the oblique cut does not affect the principle at all, it only modifies the proportions of the figure—and we place the end thus cut upon a board and scribe around it, as mentioned in the first case, the figure drawn will be an ellipse. We have thus, by rough mechanical means, produced what is technically known as an oblique projection of a circle, and which by our definition is the figure to which an ellipse corresponds. What we have here done

mechanically may be also accomplished upon the drawing board in a very simple and expeditious manner. The demonstration which follows is of especial interest to the pattern cutter, because the principles involved in it lie at the root of many practical operations which he is called upon to perform. For example, the shape to cut a piece to stop up the end of a pipe or tube which is not cut square across, and the shape to cut the hole in a piece which is to fit around a pipe passing through it at other than a right angle, like a flange to fit a pipe passing through the slope of a roof and other similar requirements of almost daily occurrence, depend entirely upon the principles which we shall here explain. With reference to such problems, an ellipse may be defined as an oblique section of a cylinder, the method of drawing the shape of which is given below.

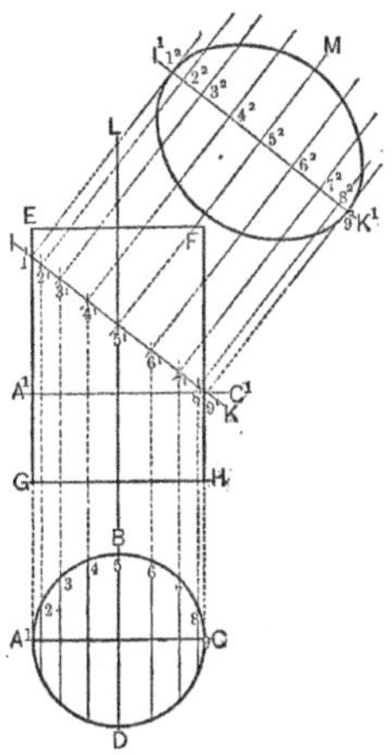

*Fig. 230.—To Describe the Form or Shape of an Oblique Section of a Cylinder, or to Draw an Ellipse as the Oblique Projection of a Circle.*

389. *To Describe the Form or Shape of an Oblique Section of a Cylinder, or to Draw an Ellipse as the Oblique Projection of a Circle.*—The two propositions which are stated above are virtually one and the same so far as concerns the pattern cutter, and they may be made quite the same so far as a demonstration is concerned. We confine our explanation of the engraving to the idea of the cylinder, believing it in that shape to be of more practical service to the readers of this book than in any other. In Fig. 230, let G E F H represent any cylinder, and A B C D the plan of the same. Let I K represent the line of any oblique cut to be made in the cylinder. It is required to draw the shape of the pipe as it would appear when cut in two by the line I K, and either piece placed with the end I K flat upon paper and a line scribed around it. Divide one-half of the plan A B C into any convenient number of equal parts, as shown by the figures 1, 2, 3, 4, etc. Through these points and at right angles to the diameter A C, draw lines as shown, cutting the opposite side of the circle. Also continue these lines upward until they cut the oblique line I K, as shown by 1', 2', 3', etc. In order to avoid confusion of lines, draw a duplicate of I K to one side, as I' K', making it parallel to I K for convenience in transferring spaces. With the T-square set at right angles to I K, and brought successively against the points in it, draw lines through I' K', as shown by $1^2$, $2^2$, $3^2$, etc. With the dividers take the distance across the plan A B C D on each of the several lines drawn through it, and set the same distance off on corresponding lines drawn through I' K'. In other words, taking A C as the base for measurement in the one case and I' K' the base of measurement in the other, set off on the latter, on each side, the same length as the several lines measure on each side of A C. Make $2^2$ equal to 2, and $3^2$ equal to 3, and so on. Through the points thus obtained, trace a line, as shown by I' M K' and the opposite side, thus completing the figure.

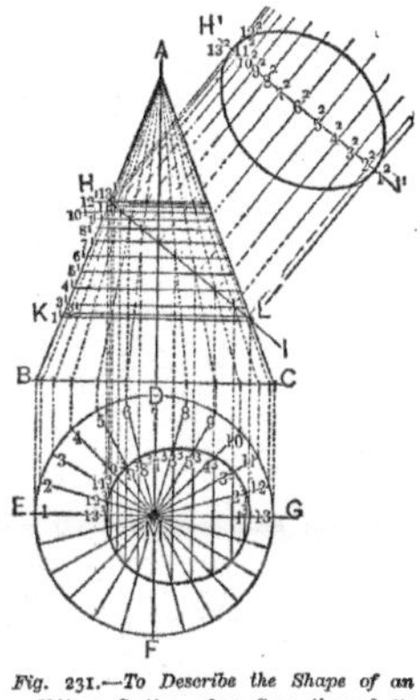

*Fig. 231.—To Describe the Shape of an Oblique Section of a Cone through its Opposite Sides, or to Draw an Ellipse as a Section of a Cone.*

390. Another definition of the ellipse is that "it is a figure bounded by a regular curve, corresponding to an oblique section of a cone through its opposite sides." It is this definition of the ellipse that classes it among what are known as conic sections. It is generally a matter of surprise to students to find that an oblique section of a cylinder, and an oblique section of a cone through its opposite sides, produce the same figure, but such is the case. The method of drawing an ellipse upon this definition of it is given in the following demonstration. The principles upon which this rule is based, no less than those referred to in the last demonstration, are of especial interest to the pattern cutter, because so many of the shapes with which he has to deal owe their origin to the cone.

391. *To Describe the Shape of an Oblique Section of a Cone through its Opposite Sides, or to Draw an Ellipse as a Section of a Cone.*—In Fig. 231, let B A C represent a cone, of which E D G F is the plan at the base. Let H I represent any oblique cut through its opposite sides. Then it is required to draw the shape of the section represented by H I, which will be an ellipse. In order to avoid confusion of lines, at any convenient place outside of the figure draw a duplicate of H I parallel to it, upon which to construct the figure sought, as H' I'. Divide one-half of the plan, as E D G, into any convenient number of equal parts, as shown by 1, 2, 3, 4, etc. From the center of the plan M draw radial lines to these points. From each of the points also erect a perpendicular line, which produce until it cuts the base line B C of the cone. From the base line of the cone continue each of these lines toward the apex A, cutting the oblique line H I. Through the points thus obtained in H I, and at right angles to the axis A D of the cone, draw lines, as shown by 1', 2', 3', 4', etc., cutting the opposite sides of the cone. From the same points in H I drop lines vertically across the plan, as shown by 1', 2', 3', 4', etc., and also from the same points in H I, at right angles to it, draw lines cutting H' I', as shown by 1', 2', 3' 4', etc., thus transferring to it the same divisions as have been given to other parts of the figure. After having obtained these several sets of lines in different portions of the figure, all of which correspond with each other, the first step is to obtain a plan view of the oblique cut, for which we proceed as follows: With the dividers take the distance from the axial line A D to one side of the cone, either A B or A C, on each of the lines 1', 2', 3', 4', etc., and set off like distance from the center of the plan M on the corresponding radial lines 1, 2, 3, 4, etc. A line traced through the points thus obtained will give a plan view of the oblique cut, as shown by the inner line in the plan. Having thus obtained the shape of the oblique cut in plan, and having previously drawn lines across the plan representing the divisions in H I, the next step is to set off the width of the plan at the several points represented by these cross lines upon the lines drawn through H' I'. With E G as a basis of measurement, with the dividers take the distance on each of the several cross lines 2', 3', 4', 5', etc., from E G to one side of the plan of the oblique cut just described, and set off the same distance on each side of H' I' on the corresponding lines. A line traced through the points thus obtained will be an ellipse.

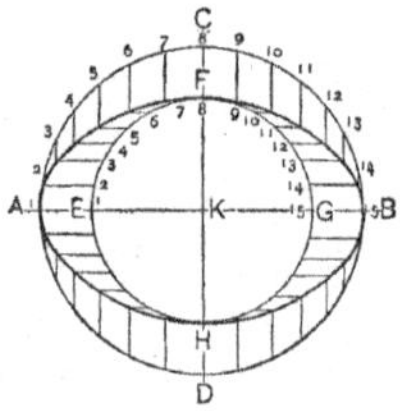


*Fig. 232.—To Construct an Ellipse to Given Dimensions by the Use of Two Circles and Intersecting Lines.*

392. *To Construct an Ellipse to Given Dimensions by the Use of Two Circles and Intersecting Lines.*—In Fig. 232, let it be required to construct an ellipse, the length of which shall equal A B and the width of which shall equal H F. Draw A B and H F at right angles, intersecting at their middle points, K. From K as center, and with one-half of the length A B as radius, describe the circle A C B D. From K as center, and with one-half of the width H F as radius, describe the circle E F G H. Divide the larger circle into any convenient number of equal parts, as shown by the small figures 1, 2, 3, 4, etc. Divide the smaller circle into the same number of equal and corresponding parts, as also shown by figures. By means of the T-square, from the points in the outer circle draw vertical lines, and from points in the inner circle draw horizontal lines, as shown, producing them until they intersect the lines first drawn. A line traced through these points of intersection will be an ellipse.

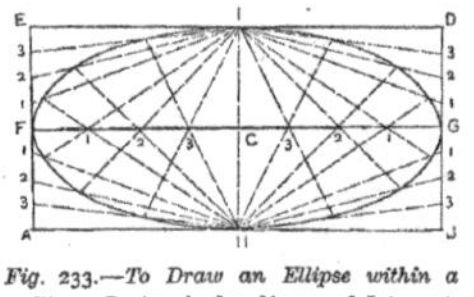


*Fig. 233.—To Draw an Ellipse within a Given Rectangle by Means of Intersecting Lines.*

393. *To Draw an Ellipse within a Given Rectangle by Means of Intersecting Lines.*—In Fig. 233, let E D B A be any rectangle within which it is required to construct an ellipse. Bisect the end A E, obtaining the point F, from which erect the perpendicular F G, dividing the rectangle horizontally into two equal portions. Bisect the side A B, obtaining the point H, and draw the perpendicular H I, dividing the rectangle vertically into two equal portions. The lines F G and H I are then the axes of the ellipse. F G represents what may be familiarly termed the length of the figure, and H I what may be called the breadth of the figure. Divide the spaces F E, F A, G D and G B into any convenient number of equal parts, as shown by the figures 1, 2, 3. From these points in F E and G D draw lines to I, and from the points in F A and G B draw lines to the point H. Divide F C and G C also into the same number of equal parts, as shown by the figures, and through each of these points draw lines to both I and H, as indicated. A line traced through the several points of intersection between the two sets of lines, as shown in the engraving, will be an ellipse.

394. *To Draw an Approximate Ellipse in a Given Rectangle by Means of Intersecting Lines.*—In Fig. 234, let F G H E be any rectangle, within which it is required to draw a figure which shall approximate an ellipse in shape, and which shall give the largest surface within the boundary of the figure consistent with easy curves. Divide the rectangle into four equal portions by the lines A B and C D, as shown. Divide each half of each end into any convenient number of equal parts, as shown by the figures. Divide each half of each side into the same number of equal parts. Then draw the intersecting lines, as shown. Commencing at D, connect 0 with 9, 1 with 8, 2 with 7, 3 with 6, 4 with 5, and so on. A line traced through the several points of intersection will be the figure sought.

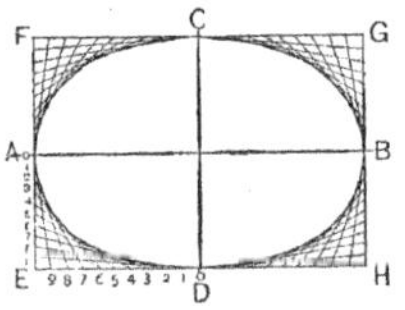


*Fig. 234.—To Draw an Approximate Ellipse in a Given Rectangle by Means of Intersecting Lines.*

395. *To Draw an Elliptical Figure with the Compasses, the Length only being Given.*—In Fig. 235, let A C be any length to which it is desired to draw an elliptical figure. Divide A C into four equal parts. From 3 as center, and with 3 1 as radius, strike the arc B 1 D, and from 1 as center, and with the same radius, strike the arc B 3 D, intersecting the arc first struck in the points B and D. From B, through the points 1 and 3, draw the lines B E and B F indefinitely, and from D, in like manner, draw the lines D G and D H. From the point 1 as center, and with 1 A as radius, strike the arc E G, and from 3 as center, with the same radius, or, what is equivalent, with 3 C as radius, strike the arc H F. From D as center, with radius D G, strike the arc G H, and from B as center, with the same radius, or, what is equivalent, with B A as radius, strike the arc E F, thus completing the figure.

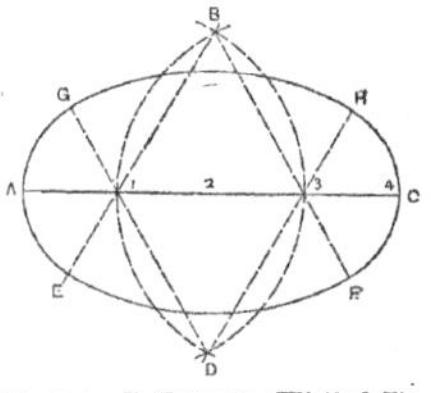


*Fig. 235.—To Draw an Elliptical Figure with the Compasses, the Length only being Given.*

396. A figure of different proportions may be drawn in the same general manner as follows: Divide the length A C into four equal parts, as indicated in Fig. 236. From 2 as center, and with 2 1 as radius, strike the circle 1 E 3 F. Bisect the given length A C by the line B D, as shown, cutting the circle in the points E and F. From E, through the points 1 and 3, draw the lines E G and E H indefinitely, and from F, through the same points, draw similar lines, F I and F K. From 1 as center, and with 1 A as radius, strike the arc I A G, and from 3 as center, with equal radius, strike the arc K C H. From E as center, and with radius E G, strike the arc G D H, and from F as center, with corresponding radius, strike the arc I B K, thus completing the figure.

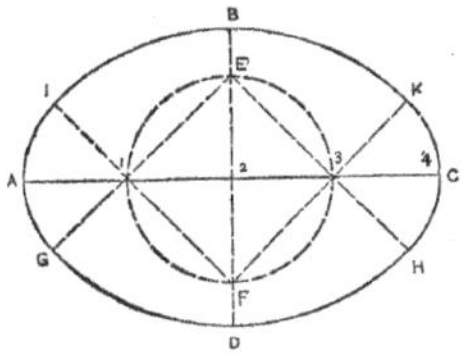


*Fig. 236.—To Draw an Elliptical Figure with the Compasses, the Length only being Given.—Another Method.*

397. *To Draw an Approximate Ellipse with the Compasses to Given Dimensions, Using Two Sets of Centers.—First Method.*—In Fig. 237, let A B represent the length of the required figure and D E its width. Draw A B and D E at right angles to each other, and intersecting at their middle points. At the point A erect the perpendicular A F, and in length make it equal to C D. Bisect A F, obtaining the point N. Draw N D. From F draw a line to E, as shown, cutting N D in the point G. Bisect the line G D by the line H I, perpendicular to G D and meeting D E in the point I. In the same manner draw lines corresponding to G I, as shown by L I, M O and R O. From I and O as centers, and with I G as radius, strike the arcs G D L and M E R, and from K and P as centers, with K G as radius, strike the arcs G A M and L B R, thus completing the figure.

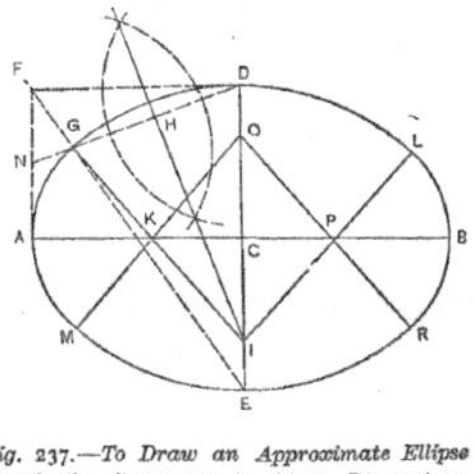


*Fig. 237.—To Draw an Approximate Ellipse with the Compasses to Given Dimensions, Using Two Sets of Centers.—First Method.*

398. *To Draw an Approximate Ellipse with the Compasses to Given Dimensions, Using two Sets of Centers.—Second Method.*—In Fig. 238, let C D represent the length of a required ellipse and A B the width.

Lay off these two dimensions at right angles to each other, as shown. On C D lay off a space equal to the width of the required figure, as shown by D E. Divide the remainder of D C, or the space E C, into three equal parts, as shown in the cut. With a radius equal to two of these parts, and from R as center, strike the circle G S F T. Then with F as center, and F G as radius, and with G as center, and G F as radius, strike the arcs, as shown, intersecting upon A B prolonged at O and P. From O, through the points G and F, draw O L and O M, and likewise from P, through the same points, draw P K and P N. From O as center, with O A as radius, strike the arc L M, and with the same radius, and P as center, strike the arc K N. From F and G as centers, and with F D and G C as radii, strike the arcs N M and K L respectively, thus completing the figure.

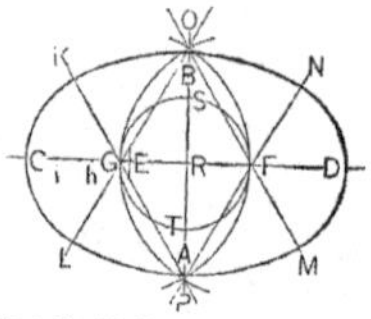

*Fig. 238.—To Draw an Approximate Ellipse with the Compasses to Given Dimensions, Using Two Sets of Centers.—Second Method.*

399. *To Draw an Approximate Ellipse with the Compasses to Given Dimensions, Using Three Sets of Centers*—In Fig. 239, let A B represent the length of the required figure and D E the width. Draw A B and D E at right angles to each other, intersecting at their middle points, as shown at C. From the point A draw A F, perpendicular to A B, and in length equal to C D. Join the points F and D, as shown. Divide A F into three equal parts, thus obtaining the points Z and I, and draw the lines Z D and I D. Divide A C into three equal parts, as shown by Y and G, and draw E G and E Y, prolonging them until they intersect with Z D and I D respectively, in the points H and J. Bisect J D, and draw K L perpendicular to its central point, intersecting D E prolonged in the point L. Draw J L and H J. Bisect H J, and draw M N perpendicular to its central point, meeting J L in N. Draw N H, cutting A B in the point O. L then is the center of the arc J D P, N is the center of the arc H J, and O is the center of the arc H A R. The points S and U, corresponding to N and O, from which to strike the remainder of the upper part of the figure, may be obtained by measurement, as indicated. Having drawn so much of the figure as can be struck from these centers, set the dividers to the distance L P or L J. By placing one point at E, the remaining center will be at the other point of the dividers, in the line E D prolonged, as shown by X.

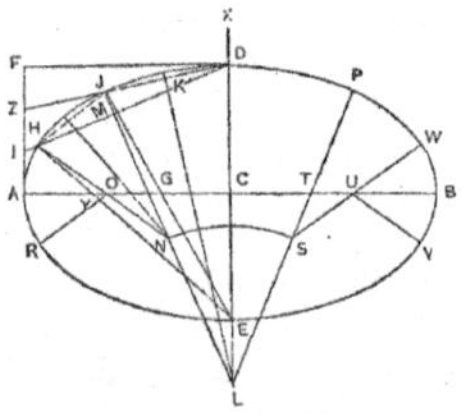

*Fig. 239.—To Draw an Approximate Ellipse with the Compasses to Given Dimensions, Using Three Sets of Centers.*

400. *To Find the Centers and True Axes of an Ellipse.*—In Fig. 240, let N B O R be any ellipse, of which it is required to find the center and the two axes. Through the ellipse draw any lines, A B and D E, parallel to each other. Bisect these two lines and draw F G, prolonging it until it meets the sides of the ellipse in the points H and I. Bisect the line H I, obtaining the point C. From C as center, with any convenient radius, describe the arc K L M, cutting the sides of the ellipse at the points K and M. Join K and M by a straight line, as shown. Bisect M K by the line N O, perpendicular to it. Through C, which will also be found to be the center of N O, draw P R, perpendicular to N O and parallel to K M. Then N O and P R are the axes of the ellipse and C the point of intersection or center.

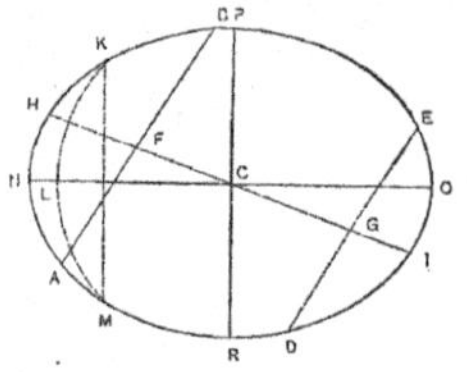

*Fig. 240.—To Find the Centers and True Axes of an Ellipse.*

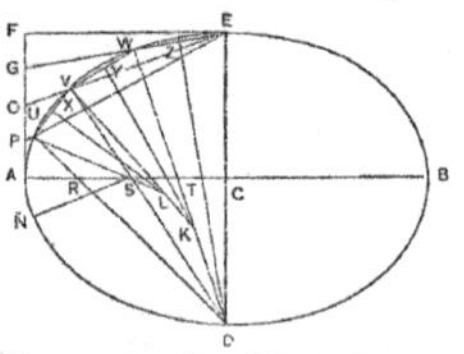

*Fig. 241.—In a Given Ellipse, to Find Centers by which an Approximate Figure may be Constructed.*

401. *In a Given Ellipse, to Find Centers by which an Approximate Figure may be Constructed.*—In Fig. 241, let A E B D be any ellipse, in which it is required to find centers by which an approximate figure may be drawn with the compasses. Draw the axes A B and E D. From the point A draw A F, perpendicular to A B, and make it equal to C E. Join F E. Divide A F into as many equal parts as it is desired to have sets of centers for the figure. In this instance we have determined upon

four. Therefore, A F is divided into four equal parts, as shown by P O G. Divide A C into the same number of equal parts, as shown by R S T. From the points of division in A F draw lines to E. From D draw lines passing through the divisions in A C, prolonging them until they intersect the lines drawn from A F to E, as shown by D U, D V and D W. Draw the chords U V, V W and W E, and from the center of each erect a perpendicular, which prolong until they meet other lines, as shown. Thus, commencing at the top, the perpendicular to W E reaches to the point D; that to W V intersects the line W D in the point K, and that to U V meets the line V K in the point L. Draw U L, cutting A C in the point S. Then D is the center of the arc E W, K is the center of the arc W V, L is the center of the arc V U, and S is the center of the arc U N. By these centers it will be seen that one-quarter of the figure (A to E) may be struck. By measurement, corresponding points may be located in other portions of the figure.

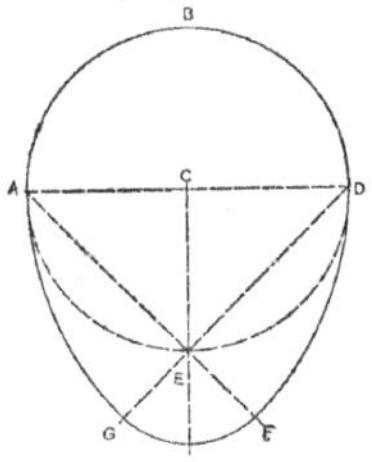

*Fig. 242.—To Draw an Egg-Shaped or Oval Figure.*

402. *To Draw an Egg-Shaped or Oval Figure.*—In Fig. 242, let A D be the required width. Upon A D describe the circle A B D E. From the center of this circle draw C E, at right angles to A D, cutting the circle in the point E. Draw D E and A E, and prolong them in the direction of G and F respectively. From A as center, and with A D as radius, describe the arc D F. From D as center, and with the same radius, describe the arc A G. From E as center, and with E G as radius, complete the figure, as shown.

# THE ART AND SCIENCE OF PATTERN CUTTING.

403. Before introducing pattern problems, it is appropriate that we should give some attention to the art and science of pattern cutting, in order that the reasons for the steps taken in the demonstrations following, and the directions for the use of tools which are occasionally introduced, may be readily understood. Underlying the entire range of problems peculiar to sheet-metal work, are certain fundamental principles, which, when thoroughly understood, make plain and simple that which otherwise would appear arbitrary, if not actually mysterious. So true is this, that we risk nothing in asserting that any one who thoroughly comprehends all the steps in connection with cutting a simple square miter, is able to cut any miter whatsoever. Since almost any one can cut a square miter, the question at once arises, in view of this statement, why is it that he cannot cut a raking miter, or a pinnacle miter, or any other equally hard form? The answer is, because he does not understand how he cuts the square miter. He may perform the operation just as he has seen some one else do it, or as laid down in some book or paper. He may produce results entirely satisfactory from a mechanical standpoint, but after all is finished he is not intelligent as to what he has done. He does not comprehend the why and wherefore of the steps taken. Hence it is, when he undertakes some other miter, that he finds himself deficient. Similar statements with reference to patterns of shapes derived from cones, and to each and every class of problems in sheet-metal pattern cutting, might be made, all teaching the same lesson, and all illustrating the importance of a thorough understanding of ground principles. There is a wide difference between the skill that produces a pattern by rote—by a mere effort of the memory—and that which reasons out the successive steps. One is worth but very little, while the other renders its possessor independent. It is with a desire to put the student in possession of this latter kind of skill, to render him intelligent as to every operation to be performed, that the present chapter is written.

404. The forms with which the pattern cutter has to deal, for convenience of description, may be divided into two general classes. The first of these we will call forms of parallel lines. It embraces moldings, pipes, flat surfaces, &c. The second we will call tapering forms. It comprehends all the shapes derived from cones, pyramids, &c. We might introduce a third class, embracing forms which in their characteristics belong to both of the other two, but since in pattern cutting such forms are treated as belonging to one or the other of the classes named, all necessary analysis is obtained by the divisions specified. For example, a vase, the plan of which is octagonal, viewed from one standpoint, belongs to the first class, because the lines of molding running around it are parallel, while viewed from another standpoint it seems to belong to the second class, because it is pyramidal in shape. It rightfully belongs to the first class, because in developing the patterns the form is treated as a molding in which octagon miters occur.

405. The patterns which arise in forms of the first class are, for the most part, what are known as miters, and, so far as principles and methods of developing are concerned, are among the simplest and easiest with which the pattern cutter has to deal. The methods of measurement, the use of tools, and the general plan of work in cutting miter patterns, are not unlike those used in developing shapes derived from cones, &c., although at first thought it would seem that they are totally distinct operations. Accordingly, an exemplification of the processes of miter cutting, provided we introduce the reason for every step taken, will also cast some light upon the second part of our subject. It is possible, moreover, to consider all shapes miters, and to treat everything in the same general way as moldings. While we shall follow this idea in part, for the sake of better explaining the

various steps taken, we shall take up the second class afterward, and give special explanations of the principles upon which its forms depend.

406. Although the shapes entering into tinware, by daily contact and long association, come to look simple, they are in reality the most difficult, in the matter of the development of their surfaces, with which the pattern cutter has to deal. On the other hand, moldings, the forms with which cornice makers deal almost exclusively, appear to those not conversant with that trade as very difficult indeed. It is necessary to divest the reader's mind of these ideas, in order to prepare him for that form of explanation which seems most desirable to introduce in this connection. We shall attempt to make clear the science of pattern cutting, first by a familiar talk about moldings, and afterward by a similar consideration of cones. The student, therefore, must cease to think that moldings are necessarily difficult forms. Although he may not be acquainted with cornice work, he will have no difficulty in understanding what we have to say. By familiarizing himself with what follows about moldings, he will be the better prepared to understand what we shall say about cones. If necessary to his comprehension of moldings and their miters, the experiments herein described should be patiently worked out. The encouragement to painstaking effort at this stage is the assurance that a thorough understanding of ground principles will make the student independent of all examples and precedents. It will enable him to formulate his own rules as occasion may require.

*Fig. 243.—Profile of a Molding.*

407. Since in sheet-metal work a molding is made by bending the sheet until it fits a given stay, a molding may be defined as a succession of parallel forms or bends made to a given stay, and, so far as the mechanic is concerned, any continuous form or arrangement of parallel continuous forms, made for any purpose whatever, may be considered a molding and treated as such in all the operations of pattern cutting. Keeping in mind, therefore, this fact, that almost any shape may be considered a molding so far as the method of obtaining its pattern is concerned, let us examine the nature of moldings and the joints occurring in them, commonly called miters.

*Fig. 244.—A Stay.*

408. A molding may be described as a form or surface generated by a profile passed in a straight or curved line from one point to another, this profile being the shape that would be seen when looking at the end if the molding were cut off square. Let us consider this definition in the light of a familiar illustration. In Fig. 243, let the form shown be the profile of some molding. If we cut the shape out of tin plate or sheet iron, as shown in Fig. 244, it is called a stay. For our purpose, as will appear further on, we require the reverse of the stay shown in Fig. 244, or, in other words, the piece cut from the face of the shape represented in that figure, which is shown in Fig. 245.

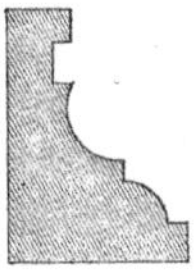

*Fig. 245.—A Reverse Stay.*

409. Having provided ourselves with a reverse stay, or "outside stay," as it is sometimes called, as shown in Fig. 245, let us take some plastic material—as, for instance, wax or potter's clay—and, placing it against a smooth surface, as of a board, move this reverse stay along its face until we obtain a continuous form in the clay corresponding to the reverse stay, all as illustrated in Fig. 246. By this operation we will have produced a molding in accordance with our definition. Our purpose in introducing this illustration is to show more clearly than we are able otherwise the principles upon which moldings—and, for that matter, all irregular surfaces—are measured in the process of pattern cutting; therefore, let us carry this same operation a step further.

410. Suppose that the form illustrated in Fig. 246 be completed, and that both ends of the molding be cut off square. It is evident, upon inspection, that the length of a piece of sheet metal necessary to form a covering to this molding will be the length of the molding itself, and that the width of the piece will be equal to the distance obtained by measuring around the face of

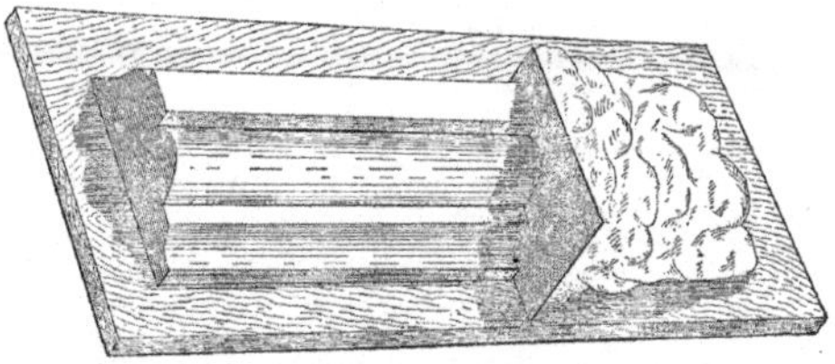

*Fig. 246.—Developing a Molding in a Plastic Material, like Clay, by Means of a Reverse Stay.*

the stay which was used in giving shape to the molding. Keeping this in mind, let us see what we must do in order to obtain a covering for it if one end is cut off obliquely. With a thin-bladed knife, or by means of a piece of fine wire stretched tight, let us cut off, at any angle, one end of the clay molding which we have constructed. By inspection of the form when thus cut, as clearly shown in the upper part of Fig. 247, it is evident that we must have such a shape to the end of the pattern as will make it correspond to the oblique end of the molding.

411. To cut such a pattern as we have just described by a straight line drawn from a point corresponding to the end of the longer side of the mold, to a point corresponding to the end of the shorter side of it, would not be right, evidently, because certain parts of the covering, when formed up, fold down into the angles of the molding, and therefore would require to be either longer or shorter, as the case might be, than if cut straight, as we have supposed. It is plain, then, that we must devise some plan by which measurements can be taken in all these angles, and at as many intermediate points as may be necessary, in order to obtain the right length at all points throughout its width. It is easy to measure the length of the molding in the lines of the several angles, and we can also readily obtain measurements at as many intermediate points as we require, by a simple plan.

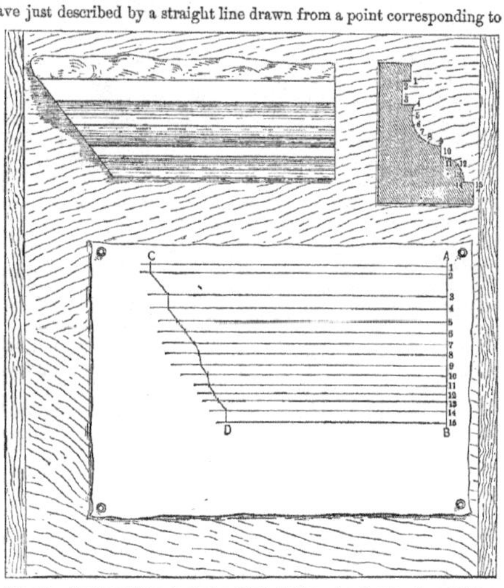


*Fig. 247.—The Use of Lines in Laying Off the Pattern of a Covering for a Molding.*

412. Divide the curved parts of the stay into any convenient number of equal parts, and at each division cut a notch, or affix a point to it. Replace the stay in the position it occupied in producing the molding, and pass it over the entire length of the molding. The points fastened to the stay will then leave tracks or lines upon the surface of the molding. Now, by measurements upon these lines, the length of the molding at all of the several points established in the stay may be obtained. All this is clearly illustrated in Fig. 247. In the upper right hand corner of the illustration is shown the stay prepared with points. By moving it as described, lines are left upon the face of the molding, as shown to the left.

413. Now, if we take a sheet of paper, and upon any part of it draw a straight line, as shown by A B in Fig. 247, and upon that line set off with the dividers the width of each space or part of the profile of the stay —that is, make the space 1 2 in the line A B equal to the space 1 2 in the profile, and 2 3 in the line A B equal to 2 3 of the profile, and so continue until all the spaces are transferred—and from the points thus obtained in A B draw lines at right angles to it indefinitely, we shall have lines upon the paper corresponding to the lines upon the clay molding made by the points fastened to the stay. Next, if we measure the molding upon each of the lines drawn upon it, and set off the same length upon the lines drawn upon the paper, we shall obtain points through which a line may be traced which will correspond to the oblique end of the molding. Therefore we set off, on the line 1 from A B, the length of the molding, measured from its straight end to its oblique end, upon the corresponding line upon its face, and upon each of the other lines on the paper the length of the molding on the corresponding line on its face. By this means we obtain points, through which, if a line be traced, as shown by C D, the pattern of the covering will be described. The line A B, laid down by measuring

from the profile, is called the "stretchout line," and the lines drawn through the points in it at right angles to it are called "measuring lines."

414. Now, what we have done in Fig. 247 illustrates what is called "miter cutting." The strict definition of the word miter, is the joint between two moldings of like profile at any angle; but in sheet-metal work it has come to mean the shape of the end of a molding or other form required to make it fit against any surface, regular or irregular, at any angle. Miter cutting, then, consists of describing in the flat the shape of a given form required to fit against a given surface at a given angle. In this sense almost all patterns are miter patterns.

415. What we have obtained in Fig. 247, by means of a clay model—that is, what we have obtained in the way of the pattern shown in the lower part of the figure, measurements for which were obtained from the lines drawn on the surface of the clay model—may be obtained just as well by a drawing. The question then is, how can we obtain, by lines drawn upon a flat surface, the same results as are obtained by measurements on lines drawn along the surface of a molding?

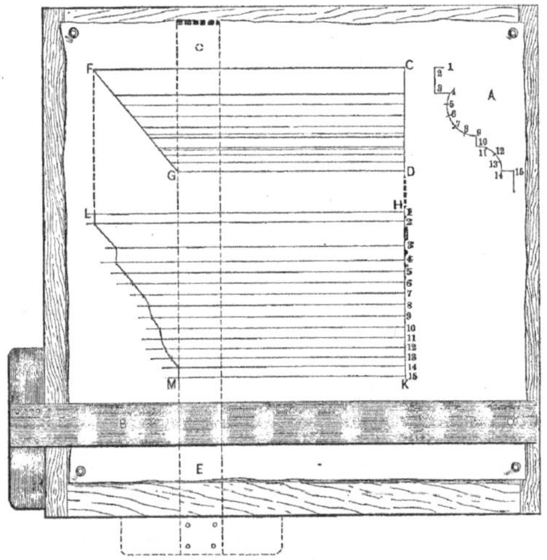


*Fig. 248.—Obtaining the Lines of Measurement for the Covering of a Molding by Means of a Drawing. Also Illustrating the Use of the T-Square in Miter Cutting.*

416. In moving the profile along the clay molding, certain lines were made by means of the points affixed. If the reader will carefully examine Fig. 247, he will doubtless notice that the lines upon the molding made by this means corresponded in number and position with the points in the profile when it is laid flat on its side. Hence, if we draw the stay or profile, and also represent the molding by lines, we are able to accomplish the same ends, care only being necessary that the relative positions of the parts be correctly maintained. This is clearly illustrated in Fig. 248, which is to be compared with Fig. 247.

417. Let us examine Fig. 248, in order to see just what is done to obtain the points of measurement and the dimensions required. First, the profile A is drawn in position, as shown. Next, from it a drawing of the required molding is made, as shown by F C D G. The rule for drawing the molding and profile may be stated as follows: Place the profile A—which, for the sake of comparison, may be a duplicate of the stay used in the preceding illustration, including all the intermediate points—in line with the space it is desired the elevation of the molding shall occupy. For the lines of the molding, use the T-square in the general position shown by B in the engraving, bringing it against the several points in A in order to draw the lines. Draw a line for each of the angles in A, and also one corresponding to each of the intermediate points in the stay. Draw the line F G, representing the oblique cut, and the line C D, representing the straight end. Then it will be seen that F C D G of Fig. 248, so far as lines are concerned, is exactly the same as the molding we made of clay, shown in Fig. 247. The line F G, by the definition of a miter, is the "miter line" of this molding. It represents the surface against which the molding is supposed to fit. Next lay off a stretchout of the profile A, in the same manner as described in connection with Fig. 247, all as shown by H K in Fig. 248, through the points in which draw measuring lines at right angles to it, or, what is the same, parallel to the lines of the moldings. In length make them equal to the length of the molding measured upon the corresponding lines in C D G F.

418. Now, if we proceed as suggested in the previous illustration—that is, by using a pair of dividers to measure the length of the molding on the several lines, from C D to F G—and if we set off like lengths on corresponding lines drawn from the stretchout H K, we will obtain a pattern in all respects corresponding to the pattern shown in Fig. 247, already referred to. By inspection of the result thus obtained, however, it will be seen that the same thing may be accomplished by using the T-square, as shown by the dotted lines in Fig. 248. Therefore proceed as follows: Place the T-square as shown at E, and, bringing it successively against the points in F G, established by the lines drawn from the profile A, cut corresponding measuring lines drawn from the stretchout H K. Then a line traced through the points of intersection thus obtained, as shown by L M, will be the shape of the pattern corresponding to the miter line F G. By this illustration it is evident that the T-square may be used with great advantage in transferring measurements under almost all circumstances.

419. Since we no longer use the dividers to locate the points in the patterns, the position of the stretchout line may be taken at will. For convenience, it should be placed as near to the miter line as possible. Hence, in practical work, supposing that the molding represented by F C D G is not a very short piece, the stretchout line, instead of being opposite the end C D, would be placed somewhere near the line of the blade of the T-square when in the position shown by E. We purposely except short pieces of moldings, for the advantage of describing the pattern at one operation in such cases sometimes overcomes the advantage of placing the stretchout close to the miter line.

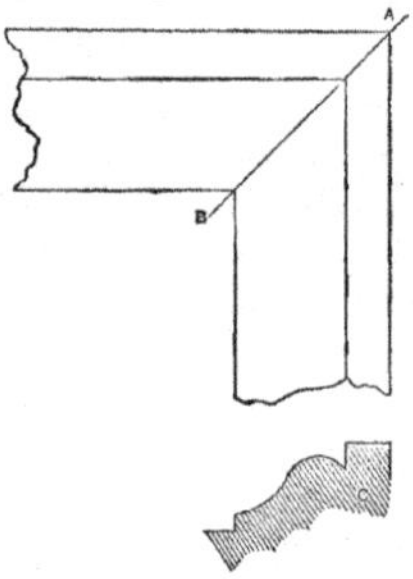

*Fig. 249.—The Cut in each Arm of the Molding required to Unite them in the Joint A B, is the same as though each piece was Calculated to Fit against a Plane Surface represented by A B.*

420. By further inspection of Fig. 248, it will be seen that, instead of drawing the lines from the points in the profile A the entire length of the molding, as there shown, all that is necessary to the operation is short lines corresponding to the points of the profile, and extending across the miter line F G. The use of these lines, it is evident, is only to locate intersections upon the miter line. In other words, all we need is the points in the profile A transferred to the miter line F G. The operation of transferring these points by short lines, as above described, is termed "dropping the points" from the profile to the miter line.

421. If, instead of the molding terminating against a plane surface, as shown by F G in Fig. 248, it be required to develop the pattern to fit against an irregular surface, we proceed in exactly the same manner, simply substituting for the straight line F G a representation of that surface. From this it will be seen that all that is required to develop the pattern of any miter, is that a correct representation of the molding be made, showing the angle of the miter, and that a profile be so drawn that it shall be in line with the elevation of the molding—its face being so placed as to agree with the face of the molding—and that points from the subdivisions of the profile be carried parallel to the molding, their intersections with the miter line being marked by short lines.

422. In order to more clearly indicate the point we desire to make by this summary of requirements, let us suppose that we have two pieces of molding made of wood, and that we cut the required miter on them by means of a saw, and then place them together, as shown in Fig. 249. Now, if we take a piece of sheet iron, for example, and slip it into the joint, as shown by A B, and then remove one arm of the miter, we readily see that what we have left is exactly what we had in Fig. 248. In other words, it amounts to a molding fitting against a plane, and, hence, the operation of cutting the pattern in such a case as shown in Fig. 249 is identical with that described in Figs. 247 and 248.

423. From all this it is plain to be seen that the central idea in miter cutting is to bring the points from the profile against the miter line, no matter what may be its shape or position. Inasmuch as all moldings, if they do not member or miter with duplicates of themselves, must either terminate square or against some dissimilar profile, it follows that the two illustrations given cover in principle the entire catalogue of miters.

424. As we remarked at the outset, all patterns may be, in one sense or another, considered miter patterns. The principles we have here explained are the fundamental principles in the art of pattern cutting, and their application is universal in sheet-metal work. It would be difficult to compile a complete list of miter problems. New combinations of shapes and new conditions are continually arising. The best that can be done, therefore, in a book of this character, is to present a selection of problems calculated to show the most common applica-

tions of principles which, carefully studied, will so familiarize the student with them that he will have no difficulty afterward in working out the patterns for whatever shapes may come up in his practice, whether they be of those specifically illustrated or not.

425. From what has preceded we derive the following summary of requirements, together with a general rule for cutting all patterns whatsoever: *Requirements.*—There must be a plan, elevation or other view of the shape, in line with its profile, showing the line of the surface against which it miters.

426. *Rule.*—1. Place a stretchout of the profile on a line at right angles to the direction of the molding or other shape, as shown by the plan, elevation or other view, and draw measuring lines parallel to the molding. 2. Drop points from the profile to the miter line or line of joint, carrying them in the direction of the molding or other surface. 3. Drop the points thus obtained from the miter or joint-line on to the measuring lines of the stretchout, at right angles to the direction of the molding or surface.

427. The student who gives careful attention to these rules will at once remark that the operation of cutting a common square miter—that is, a square miter between the moldings running across two adjacent sides of a building, for example—does not employ a miter line, and therefore appears to be an exception. Yet we have remarked (Section 403) that a thorough understanding of how a square miter is cut comprehends within itself the entire science of pattern cutting. It is because a square return miter—for such is the distinctive name applied to the kind of square miter in question—is in one sense an exception to the general rule, that it is so valuable for the purposes of illustration. A miter of this kind admits of an abbreviated method. The short rule for cutting it is usually the first thing a pattern cutter learns, and the operation is very generally explained to him without any reason being given for the several steps taken. In many cases it would bother him to cut the pattern by any other than the short method, even after he has obtained considerable proficiency in his art. Hence it is that, to all who have any previous knowledge of pattern cutting, the rules above set forth seem inadequate, or, to put it otherwise, a formula to which there are exceptions.

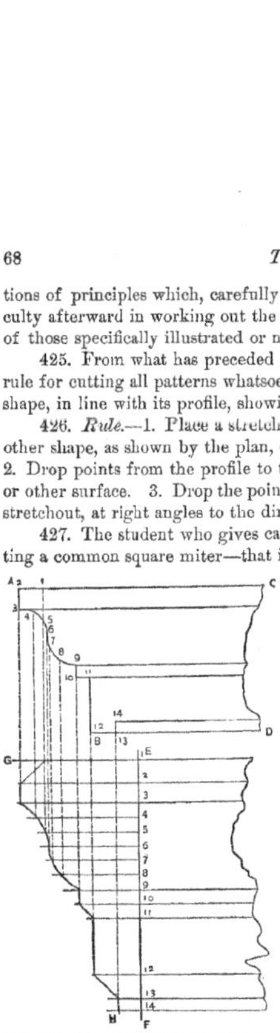

*Fig. 250.—The Usual Plan of Cutting a Square Miter in which no Joint Line is used.*

428. To clear up these doubts in the mind of the student, we will first introduce an illustration of the short method of cutting a square return miter, and afterward we will show the long method, or the plan which is in strict accordance with the rule above given, combined with the short method, thus showing the relationship and correspondence between the two.

429. Fig. 250 shows the usual method of developing a square return miter, being that in which no plan line is employed. The profile A B is divided into any convenient number of spaces, as indicated by the small figures in the engraving. The stretchout E F is laid off at right angles to the lines of the moldings, and, through the points in it, measuring lines are drawn parallel to the lines of moldings. From the points established in the profile, lines are dropped cutting corresponding measuring lines. Then the pattern is obtained by tracing a line through these points of intersection.

430. In this operation it will be noticed that we have fully complied with the stipulations of the first rule given. We have placed the stretchout at right angles to the lines of the molding, and have drawn measuring lines parallel to those lines, but when it comes to the second and third parts of the rule it would seem that we have done something else than anticipated therein. We have, apparently, employed no joint line or plan, but have dropped points directly from the profile on to the measuring lines.

431. Let us now examine Fig. 252, which in its upper part contains the short rule just described, and which, by G F, shows the use of the plan line of the joint or miter line. The pattern, as developed by the long method, is shown on the lower portion of the cut to the right. Referring to Section 425, it will be seen that we have complied with the requirements therein recited. We have a plan of the shape (F G) in line with the profile A B. By spacing the profile in the usual manner, and drawing lines from the points in it toward the miter line, we have the lines of the molding in plan, at right angles to which, by the first part of the rule in Section 426, the stretchout is to be placed. Therefore, we lay off C D at right angles to H F, and draw measuring

lines perpendicular to it, or, what is the same, parallel to the lines of the molding in the plan, as stipulated in the rule. We have already dropped points from the profile on to the miter line, as recited in the second part of the rule. So there remains only the third part to be complied with. Placing the blade of the T-square at right angles to the lines of the molding in the plan, and bringing it successively against the several points in F G, we cut corresponding measuring lines drawn through the stretchout. Then a line traced through these points of intersection will be the pattern sought.

432. Laying off a stretchout below the profile and at right angles to it, as shown by C D, through the points in which measuring lines are drawn, and tracing a line through the points of intersection between corresponding measuring lines and lines dropped from the profile, also produces the pattern, as shown by C E. This last operation is the short method, or the same as shown in Fig. 250. By comparison it will be seen that the two patterns C E and C' E' are identical.

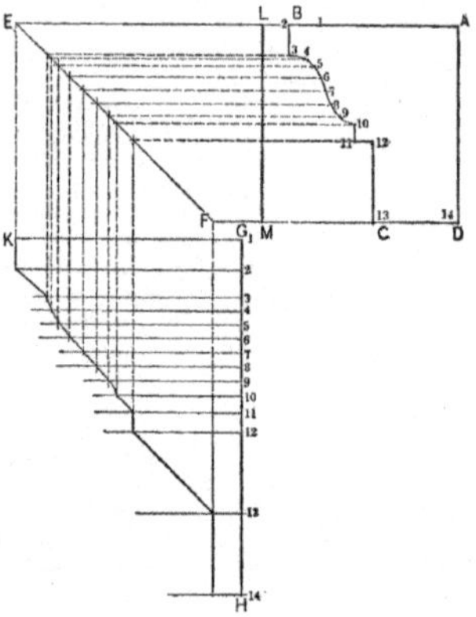


*Fig. 251.—A Square Face or Panel Miter. This Cut Illustrates a Mistake often made by Students in Attempting to Employ a Miter Line in Cutting Square Return Miters.*

433. Since the miter line F G bisects the right angle H F K, the two arms of the miter must be identical. Hence, all the operations in connection with the patterns may be performed on one side of the line. By comparison it will be seen that the relationship between C E and the miter line, and C' E' and the line, are the same. Dropping points from a profile against a line inclined 45 degrees, as F G, and thence on to a stretchout, gives the same result as dropping them on to the stretchout in the first place. Hence it is that the portion of the operation shown in the lower part of the engraving may be dispensed with.

434. A very common mistake made by beginners in attempting to apply the general rule for cutting miters given in Section 426, is that of getting the miter line in a wrong position with reference to the profile. For example, instead of drawing a complete plan, as shown by L H F K M in Fig. 252, by which the miter line is located to a certainty, and in connection with which it is a simple matter to correctly place the profile, it is very customary to attempt the operation by drawing the miter line only, placing it either above, below or to one side of the profile. The mistake is made by having the line to the side of the profile when it should be either above or below it, and *vice versa*. Fig. 251 illustrates a case in point. The engraving was made from the drawing of a person who attempted to cut a square return miter by the rule, using a miter line. By placing the line E F to the side instead of below the profile, as shown in Fig. 252, a square face miter—for example, such as would be used in the molding running around a panel or a picture frame—was produced in place of what was desired.

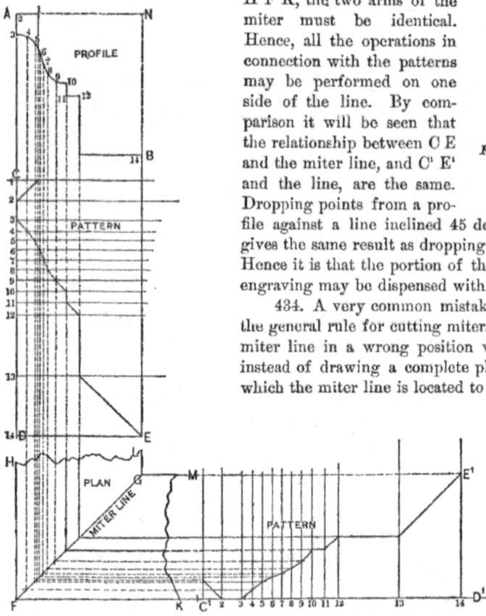


*Fig. 252.—A Comparison of the Short or Usual Method of Cutting a Square Miter, with the Long Method, or that which takes all the Steps laid down in the Rule.*

435. No better rule for avoiding errors of this kind can be given than the exercise of the greatest thoughtfulness and care. It is better to draw a complete plan, as shown in Fig. 252, thus demonstrating to a certainty the correct relationship of the parts, than to save a little labor and run the risk of error. So far as it is possible to formulate a rule for such operations, it may be presented thus: Place the profile, with reference to the plan or elevation, so that lines drawn from the points in it will correctly represent the molding in plan or elevation, as the case may be. Thus, in Fig. 252, the lines dropped from the profile to the miter line, and thence carried to the right, represent the members of the molding as they would appear if we were above it and looked down upon and through it. The relative position of the parts is evidently correct for the end in view. Applying the same test to Fig. 251, it will be observed that the lines drawn in the plan, or elevation, whichever it may be considered, are correct for a square panel miter, but are incorrect for the plan of a square return miter, which it was the design of the draftsman to produce when he made the drawing.

436. Always bear in mind that miter cutting, and for that matter all pattern cutting whatsoever, is simply a system of measurements upon surfaces. Of necessity, the surfaces are represented by diagrams in the flat. Two or more views are required to obtain the same dimensions from a drawing of an object as would be got in one operation from the object itself. The two or more views are to be so arranged that different portions are presented at the same time in proper combination. We have already seen (Section 417) how, by means of a profile and a drawing properly placed, the same results were accomplished as were obtained by measurements upon the molding itself. Keep such comparisons in mind, and think out what is wanted to be done before the drawing is commenced.

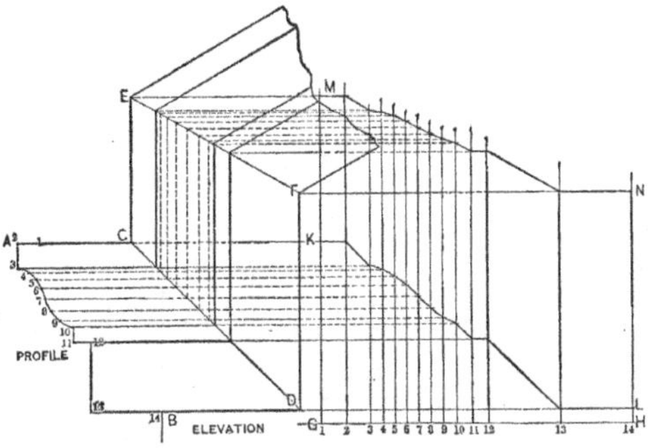


*Fig. 253.—Patterns in a Common Form of Window Cap, Introduced in Further Elucidation of Principles.*

437. In further elucidation of the principles of miter cutting, and as illustrating the directions just given, we show in Fig. 253 some of the patterns in a very common form of window cap. The miters illustrated are of the kind called "face miters," the one represented by the line C D being a square miter, while that at E F is at some other than a right angle. The profile A B is spaced in the usual manner, and lines from the points are carried through the various parts composing the cap, parallel to the lines of molding. The stretchout G H is laid off at right angles to the lines of that portion of which the pattern is required. The measuring lines being drawn in accordance with directions already given, the T-square is placed with the blade at right angles to the lines of the molding, and being brought successively against the points in the two miter lines, the measuring lines of corresponding number are cut. Then lines traced through these points of intersection complete the pattern. Had it been desired to obtain the pattern for that portion shown by A C D B in the elevation, a stretchout line would have been drawn at right angles to it. The square return miter would be dropped from the profile, while the opposite end of the piece would be obtained by dropping points from the miter line D C.

438. Having now, as we think, made clear the principles of pattern cutting, at least so far as they can be illustrated by simple miters, we desire to return again to the rule laid down in Section 426, in order to present another conception of a square return miter, which will show that the short method we have taken so much pains to explain by comparing it with the long rule, is not so much an exception to the general rule as would at first be supposed. Referring to Fig. 250 for illustration, we will apply the rule to the operations there shown. In the first place, we place a stretchout, E F, at right angles to the direction of the molding, as shown by the elevation, for A C D B represents the molding in elevation. We next draw measuring lines parallel to the molding. Thus E G and the lines below it are parallel to A C and B D. The second part of the rule

says: "Drop points from the profile to the miter line or line of joint, carrying them in the direction of the molding." A B is evidently the profile, from which points are to be dropped on to the miter line, or line of joint. A moment's investigation will show that A B is also the miter line in this case. What we really want to do is to cut the pattern to such a shape that, when it is formed up, one end will be straight and the other end present the profile shown by A B. This view of the case makes A B the miter line. In connection with our description of Figs. 247 and 248, we remarked that if the shape required to be given to the end of the molding were other than that represented by a straight line in the elevation (F G, Fig. 248), the operation would still be the same, the only change to be made being the substitution of a curved or mixed line in place of the straight line. Now, A B of Fig. 250 may be considered a mixed line, substituted for the straight line F G in Fig. 248.

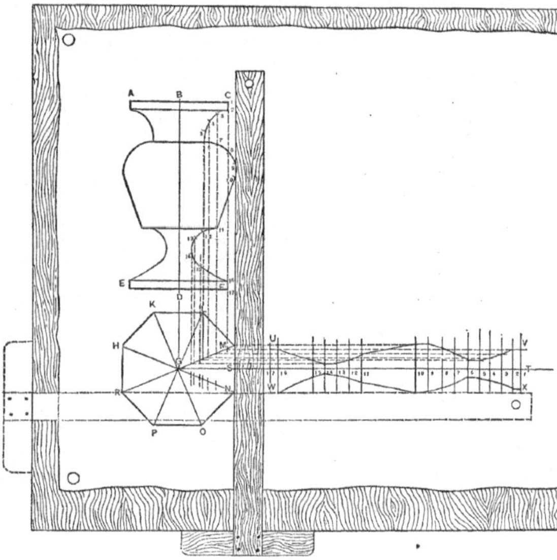


*Fig. 254.—The Patterns of an Octagonal Vase Developed by the T-Square, Introduced to Illustrate the Use of that Instrument.*

Therefore, in spacing the profile we also dropped points upon the miter line. Our compliance with the third part of the rule is evident without special explanation. By investigations and comparisons of this kind it becomes evident that there is a unity of principle underlying all the operations in pattern cutting. If the student is able to grasp and master this central idea, his success as a pattern cutter is assured.

439. In order to make the use of the T-square for transferring distances and dropping points better understood, we present a diagram of the patterns of an octagonal vase, in the development of which this instrument plays an important part. Referring to Fig. 254, it will be seen that the profile C E is drawn directly over the plan, and that points from the profile are dropped across so much of the plan as it is necessary to use in developing the pattern for one section. For this purpose the T-square is employed in the position shown in the engraving. Thus the miter lines G M and G N represent the boundaries of one of the sections in the plan. Points from the profile are dropped so as to cut these two lines. It is not necessary to continue them entirely

across the plan. Simply crossing the miter lines with short fine marks answers every requirement. At right angles to the side of the vase, as shown in the plan, the stretchout S T is laid off, using the T-square as shown by the dotted lines in the engraving. Through the stretchout measuring lines are drawn in the usual manner. By bringing the T-square against the several points in the miter lines G M and G N, and thus cutting measuring lines of corresponding numbers, points of intersection are obtained, through which, if lines be traced, the form of the pattern, as shown by U V X W, will be obtained.

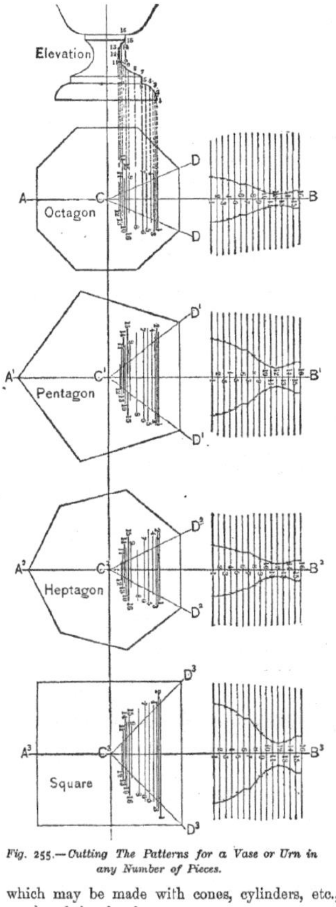


*Fig. 255.—Cutting The Patterns for a Vase or Urn in any Number of Pieces.*

440. Before taking up the subject of tapering surfaces, we will introduce Fig. 255, which shows patterns of vases, the plans of which are various regular polygons, all developed from the same profile. This diagram serves to illustrate several points. It shows the relationship of profile and plan; the use of miter lines in the plan, and the application of one general rule to what are ordinarily considered separate and distinct problems. It further shows, in part, the reason for the assertion made at the commencement of this chapter, that proper knowledge of a square miter is adequate for cutting any miter. Detailed description of the steps shown is not necessary, because they are the same as described in connection with the last figure. Problems illustrating the same miters are also to be found in their proper places in another portion of the book.

441. A term of somewhat frequent occurrence in geometrical works is "a solid of revolution," the meaning of which, as defined by Webster, is as follows: "A solid generated by the motion of a surface about a line as its center or axis." Defined in more familiar terms, it may be described as a solid whose outline corresponds to the form described by the rotation of a plane of some defined shape around one of its sides. A right cone (see Section 98) is one of the most common examples of a solid of revolution. Thus, if a right-angled triangle, C E D, Fig. 256, be revolved about its altitude, C E, as an axis, the form described by its hypothenuse will be a cone. A cylinder, Fig. 257, is another example in point. If a rectangle, as shown by C D F E, be revolved about one of its sides, C E, as an axis, the form generated will be a right cylinder.

442. Our purpose in introducing this term and these illustrations in this connection, is to make clear by contrast what cannot be so well shown by other means. We have already explained that in sheet-metal pattern cutting all objects are treated as solids—that the shell, with which we really deal, is considered the envelope stripped from a solid. Keeping this in mind, and examining the nature of cones and cylinders in the light of the definition above given, the reasons for some of the steps taken in developing patterns for them at once become apparent. Perhaps, however, we can show this better by describing one or two experiments which may be made with cones, cylinders, etc., just as was shown in connection with moldings where we employed the clay form.

443. Keeping in mind how solids of revolution are generated, let us investigate their properties by some

experiments in the revolution of solids. Let us suppose that we have a cone, a cylinder, a cube, a prism and a pyramid, all with their surfaces blackened in such a way as to make an impression or print when they are revolved or rolled over a sheet of paper. Commencing with the cone, as shown by A B in Fig. 258, we will mark some point in its base by which to note how far it has revolved, and will turn it so as to make one complete revolution. The resulting figure, which of course corresponds to its surface, is as shown by C A' D. The point A, having no diameter, remains in one spot during the operation, but the base travels the distance shown by the curved line C D, or, in other words, a distance equal to the circumference of the base. From this simple experiment it is easy to see what is to be done to work out by lines the pattern of a right cone. In the first place, one end of the pattern will be a point like A'; the other end, evidently, will be a curve, all points in which are equally distant from A', and the length of which is equal to the circumference of the base. Therefore we set the dividers to a radius equal to the hight of the cone, and from any point, as A', describe an arc, making its length equal to the circumference of the base.

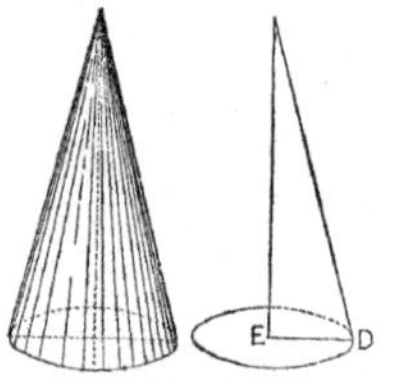


*Fig. 256.—A Right Cone, Generated by the Revolution of the Right-Angled Triangle C E D about its Axis, C E.*

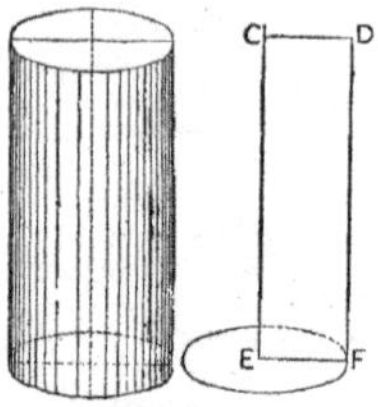


*Fig. 257.—A Right Cylinder, Generated by the Revolution of the Rectangle C D F E about C E, one of its Sides.*

444. Since the distance from the apex of a right cone to its base is the same at all points, the system of measurements which we described in connection with the clay molding would not seem to apply in problems relating to cones. This, however, is not the case. The peculiarity mentioned is incidental to one form of the cone alone, and makes abbreviated methods possible with it. We could obtain the same results with a right cone by measuring on the surface instead of revolving it. For example, we might have drawn a straight line from the apex to the base of the cone, and then, measuring one inch along the base in the direction of its circumference, drawn another line to the apex, and so continue around it. For the pattern shape in this case we would have drawn a line equal in length to the first line above described, and then, measuring from one end of it, we would have laid off an inch space, drawing another line to the end, representing in the pattern the apex of the cone, and so on. Or, to describe the operation in another way, for the pattern we would have proceeded to construct, side by side, a number of triangles corresponding to the triangles drawn on the face of the cone. By this plan the similarity between the measurements necessary to the development of a cone pattern and those employed in miter patterns is at once perceived. The right cone, like the square return miter, admits of a short method, but patterns of other forms of cones require measurements somewhat after the plan above outlined, though in many cases much more complicated. To these we shall give attention further on.

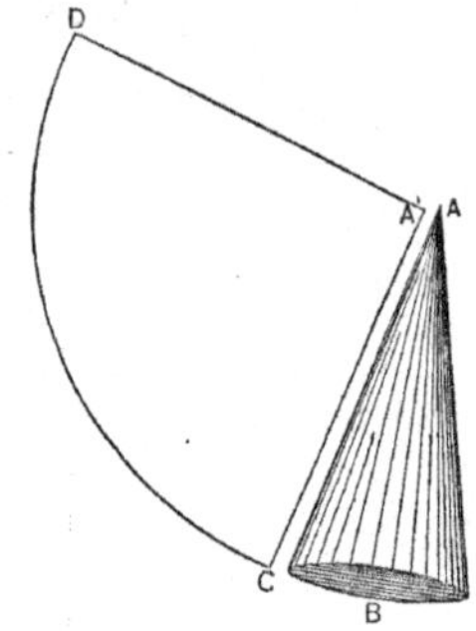


*Fig. 258.—The Revolution of a Right Cone by which the Shape of its Envelope is Described.*

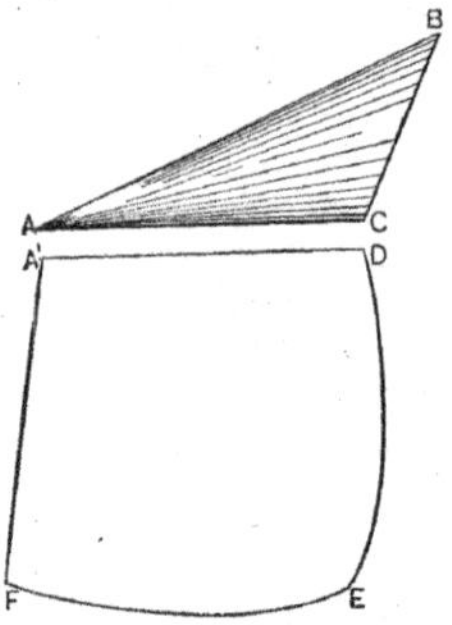


*Fig. 259.—A Scalene Cone, Revolved in such a Way as to Show the Shape of its Envelope.*

445. Fig. 259 represents a similar experiment performed with a scalene cone. In this case the revolution

of the cone is made to begin with the shortest point, so that its longest length falls in the middle of the pattern surface. By comparing this shape with that last described, it is evident that some such system of measurement as alluded to above will be necessary to determine the shape represented by D E F. We shall not stop here to describe how measurements are applied in this case, because it will be necessary to take up the same subject further on.

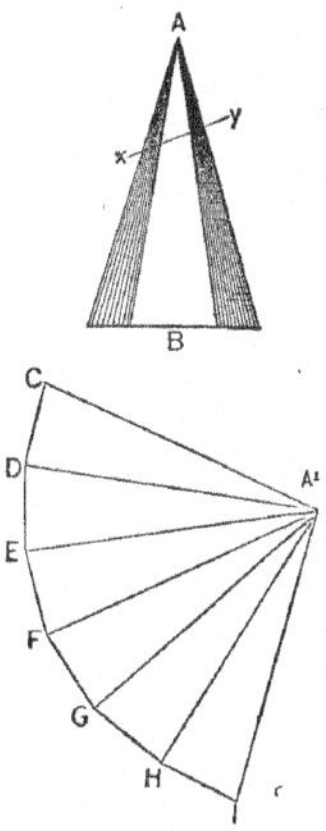


*Fig. 260.—A Hexagonal Pyramid, Revolved in such a manner as to Describe the Shape of its Covering.*

446. In Fig. 260 we show a similar experiment with a six-sided pyramid. Here it will be seen that the pattern is a succession of triangles, each of which is equal to one of the faces of the pyramid. The manner of developing a pattern of this kind is almost self-evident. By describing an arc of a circle from the center A, with a radius equal to the length of one of the faces, and then stepping off in this arc spaces equal to the width of the faces measured at the base, the shape indicated by A' I C will be obtained. Supposing that this pyramid were cut on the line $x$ $y$, as shown in Fig. 260, the revolution of the solid would give the shape indicated in Fig. 261. The pattern would be obtained by lines and measurements in the same general manner. Having established that part of the pattern corresponding to the base, as described in connection with the previous illustration, and as indicated by I, H, G, F and C, and drawn lines to the center, it is a simple matter to measure up each of the several angle lines a distance equal to the hight of the corresponding angle in the solid itself, thus determining the broken line which in Fig. 261 represents the top of the pattern. Several examples illustrating this principle will be found among the pattern problems.

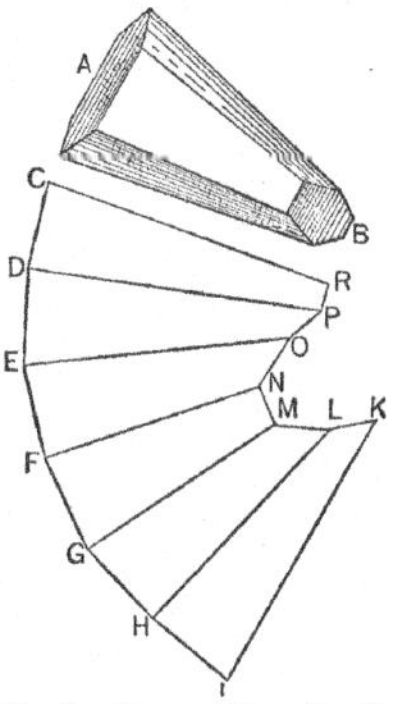


*Fig. 261.—Hexagonal Pyramid, with so much of its Apex Removed as is indicated by x y in Fig. 260, Revolved so as to show the Shape of its Covering.*

447. Leaving the cone for a moment, let us revolve a cylinder in the same general manner as we have been describing. The shape produced is shown in Fig. 262 by D E G F. It is evident that the length D E must be equal to the length A B of the cylinder, and that the width E G must be equal to the circumference of the cylinder. Suppose one end of the cylinder to be cut off obliquely, as shown in Fig. 263, and that the solid is then revolved in the same manner. Here one end of the pattern shape is irregular, as shown by E F G, and becomes what we have already described as a miter pattern. Referring again to Fig. 262, if an opening be cut in the cylinder, as indicated by C of the elevation, and it be revolved, a form similar to the shape indicated by C' in the pattern will be produced. Without describing in detail all the features of these experiments, it is evident, we think, that the system of measurement on the sur-

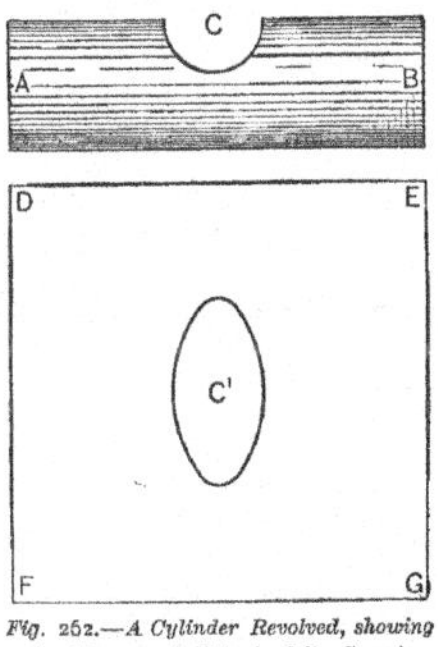


*Fig. 262.—A Cylinder Revolved, showing the Shape and Extent of its Covering. Also showing the Shape in the Pattern of an Opening made in its Side.*

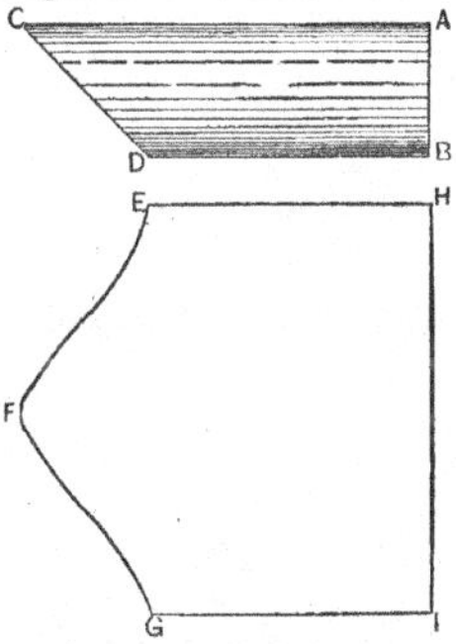


*Fig. 263.—A Cylinder with One End Cut off Obliquely and Revolved so as to show the Shape of its Envelope.*

face of solids, by which the shape of various parts is determined, is the same in all cases, and in principle is identical with that described in connection with the clay molding and miter patterns at the commencement of this chapter.

448. Fig. 264 shows the covering of a triangular prism, obtained in the same general manner as we have been describing, but which, it must be evident to the reader, can be just as well obtained by lines and measurements. Fig. 265 shows the same thing applied to a cube. If the student will keep in mind these experiments, and when puzzled over difficult problems will picture in his mind the form that would be produced by the revolution of the solids with which he has to deal, he will find it of great help to him in determining the best method of obtaining the lines and measurements required.

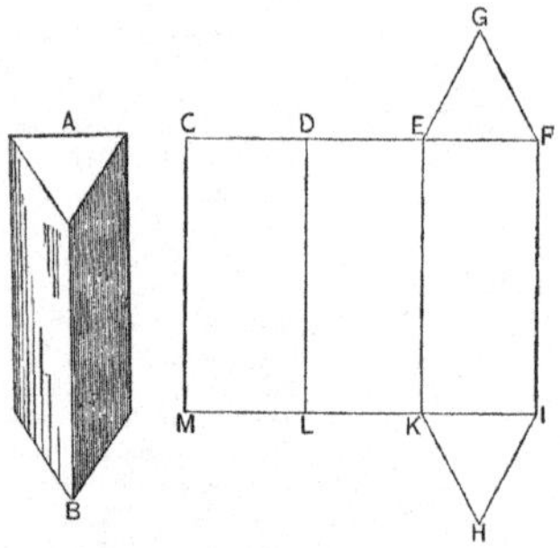


*Fig. 264.—The Covering of a Triangular Prism, obtained by Revolving it as Before Described.*

449. We have remarked that the most difficult problems in pattern cutting relate to conical shapes. Besides the right cone, some of the properties of which we have just illustrated, there are conical forms whose bases are elliptical instead of circular. With such figures the steps necessary to develop the shape of the covering are, of course, very much more complicated than those employed for the simple form we have named. Although it is possible, in some instances at least, to revolve the solids we have just referred to, the shapes thus produced are so irregular in outline as to show at once that quite different means from any so far described are necessary to obtain the requisite lines and measurements for developing the pattern. In the preceding chapter we referred to some of the properties of the ellipse, showing how it may be produced by string and pencil (Section 377); also how approximate figures may be drawn by the compasses from two or more sets of centers (Sections 395 to 399), and how an oblique section of a cone through its opposite sides and an oblique section through a cylinder both produce this figure. (Section 390.) An infinite variety of ellipses is possible, the range being from a close resemblance to a circle on one extreme, to a figure suggesting a straight line on the other, and the solids which may be erected on the ellipse for a base vary quite as much. Since an oblique section through a solid, the base of which is a circle, as, for example, a cylinder or a right cone, gives an ellipse, it follows that oblique sections through solids whose bases are elliptical may produce circles. Careful attention as to the nature of the forms with which he has to deal, is always required upon the part of the pattern cutter. Sometimes it is necessary for him to resolve a form into its simple component parts before it is possible to develop the patterns at all. It is only by thoroughly understanding the nature and properties of the ellipse and of elliptical solids, so that they are recognized in whatever form encountered, that he is enabled to develop the intricate shapes peculiar to work of this kind.

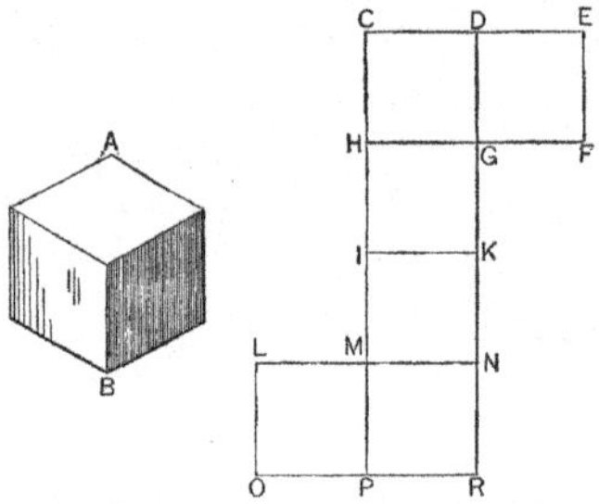


*Fig. 265.—The Covering of a Cube, developed by Revolving it so that its Several Sides come in Contact with the Paper.*

450. We described in Section 443 a method by which the envelope of a right cone may be drawn, deriving the rule from the experiment that had just been made, of revolving the solid in a way to show the form of its covering by an impression or print. The pattern of a frustum of a cone is much more frequently required in sheet-metal work than that of the complete cone. The method of proceeding in such cases is very similar to that employed with the complete figure. It is simply necessary to restore that portion of the cone cut away, as shown in Fig. 266, and employ two radii of different

length. If a pin be fastened at the apex C of a right cone, Fig. 267, and a piece of thread be tied thereto, carrying points B and A, corresponding in position to the upper and lower faces of the frustum, and the thread being drawn straight be passed around the cone, the points will follow the line of the faces of the frustum throughout its course. If we then take the thread and pin from the cone, and fastening the pin as a center upon a sheet of paper, as shown in Fig. 268, carry it around the pin, keeping it stretched all the time, the track of the points fastened to the thread will describe the shape of the envelope of the frustum. By omitting the line produced by the upper of the two points, the envelope of the complete cone will be described. In both cases the lengths of the arcs described by the thread and the points attached to it are to be governed by the circumference of the base of the cone, as we have already described. Our object in introducing this experiment is to show a method applicable in common to right and elliptical cones, by which the correspondence between these two shapes may be the more readily discerned.

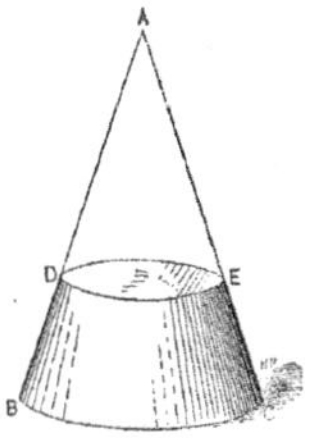

*Fig. 266.—Frustum of a Right Cone. The Dotted Lines show the Cone Restored for the purpose of Pattern Cutting.*

451. Since all the points in the boundary line of an elliptical figure are not equally distant from one common central point, it follows that the distance from the apex of an elliptical cone to the points in its base is a constantly varying one. Therefore quite different means are necessary in developing the shape of the envelope of an elliptical cone from those employed with the right cone. We will first describe the method of performing this operation with lines, and will afterward refer to the parts into which certain elliptical cones may be resolved by analysis of their shape, showing in that connection the pin and thread method applied to their development, by which comparison can be made with what we have already presented.

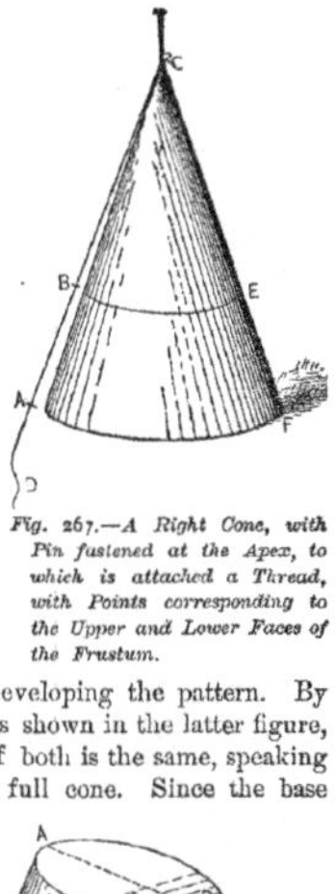

*Fig. 267.—A Right Cone, with Pin fastened at the Apex, to which is attached a Thread, with Points corresponding to the Upper and Lower Faces of the Frustum.*

452. In Fig. 269 we show the frustum of an elliptical cone, the envelope of which is required. The elevations and plan of the same figure are shown in Fig. 270, with the necessary lines for developing the pattern. By inspection of the two elevations shown in the latter figure, it will be seen that the hight of both is the same, speaking now more particularly of the full cone. Since the base width of the side is greater than the base width of the end, it follows that the angle of inclination or slant of the side is greater than that of the end. Or, by inspecting the plan, it will be seen that the distance B A, representing the flare at the ends, is greater than C D, which shows the flare at the sides. By reason of this irregularity the outline of the pattern will be a broken line instead of a regular curve, and is to be obtained by constructing a number of triangles. We can make this plainer, perhaps, by referring to the definition of a right cone, which, as already given, is "a solid generated by the revolution of a right-angled triangle around its altitude as an axis." Now, an elliptical cone, it is evident, cannot be generated by the revolution of a right-angled triangle around an axis, because the points in its base constantly vary in their distance from the center, and yet it is evident that a right angle may be constructed whose altitude shall be equal to the axis of the elliptical cone, and whose base shall be equal to half of the length of the base, and also that a similar triangle may be drawn, the altitude of which shall likewise be equal to the hight

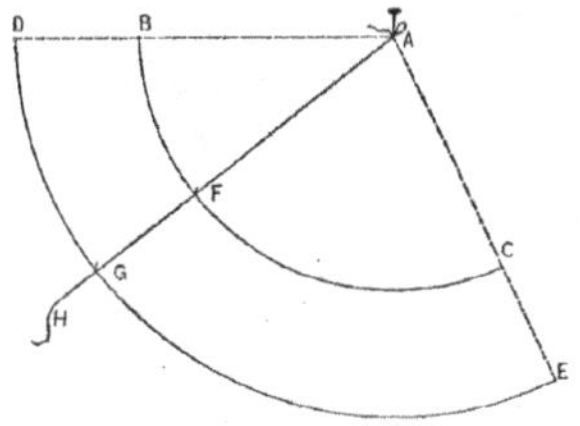

*Fig. 268.—Describing the Pattern of the Envelope of a Frustum of a Right Cone by means of a Pin and Thread.*

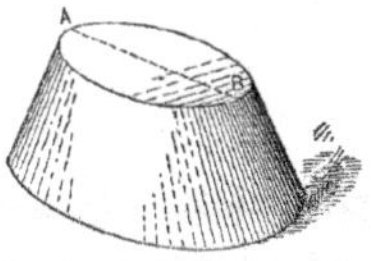

*Fig. 269.—The Frustum of an Elliptical Cone, the Envelope of which is to be Described.*

of the axis of the cone, and the base of which shall be equal to half of the width of the base of the cone. Such a right-angled triangle as we have been describing is indicated in Fig. 270 by K R F, in which K R, the hight, is the axis of the cone and R F is one-half of the length of the base, equal to X A of the plan. A similar triangle would be composed of K R for altitude and X D of the plan for base. Herein is a suggestion of the means which may be employed for describing the envelope of this shape. This solid cannot be generated by the revolution of a single triangle, but we can construct a number of triangles, having varying bases but one common altitude, by measurements on which all necessary dimensions may be obtained. In Fig. 270, divide one-quarter of the plan, as P D, into any convenient number of equal spaces, as shown by the figures 1, 2, 3, 4, etc. From the points thus established draw lines to the center of the plan X, as shown. At any convenient place draw a straight line, M X, as shown in Fig. 271, which in length make equal to the hight of the cone. At right angles to the base of this line lay off X 1, in length equal to X 1 of the plan, Fig. 270, and in like manner set off distances equal to the length of the several lines drawn from X to the boundary of the figure in the plan, and from these points draw lines to the point M. By this operation we have in one diagram a set of triangles corresponding to the lines drawn in the plan. The next step is to apply the dimensions we have now obtained to the development of the pattern. By inspection it is evident that the points in the boundary of the pattern corresponding to the points 1, 2, 3, 4, etc., in the plan, will be the same distance from one common center as these points in base or plan are from the apex of the cone. The distance of these points from the apex of the cone is indicated by the hypothenuses of the triangles constructed in Fig. 271. Therefore, taking M as a center, we set the compasses to the several lengths M D, M 7, M 6, etc., as radii, and describe arcs as shown to the left. At any convenient point in the arc corresponding to D, as D', we draw a line to M, which will represent one side of the pattern. From D' we then lay off the stretchout of the base, as shown by the divisions 1, 2, 3, 4, etc., in the plan, taking the distance in the dividers and stepping from one arc to another, as shown in Fig. 271. A similar set of arcs is to be drawn from the intersections of the line representing the top of the frustum with the hypothenuses of the triangles. Then lines drawn from the points established on the lower set of arcs, will intersect the last arcs drawn at points representing the upper line of the pattern.

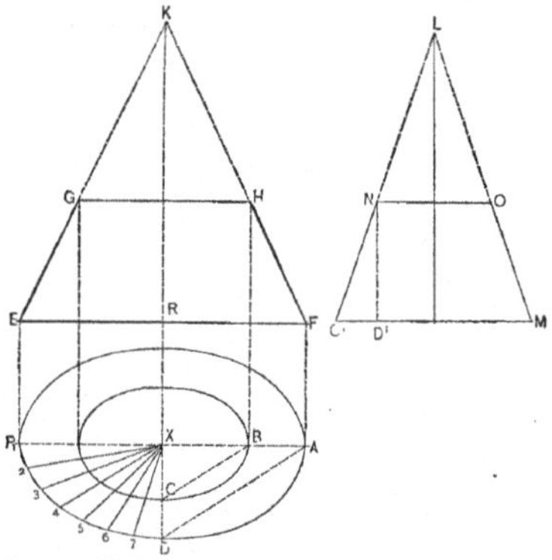


*Fig. 270.—Side Elevation, End Elevation and Plan of the Frustum shown in the Preceding Figure.*

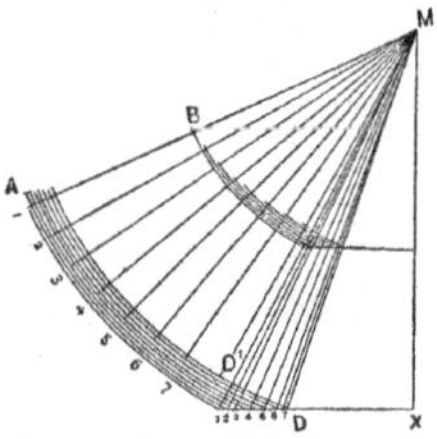


*Fig. 271.—Diagram of Triangles constructed from Measurements upon Elevations and Plan in Fig. 270, showing how they are Spread in Describing the Pattern.*

453. The method just explained for obtaining the pattern of the envelope of an elliptical cone applies to what we may term perfect elliptical cones only. By a perfect elliptical cone we mean a solid whose base is an ellipse and whose apex is a point. Such a solid as we have shown in Fig. 270, gives frustums of which the flare at sides and ends is unequal. Patterns of this kind, however, are less frequently required in practice than those in which the flare is alike throughout. If the flare of ends and sides is made the same, the resulting solid, if we attempt to complete it, will not be a perfect elliptical cone, but rather an irregular form, which it will be found, upon careful inspection, can be resolved into several simple parts. In explanation of this form we will first describe the usual rule for developing patterns of regular flaring ware, and will afterward undertake to explain the reasons for the several steps taken.

454. In Fig. 272 is shown the usual method employed for describing the patterns of regular flaring ware.

K L N M represents the frustum of which the envelope is required. A B C D is the plan of the same on the base line, while the inner curve represents the plan of the upper surface of the frustum. The ellipses representing the plan of the article, from the requirements of succeeding operations, are struck from centers, or, if true elliptical curves are employed, they are to be resolved into arcs of circles by the method explained in Section 401. In this case, in order to simplify the explanation, we have employed a plan described from two sets of centers. Having determined these centers, we draw the lines C X, D E, etc. The next operation is to construct the diagram shown in the upper part of the figure, which determines the radii to be employed in developing the pattern. Lay off O P equal to D E of the plan, the latter being the radius of the arc E C W. Upon O erect the perpendicular O J, continuing the same in the direction of J indefinitely. Make O S equal to the straight hight of the frustum, and draw S U parallel to O P, making it equal in length to D H of the plan, or the radius of the arc H G V. Now, if we draw a straight line through the points P and U thus established, and continue the same upward until it meets the perpendicular O J in the point J, we shall have a triangle which, if revolved upon its side J O as an axis, will generate so much of the conical shape of which the frustum in question is a part as corresponds to the arcs in the plan struck from D as center. Next, if we locate the point R in the base of the diagram by making O R equal to D F of the plan, and upon R erect the perpendicular R Z, producing the same until it meets the line P J in the point Z, we shall have in Z R P a triangle, which, if revolved upon Z R as an axis, would generate so much of the shape of which the frustum K L N M is a part as in plan is struck from F as center. The succeeding steps are self-evident, and we shall describe them without a diagram, because it will be necessary to show the same thing in its proper place among the pattern problems. From any convenient point as center, with J U and J P as radii, strike parallel curves, in length equal to E C W and H G V of the plan respectively. This will represent so much of the envelope as belongs to the cone which we said would be generated by the revolution of the large triangle J O P. From the terminal points in these curves draw lines to the center from which they were struck, and then with radii Z U and Z P, from a center in one of the lines just drawn, continue the curves, making the arcs in this case equal in length to X A E and the corresponding inner line of the plan struck from F as center. In this operation we have described the envelope of that part of the frustum which, as explained above, is a part of the cone that would be generated by the small triangle revolving about its altitude as an axis. These steps will give one-half of the pattern, and a repetition of them will produce the other half.

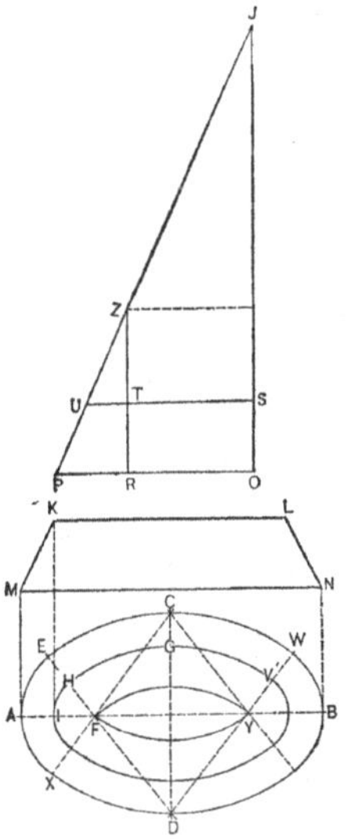


*Fig. 272.—The Usual Method of Developing the Patterns of Regular Flaring Shapes.*

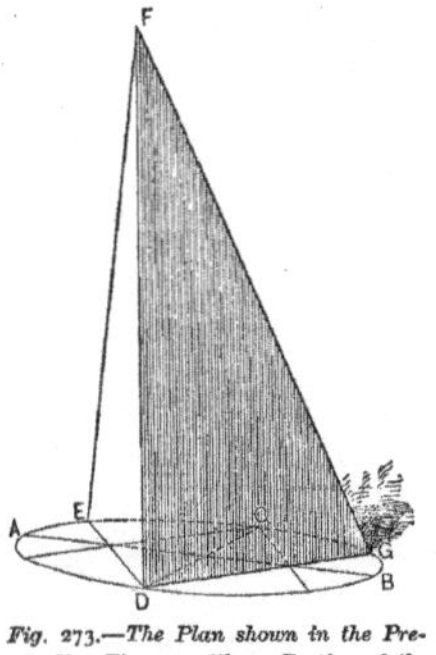


*Fig. 273.—The Plan shown in the Preceding Figure, with a Portion of the Solid, of which the Frustum K L N M is a Part, in Position.*

455. We have indicated, by the method of explanation above employed, that the solid of which the frustum K L N M of Fig. 272 is a part, is composed of portions of right cones of different diameters and altitudes joined together. The reasons of the several steps taken will be better understood if we show just what such a solid looks like, and how it is resolved into its component elements in the process of pattern cutting. We remarked that it was necessary, on account of subsequent operations, to employ a plan struck from centers, and then, having determined the centers, we constructed the triangle J O P (Fig. 272), which we said if revolved upon J O as an axis would generate a cone, a

part of which would correspond to a portion of the frustum in question. In Fig. 273, the plan A C B D corresponds to the plan represented by the same letters in Fig. 272. The triangle F D E corresponds to J O P of the preceding figure, while the portion of the solid represented in position on the plan is a part of the cone, as already explained, that would be generated by the revolution of this triangle about its altitude as an axis.

456. Fig. 274 shows the same plan, with portions of the small cone, corresponding to the end sections of the plan, in position. The triangles L F G and M H K correspond with the triangle Z R P of Fig. 272, and the small cones are the same as would be generated by the revolution of Z R P about its altitude Z R as an axis. As already remarked, we have, for the purpose of simplifying the operation, employed but two sets of centers in this illustration. From what has preceded, it is evident that if more centers were used the solid of which the frustum is a section would be composed of parts of a larger number of cones, the joining together of which would be upon the same general principle as here explained.

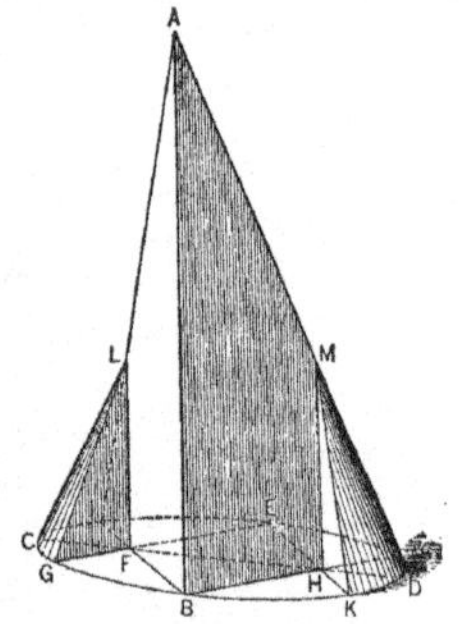


*Fig. 274.—The same Plan, with Portions of the Small Cone shown in Position, Joined to the Larger One.*

457. We will now return to the string or thread method, which we employed with the right cone, showing how it may be applied to this compound solid. A description of its use will still further explain the usual method of describing the patterns of regular flaring ware. Since all pattern cutting is, in result, a system of measurements upon the surface of the various solids, envelopes of which are required, occasional experiments in measuring the solids themselves, instead of always dealing with representations of them, are advantageous. Hence our experiments with the clay molding, the revolution of solids, and this string method of describing the envelopes of cones. Fig. 275 shows the opposite side of the parts presented in Fig. 274, with a pin fastened at the apex, and a thread attached carrying points G and H, representing the two surfaces of the frustum. Now, if we draw the string tight, and pass it along the side of the larger segment of the cone from A to B, the points will follow the upper and lower bases of the frustum. When we reach the point B, if the finger be placed upon the thread at the apex of the lesser cone, as shown by C, and the progress of the thread be continued, the points will still follow the lines of the bases of the frustum. If the pin and thread be taken from the cone and transferred to a sheet of paper, as shown in Fig. 276, the pin A being used as a center and the thread as a radius, the points will describe the envelope of the frustum. First, the radius is used full length, as shown by A L K, and arcs L M and K H are drawn, in length equal to the base of the larger segment of cone in the solid, Fig. 275. Then a second pin is put through the string, as shown at B, thus reducing the radius to the length of the side of the lesser cone, and arcs are struck in continuation of those first described, making the length of the additional arc equal to the base of the segment of the small cone.

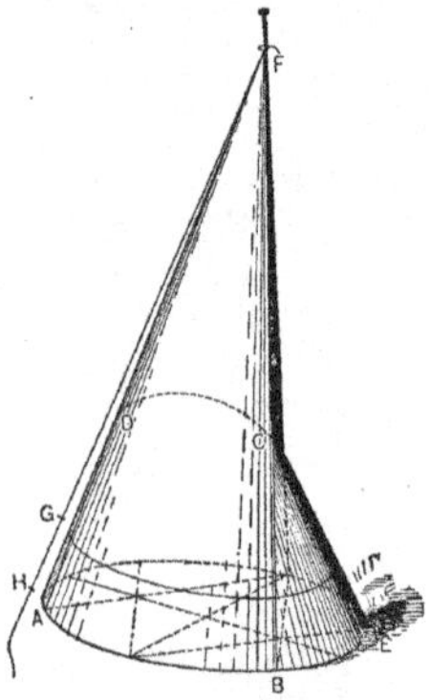


*Fig. 275.—The Opposite Side of the Parts in Fig. 274, showing a String attached to a Pin fastened at the Apex of the Larger Segment.*

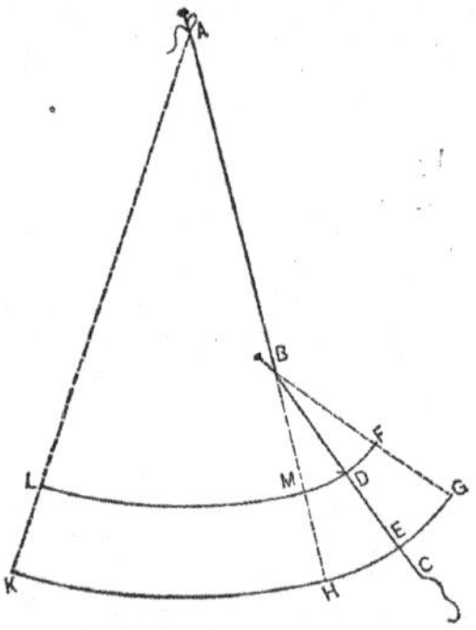


*Fig. 276.—The Pin and Thread taken from the Solid and Employed in Describing the Envelope.*

458. In our description of the solid of which the frustum that has equal flare all around is a part, we have called it a compound shape. In Figs. 273 and 274 we have shown parts of the cones corresponding to the triangles constructed in Fig. 272, which compose it. The larger cone employed has such diameter of base as causes its axis to fall upon the opposite boundary line of the plan, as shown by D in Fig. 273 and B in Fig. 274. Now, if it were desired to complete the solid—that is, to employ other portions of the cone to fill up the blank spaces in the plan—it would first be necessary to reduce the larger segment by cutting it upon a line corresponding to C D of the plan in Fig. 274. This line would pass through the top in the points L and M. By the nature of the shape with which we have to deal, the shape of the top of the solid thus cut would be a hyperbola. (See Section 120.) Completing the solid as above suggested, by adding a second section of the large cone, would produce the form shown in Fig. 277. To look at this solid, or to look at an ordinary elliptical flaring pan, affords little or no suggestions as to the possible composition of the shape and the rules for cutting the patterns which are to be deduced therefrom. Yet it is by such analyses as we have above described that the science of pattern cutting is to be understood.

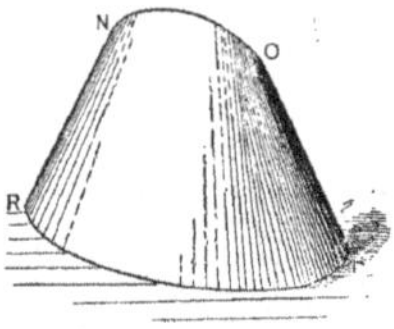

*Fig. 277.—The Solid of which a Regular Flaring Elliptical Frustum is a Part.*

459. We might extend this chapter, entering still further into the reasons and methods of sheet-metal pattern cutting, but enough has been written to afford the intelligent student such an insight as will enable him to continue investigations in other directions for himself. So we shall bring our talk about the art and science of the subject to a close at this point, adding only a few words of general advice. We would caution the student against arbitrary rules and methods. We think we have demonstrated conclusively that there are governing principles underlying all operations whatsoever. Therefore we say, search for the reason of every step to be taken. Do not be content to follow a rule because it is a rule. There should be no rules in pattern cutting, using that word in the sense in which it is ordinarily employed. There are principles and the application of principles, but not set rules. The good sense of the student must govern him in the employment of principles and in the choice of methods. There is hardly a pattern to be cut which cannot be obtained in more than a single way. Under some conditions one method is best, and under other circumstances another. Careful thought before the drawing is commenced will show which is best for the purpose in hand. To make this book of the greatest possible usefulness, we have added quite an extensive list of problems and demonstrations, but the methods we have employed are not to be taken as fixed rules. The same results in almost all cases may be reached by different methods. The student, therefore, should learn to choose between the different ways open before him and to work intelligently, otherwise he will not attain the highest degree of proficiency in his art.

# PATTERN PROBLEMS.

460. Having in the preceding chapters defined the terms most frequently employed in pattern cutting, shown how drawing tools and materials may be employed most advantageously, explained the geometrical problems of most common occurrence in practical work and the general theory of pattern cutting, we will now complete our task by presenting a selection of pattern problems, so arranged as to be convenient for reference upon the part of those who make use of this portion of the book without previous study of the other chapters. We shall attempt, therefore, to make each demonstration complete in itself and to avoid references to other parts of the book. To do this, we must assume for the reader a certain degree of familiarity with general principles and methods. If any one fails to comprehend any of the steps described, we suggest that his difficulties may be overcome by turning to those parts of the book where elementary matters are explained.

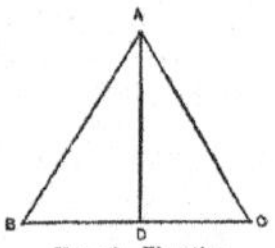

Fig. 278.—Elevation.

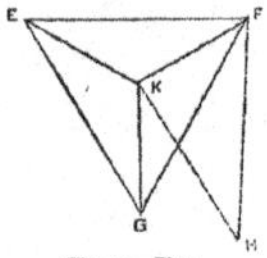

Fig. 279.—Plan.

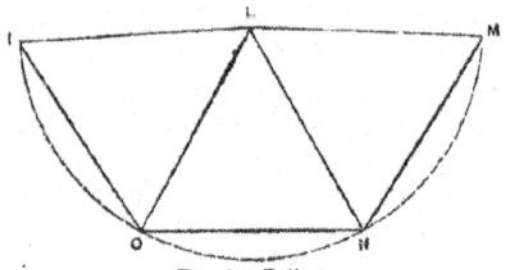

Fig. 280.—Pattern.

*The Envelope of a Triangular Pyramid.*

461. *The Envelope of a Triangular Pyramid.*—Let A B C of Fig. 278 be the elevation of the pyramid, and E F G of Fig. 279 the plan. Draw the lines E K, F K and G K in the plan, representing the angles. From the end of any one of them, as K of the line F K, erect a perpendicular, as K H, equal in length to the hight of the pyramid, as shown by A D of the elevation. Draw F H, which then represents the length of the corner lines. From any point, as L of Fig. 280, for center, with radius equal to F H, describe the arc M N O I indefinitely. Draw L M. From M set off the chord M N, in length equal to the side F G of the plan. In like manner set off N O and O I respectively, equal to G E and E F of the plan. Connect I and L, as shown, and draw L O and L N. Then L I O N M is the pattern sought.

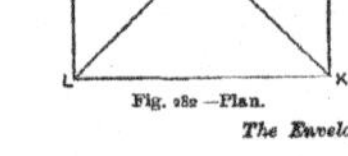

Fig. 281.—Elevation

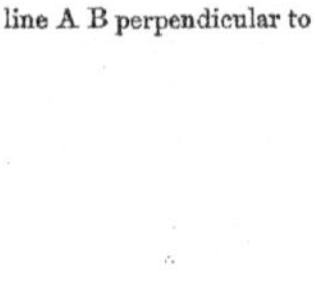

Fig. 282.—Plan.

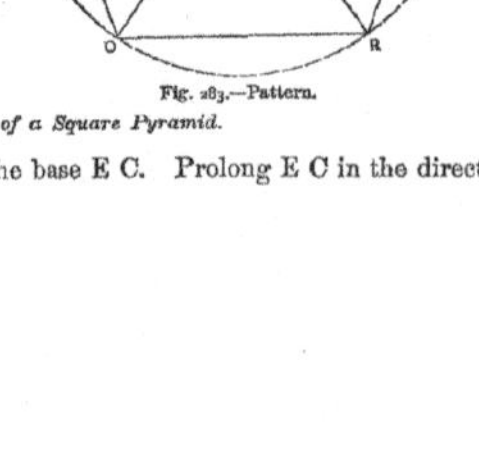

Fig. 283.—Pattern.

*The Envelope of a Square Pyramid.*

462. *The Envelope of a Square Pyramid.*—Let E A C of Fig. 281 be the elevation of the pyramid, and F H K L of Fig. 282 the plan. The diagonal lines F K and L H of the plan represent the angles or corners, and G, a point corresponding to the apex A of the elevation. From the apex A drop the line A B perpendicular to the base E C. Prolong E C in the direction of D, making B D equal to G F of

the plan. Connect D and A. Then A D will be the slant hight of the article on one of the corners, and the radius of an arc which will contain the pattern, as shown in the diagram. From any center, as M, Fig. 283, with a radius equal to A D, describe an arc, as P R O S N, indefinitely. Draw M P. From P set off a chord, P R, in length equal to one of the sides of the pyramid shown in the plan. From R set off another chord, R O, in like manner, and repeat the same operation for O S and S N. Draw M N, and likewise M S, M O and M R. Then M N S O R P will be the required pattern.

463. *The Envelope of the Frustum of a Square Pyramid.*—In Fig. 284, let G H K I be the elevation of the article, C A E D the plan of the larger end and L M O N the plan of the smaller end. Produce the miter lines C L, A M, etc., in the plan to the center P. Construct a diagonal section on the line A P as follows: Erect the perpendicular P F, making it equal to the straight hight of the article, as shown by R K of the elevation. Likewise erect the perpendicular M B of the same length. Draw F B and A B. Then P A B F is the diagonal section of the article cut on the line P A. Produce A B indefinitely in the direction of X, and also produce P F until it meets A B extended in the point X. Then X is the apex of a right cone and X B a side of the same, the base of which, if drawn, would circumscribe the plan C A E D. Therefore, from any convenient center, as $X^1$ of Fig. 285, with X A as radius, describe the arc $C^1$ $D^1$ $E^1$ $A^1$ $C^2$, and from the same center, with radius X B, draw the arc $L^1$ $N^1$ $O^1$ $M^1$ $L^2$, both indefinitely. Draw $C^1$ $L^1$. Make the chord $C^1$ $D^1$ equal to one side, C D, of the plan, and $D^1$ $E^1$ to another side, D E, of the plan, and so on. Draw $D^1$ $N^1$, $E^1$ $O^1$, etc., which will represent the lines of bend in forming up the pattern. Draw the chords $L^1$ $N^1$, $N^1$ $O^1$, etc., thus completing the pattern.

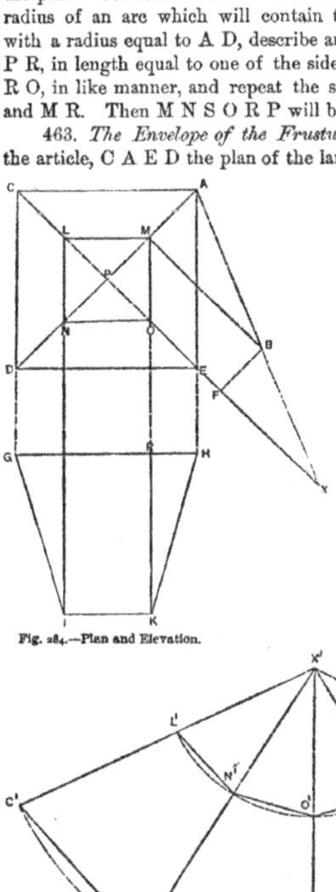


Fig. 284.—Plan and Elevation.

464. *The Envelope of a Hexagonal Pyramid.*—Let H G I of Fig. 286 represent the elevation of a hexagonal pyramid, of which D F C L B E of Fig. 287 is the plan. The first step is to construct a section on a line drawn from the center of the figure through one of its angles, as shown in the plan by A B. From the center A erect A X perpendicular to A B, making it equal to the straight hight of the article, as shown in the elevation by G K. Draw B X. Then X is the apex and X B one of the sides of a right cone, the plan of the base of which, if drawn, would circumscribe the plan of the hexagonal pyramid. From any convenient center, as $X^1$ of Fig. 288, with X B as radius, describe an arc indefinitely, as shown by the dotted line. Through one extremity to the center draw a line, as shown by $D^1X^1$. With the dividers set to a space equal to any side of the plan, commencing at $D^1$ set off this distance on the arc six times, as shown. From the several points $E^1$ $B^1$ $L^1$ in the arc thus obtained, draw lines to the center, as shown by $E^1$ $X^1$, $B^1$ $X^1$, etc. These lines represent the angles of the completed shape, and serve to locate the bends to be made in process of forming up.

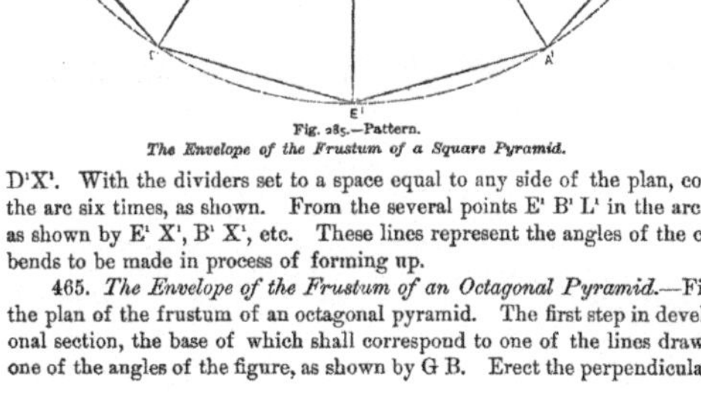


Fig. 285.—Pattern.

*The Envelope of the Frustum of a Square Pyramid.*

465. *The Envelope of the Frustum of an Octagonal Pyramid.*—Fig. 289 shows the elevation and Fig. 290 the plan of the frustum of an octagonal pyramid. The first step in developing the pattern is to construct a diagonal section, the base of which shall correspond to one of the lines drawn from the center of the plan through one of the angles of the figure, as shown by G B. Erect the perpendicular G C equal to the straight hight of the

frustum, as shown by N M of the elevation, and at *b* erect a perpendicular, *b* A, of like length. Draw B A and A C. Then G B A C is a section of the article as it would appear if cut on the line G B. Produce B A indefinitely in the direction of X, and likewise prolong G C until it intersects B A produced in X. Then X is the apex and X B the side of a right cone, the plan of which, if drawn, would circumscribe the base of the frustum. From any convenient center, as $X^1$, Fig. 291, with radius X B, describe an arc indefinitely, as shown by the dotted lines $E^1$ $E^2$ of the pattern, and from the same center, with X A for radius, describe the arc $e^1$ $e^2$ of the pattern. Through one extremity of each to the center draw a straight line, as shown by $E^1$ $e^1$ $X^1$. Set off on the arc $E^1$ $E^2$ spaces equal to the sides of the plan of the base of the article and connect the points by chords. Thus make $E^1$ $P^1$ of the pattern equal to E P of the plan, and so on. Also from these points in the arc draw lines to the center, cutting the arc $e^1$ $e^2$, as shown. Connect the points thus obtained in this arc by chords, as shown by $e^1$ $p^1$, $p^1$ $d^1$, $d^1$ $o^1$, etc. Then $e^1$ $E^1$ $E^2$ $e^2$ will be the pattern sought.

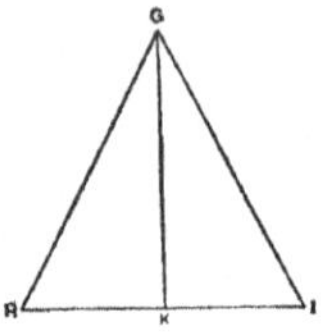

Fig. 286.—Elevation.

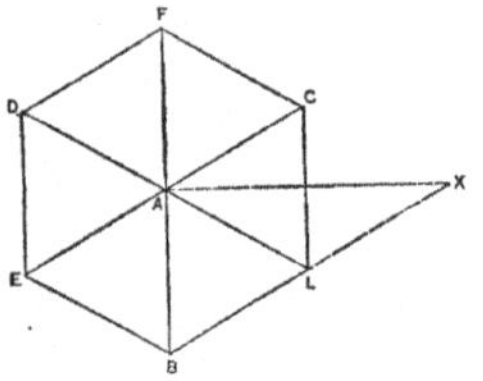

Fig. 287.—Plan.

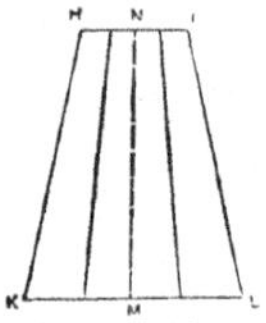

Fig. 289.—Elevation.

466. *The Envelope of the Frustum of an Octagonal Pyramid having Alternate Long and Short Sides.*—In Fig. 292, let I M B N O P K L be the plan of the article of which G H F E is the elevation. The first thing to do in describing the pattern is to construct a section corresponding to a line drawn from the center to one of the angles in the plan, as S B. At S erect the perpendicular S R, in length equal to the straight hight of the article, as shown by C D of the elevation. Upon the point *b* erect a corresponding perpendicular, as shown by *b* A. Draw R A and A B. Then B A R S is a section of the article taken upon the line S B. Produce S R and B A until they meet in the point X. Then X is the apex and X B is the side of a cone, the base of which, if drawn, would circumscribe the plan of the article. From any convenient center, as $X^1$, Fig. 293, with radius equal to X B, describe an arc, as shown by $M^1$ $M^2$. Draw $X^2$ $M^2$ as one side of the pattern. Then, starting from $M^1$, set off chords to the arc, as shown by $M^1$ $B^1$, $B^1$ $N^1$, etc., equal to and corresponding with the several sides of the article, as shown by M B, B N, etc., in the plan. From these points, $B^1$, $N^1$, etc., in the arc, draw lines to the center $X^1$.

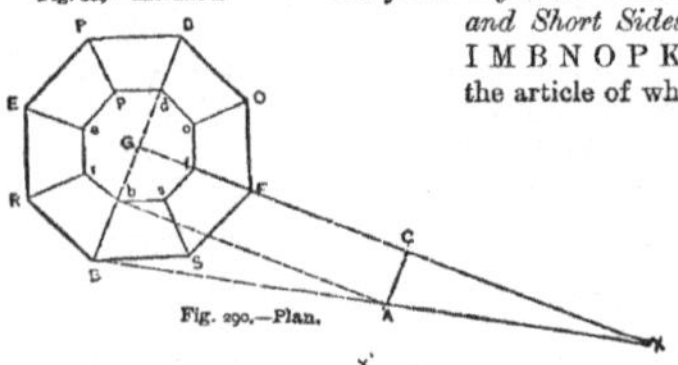

Fig. 290.—Plan.

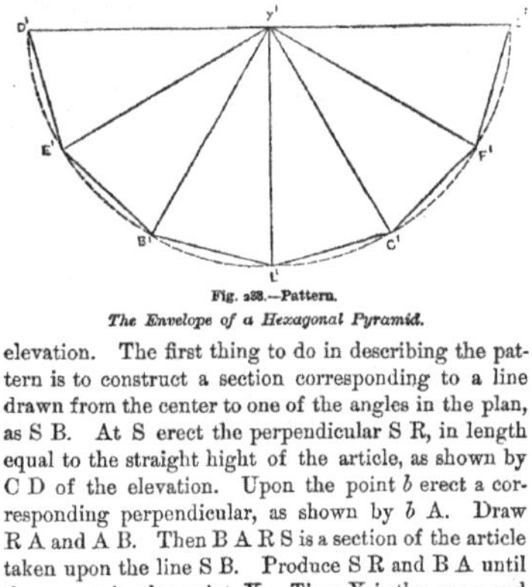

Fig. 288.—Pattern.

*The Envelope of a Hexagonal Pyramid.*

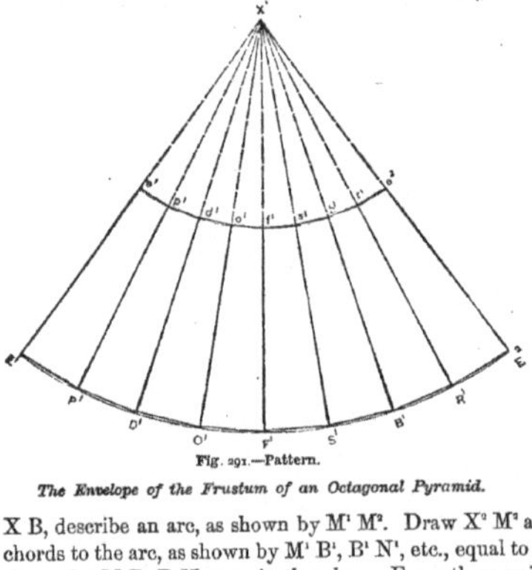

Fig. 291.—Pattern.

*The Envelope of the Frustum of an Octagonal Pyramid.*

From $X^1$, with X A as radius, describe an arc cutting these lines, as shown by $m^1$ $m^2$. Connect the points of intersection by straight lines, as shown by $m^1$ $b^1$, $b^1$ $n^1$, $n^1$ $o^1$, etc. Then $m^1$ $m^2$ $M^2$ $M^1$ will be the pattern sought, and the lines $B^1$ $b^1$, $N^1$ $n^1$, etc., will represent the lines of bends to be made in forming up the article.

467. *The Envelope of the Frustum of a Pyramid which is Diamond Shape in Plan.*—In Fig. 294, let A B D E be the elevation and K G I O the plan of the pyramid at the base. Complete the plan by drawing M L P R, the shape of the top of the article, and draw the angle lines G L, I P, O R and

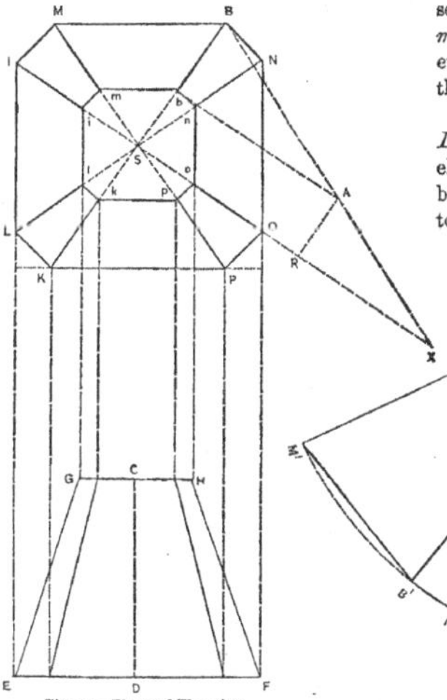


Fig. 292.—Plan and Elevation.

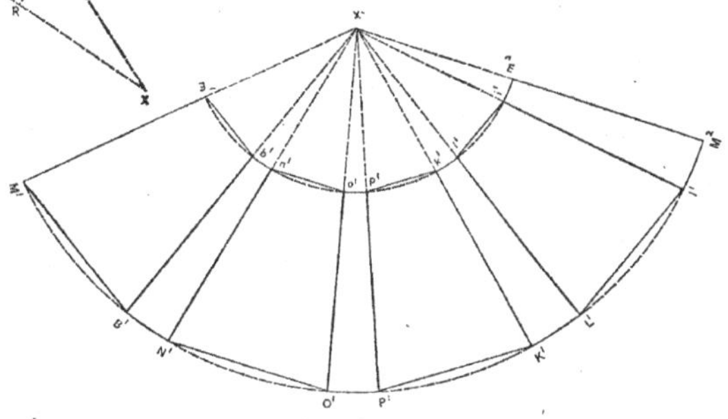


Fig. 293.—Pattern.

*The Envelope of the Frustum of an Octagonal Pyramid having Alternate Long and Short Sides.*

K M. Draw C F in the elevation, representing the angle G L of the plan. It also serves to measure the straight hight of the frustum. At right angles to M R of the plan draw S W, making its length equal to the straight hight of the frustum, as shown by C F of the elevation. Through W draw N H indefinitely, parallel to K O. At right angles to K O, through the points K and O, draw lines, K N and O H, cutting N H in the points N and H, thus establishing its length. Connect M N and R H. Then M R H N will be the pattern of one of the four sides composing the article.

468. *The Pattern of a Rectangular Flaring Article.*—In Fig. 295, let C A B E be the side elevation of the article, of which F I K M is the plan at the base and G H L N the plan at the top. Let it be required to produce the pattern in one piece, the top included. Make $H^1$ $L^1$ $N^1$ $G^1$ in all respects equal to H L N G of the plan. Through the center of it lengthwise draw R P indefinitely, and through the center in the opposite direction draw O S indefinitely. From the lines $H^1$ $L^1$ and $G^1$ $N^1$ set off T O and W S respectively, each in length equal to the slant hight of the article, as shown by C A or E B of the elevation. Through O and S respectively draw $I^1$ $K^1$ and $F^1$ $M^1$, parallel to $H^1$ $L^1$ and $G^1$ $N^1$, and in length equal to the corresponding sides in the plan I K and F M, letting the points O and S fall midway of these lengths respectively, as shown.

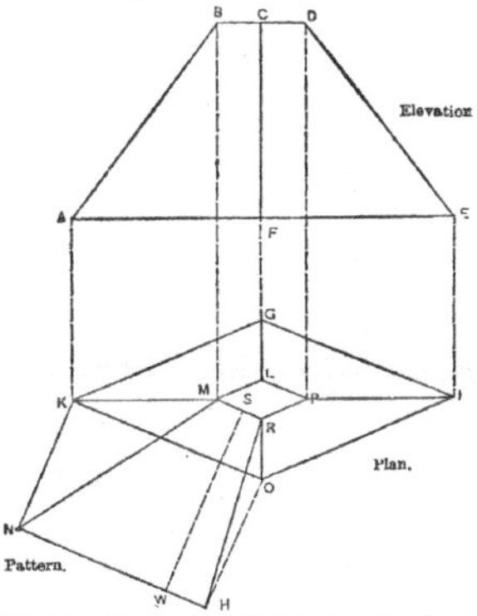


*Fig. 294.—The Envelope of the Frustum of a Pyramid which is Diamond Shape in Plan.*

In like manner set off V P and U R, and draw through R and P the lines $F^2$ $I^2$ and $K^2$ $M^2$, parallel to the ends of the pattern of the top part as already drawn, and in length equal to I F and K M of the plan. Draw $I^2$ H', K' L', $K^2$ L', $M^2$ N', M' N', F' G', $F^2$ G' and $I^2$ H', thus completing the pattern sought. In the same general way the pattern may be described, including the bottom instead of the top, if it be required that way, or the sides may be developed independent of either top or bottom.

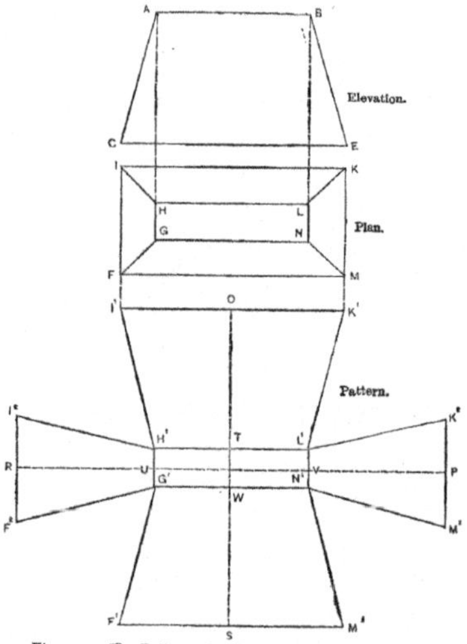


*Fig.* 295.—*The Pattern of a Rectangular Flaring Article.*

469. *The Pattern of a Rectangular Article, Three Sides of which are Vertical, the Fourth being Inclined.*—In Fig. 296, let L H I K be the elevation and C B A F E D the plan. Let it be required to describe the pattern in one piece, locating the seam at the point G in the plan. Draw L' G' indefinitely. From G', at right angles to L' G', draw G' O, in length equal to H I, the straight hight of the article in the elevation. Draw O E' indefinitely, parallel to G' L'. From G', which represents the end of the pattern, set off G' H', equal to G A of the plan. Draw H' F' at right angles to L' G', cutting O E' in F'. From H' set off H' L', in length equal to H L of the elevation, and from F' set off F' E', equal to I K of the elevation. Draw L' E', which corresponds to L K of the elevation. At right angles to L' E', from L', draw L' D', in length equal to the width of the article, as shown by C D of the plan. In like manner draw E' K' of the same length. Draw D' K'. Upon the point E' erect the perpendicular E' M. With the dividers set to the distance M L', from D' as center, strike the arc *a b*, and with the radius E' M, from K' as center, strike the intersecting arc *y x*. Then the point of intersection, or M', corresponds to M of the other arm of the pattern, and is a point through which the line of the side must pass. Therefore, from D' through M' draw D' R, and parallel to it, from K', draw K' S. Make these lines respectively equal to H L and I K of the elevation and connect their extremities, which will complete the pattern.

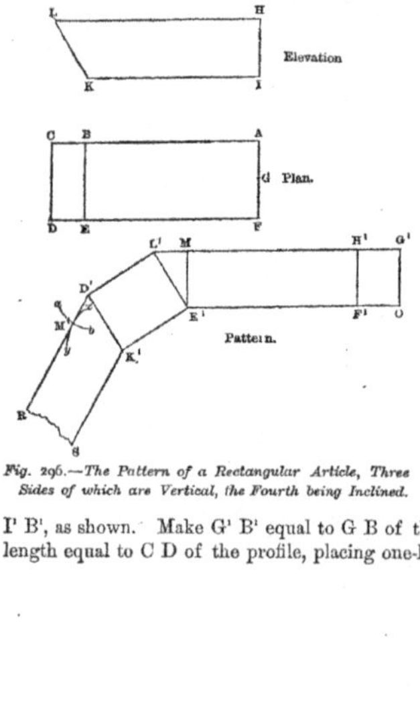


*Fig.* 296.—*The Pattern of a Rectangular Article, Three Sides of which are Vertical, the Fourth being Inclined.*

470. *The Pattern of a Rectangular Flaring Article having One End Upright.*—Let A B G H of Fig. 297 be the side elevation of the required article, and C U V D the elevation of the end, or section—both being the same in this case. Construct a plan, as shown by L M T S, Fig. 298, making L M and S T equal to A B of the elevation, and L S and M T equal to C D of the profile. Also make N P and O R equal to U V, and N O and P R equal to H G. From these three views of the article the pattern may be obtained as follows: Lay off N' O' R' P', Fig. 299, equal to N O R P of the plan, and through the center of it draw I' B', as shown. Make G' B' equal to G B of the elevation, and through B', parallel to O' R', draw C' D', in length equal to C D of the profile, placing one-half on each side from B'. Draw C' O' and D' R'. Produce

$O^1 R^1$ in the direction of $M^1$ and $T^1$, making $O^1 M^1$ and $R^1 T^1$ each equal to C U of the profile. From $M^1$ and $T^1$, parallel to the plan of the bottom already drawn, draw $M^1 L^1$ and $T^1 S^1$, in length equal to the sides, as shown by M L and T S in the plan. Connect $L^1 N^1$ and $S^1 P^1$. Make $H^1 A^1$ equal to H A of the elevation. Through $A^1$ draw $S^2 L^2$, parallel to $P^1 N^1$, and in length equal to C D of the profile, placing equal portions of it each side of $A^1$. Connect $L^2 N^1$ and $S^2 P^1$, thus completing the pattern. It is to be observed that the plan M T S L is not absolutely necessary in describing the pattern. All the necessary measurements may be obtained from the profile and the elevation. It is given in this connection, as more clearly showing the principles pertaining to patterns of this class than could be done without it.

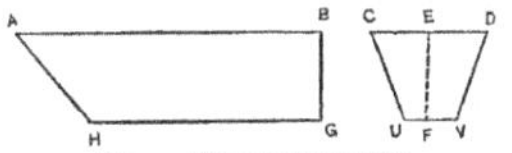

Fig. 297.—Side and End Elevation.

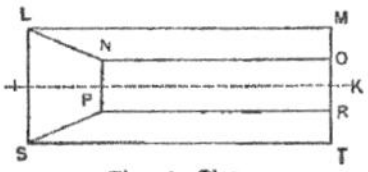

Fig. 298.—Plan.

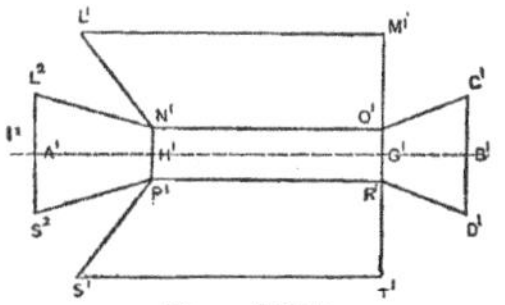

Fig. 299.—Pattern.

*The Pattern of a Rectangular Flaring Article having One End Upright.*

471. *The Pattern of a Flaring Article of which the Base is an Oblong and the Top Square.*—Let A B D E of Fig. 300 be the elevation of the article, and F N O I the plan. Let K M P L represent the top of the article. If the article is to be used as a cover, the top being solid and the bottom open, proceed for the pattern as follows: Draw $K^1 M^1 P^1 L^1$, Fig. 301, equal in all respects to K M P L of the plan. Through the center of it, and at right angles to each other, draw lines V U and S T indefinitely. Through the elevation, and perpendicular to the base and top, draw the line C G, which will measure the straight hight of the article. From G set off G H, in length equal to M R of the plan. Draw H C. Then H C will be the slant hight of the article on the side, and therefore the width of the pattern of that portion. The slant hight of the article at the ends, or the width of the pattern for the ends, is shown by A B and E D of the elevation. Upon V U of the pattern, from $K^1 M^1$ set off W V, and from $L^1 P^1$ set off X U, in length equal to H C of the elevation, and upon S T set off Z T from $M^1 P^1$, and Y S from $K^1 L^1$, in length equal to A B or D E of the elevation. Through U and V draw lines parallel to $K^1 M^1$ and $L^1 P^1$, making them in length equal to F N and I O of the plan, letting the points V and U come midway of their lengths respectively. Draw $F^1 K^1$, $N^1 M^1$ and $I^1 L^1$, $O^1 P^1$. In like manner through the points S and T draw $F^2 I^2$ and $N^2 O^2$ parallel to $K^1 L^1$ and $M^1 P^1$, and in length equal to F I and N O of the

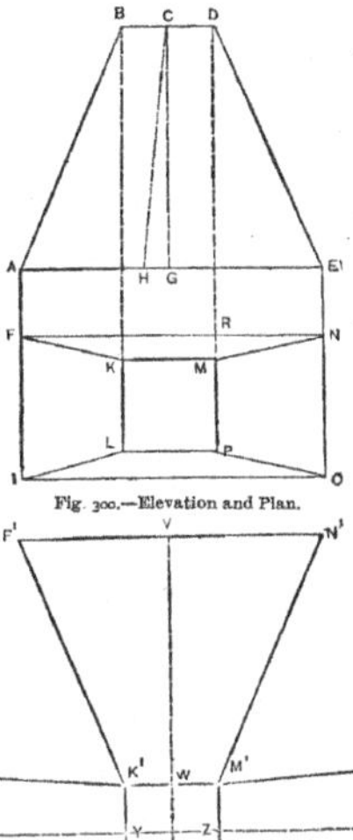

Fig. 300.—Elevation and Plan.

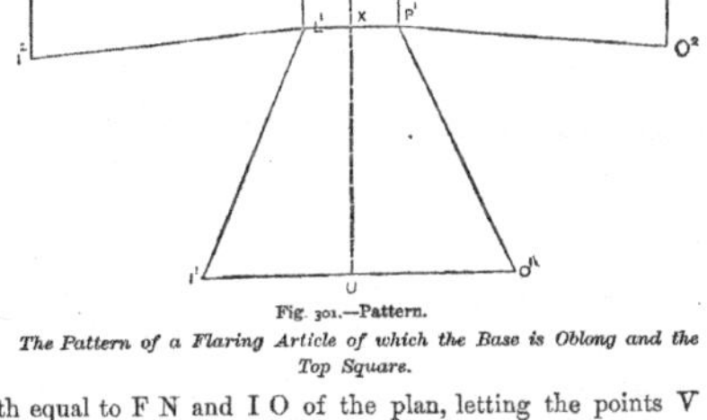

Fig. 301.—Pattern.

*The Pattern of a Flaring Article of which the Base is Oblong and the Top Square.*

plan, letting the points S and T fall midway of their lengths respectively. Draw $F^2 K^1$, $I^2 L^1$ and $N^2 M^1$, $O^2 P^1$, which will complete the pattern. If the pattern is wanted in four pieces instead of one, as above described, set off $K^1 M^1$, upon which erect $K^1 F^1 N^1 M^1$ in the same manner as explained, and likewise upon $M^1 P^1$ erect $M^1 N^2 O^2 P^1$. The other lines and parts may be dispensed with.

472. *The Patterns of a Tapering Article which is Square at the Base and Octagonal at the Top.*—A B D C in Fig. 302 shows the plan of the article at the base, I K L M H G F E represents the shape at the top, $E^2 H^2 D^2 C^2$ is an elevation of one side. Construct a diagonal elevation, as shown by $I^1 G^1 D^1 A^1$, as follows: Extend the base line $D^2 C^2$ of the elevation, as shown, making $D^1 A^1$ equal to the diagonal length across the plan, as shown by D A. In like manner extend the top line $H^2 E^2$ of the elevation, making $G^1 I^1$ equal to the distance from G to I of the plan, letting the middle point $R^1$ in $I^1 G^1$ fall directly above the middle point $C^2$ in $A^1 D^1$. Draw $I^1 A^1$ and $G^1 D^1$. Then $I^1 G^1 D^1 A^1$ is an elevation or section of the article taken upon the line A D of the plan, and therefore $A^1 I^1$ represents the length of one of the smaller sides of the article. Produce the diagonal line R C, as shown, making N $C^1$ in length equal to $I^1 A^1$ of the diagonal section. By means of the T-square, as indicated by the dotted lines, set off $E^1 F^1$ equal to E F of the plan and draw $C^1 E^1$ and $C^1 F^1$. Then $E^1 C^1 F^1$ is the pattern of one of the smaller sides of the article. From the center R of the plan draw R P perpendicular to the side A C, upon which set off O P, in length equal to $E^2 C^2$ of the elevation. At right angles to it draw $A^2 C^4$, which, by means of the T-square, as shown by the dotted lines, make equal to A C of the plan. In like manner draw $I^2 E^4$ equal to I E of the plan. Connect $A^2 I^2$ and $C^4 E^4$. Then $A^2 I^2 E^4 C^4$ will be the pattern of one of the larger sides of the article. If for any reason the pattern is desired to be all in one piece, the shapes of the different sides may be laid off adjacent to each other, the large and small sides alternating, all as indicated by *i i*¹ *a a*¹, Fig. 303.

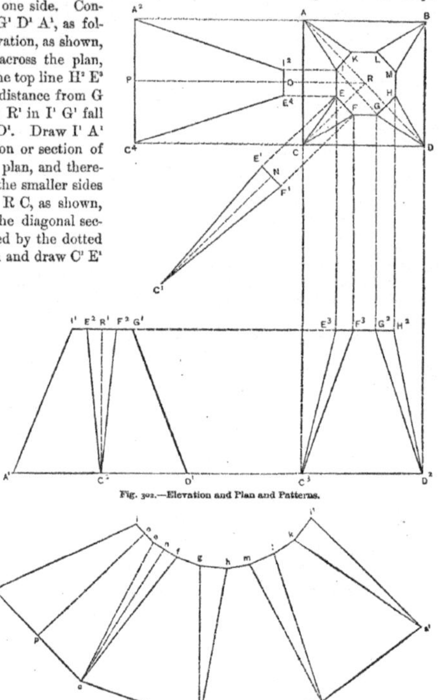


Fig. 302.—Elevation and Plan and Patterns.

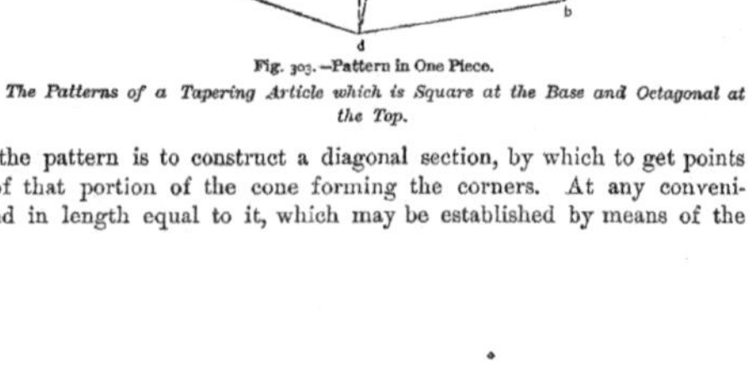


Fig. 303.—Pattern in One Piece.

*The Patterns of a Tapering Article which is Square at the Base and Octagonal at the Top.*

473. *The Pattern of a Flaring Article Square at the Base and Round at the Top.*—Let P R T S of Fig. 304 be the elevation of the article, G B K H, Fig. 305, the plan at the base, and L O N M the plan at the top. The corners, one of which is shown at M K N, are to be regarded as sections of oblique cones, the apexes of which lie in the angles of the plan of the base, or, in the case above cited, in the point K. The first step in developing the pattern is to construct a diagonal section, by which to get points from which to describe the envelope of that portion of the cone forming the corners. At any convenient place draw $A^1 B^1$ parallel to A B, and in length equal to it, which may be established by means of the

T-square, as indicated by the dotted lines A A' and B B'. From A', on A' A, set off A' A², in length equal to the straight hight of the article, as indicated by U V of the elevation. From A² draw A² C' parallel and equal to A C of the plan. Draw A² B' and C' B'. Then A' A² C' B' is a diagonal section of the article corresponding to the line A B of the plan, and A² B' C' represents a section of the oblique cone forming the corners. B' is the apex, B' A' is the axis, and B' C' is one of the sides, while N C O of the plan is a section of its base. Divide N C O into any convenient number of equal parts, and from the points thus obtained drop points perpendicular to A B on to A² C'. From A² C' carry the points in a direction perpendicular to the axis A² B' until they meet B' C',

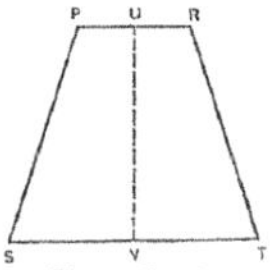

Fig. 304.—Elevation.

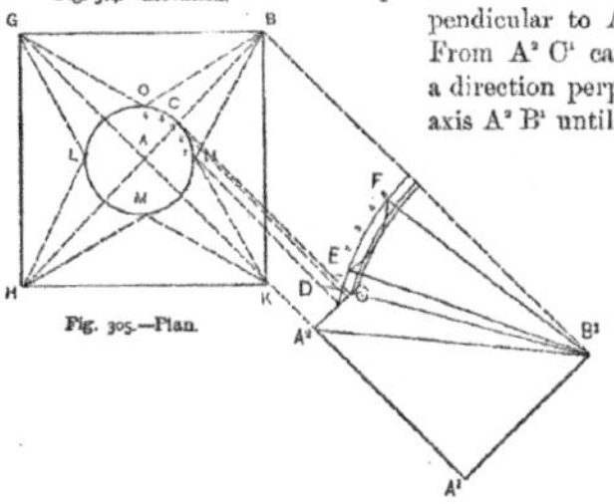
Fig. 305.—Plan.

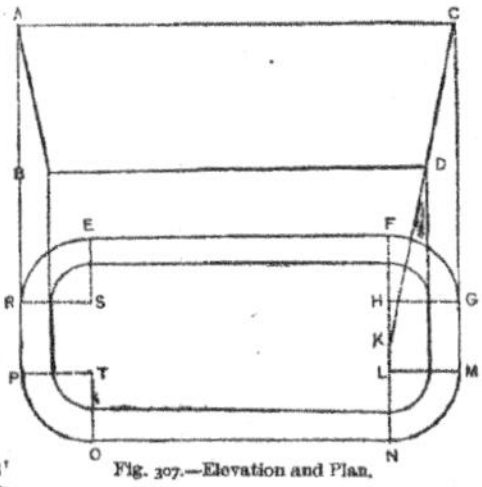
Fig. 307.—Elevation and Plan.

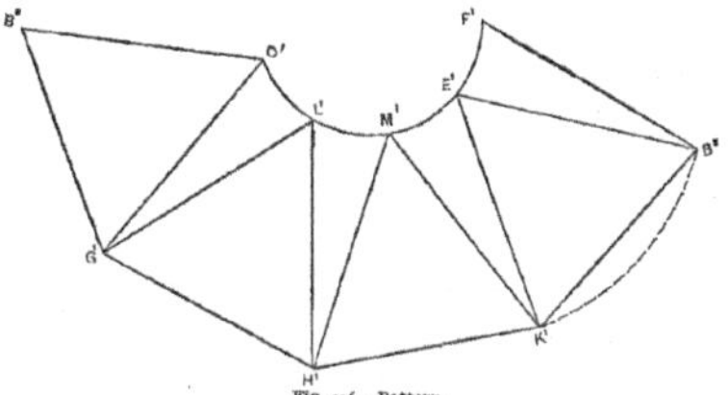
Fig. 306.—Pattern.

*The Pattern of a Flaring Article Square at the Base and Round at the Top.*

extended as shown by C' D. From B' as center, strike arcs corresponding to each of the several points in C' D, as shown. From any convenient point in the first arc, as E, draw a line to B', as shown by E B'. Set the dividers to the space used in stepping off the plan N C O and, commencing with the point E', lay off the stretchout of N C O, stepping from arc to arc as shown, the last point being F. Draw F B', and trace a line through the points in the arcs, as indicated by E F. Then E B' F is the pattern of one of the corners. For convenience in laying off the pattern in one piece, transfer this part of the pattern to any space sufficiently removed from the diagram of the plan to avoid confusion of lines,

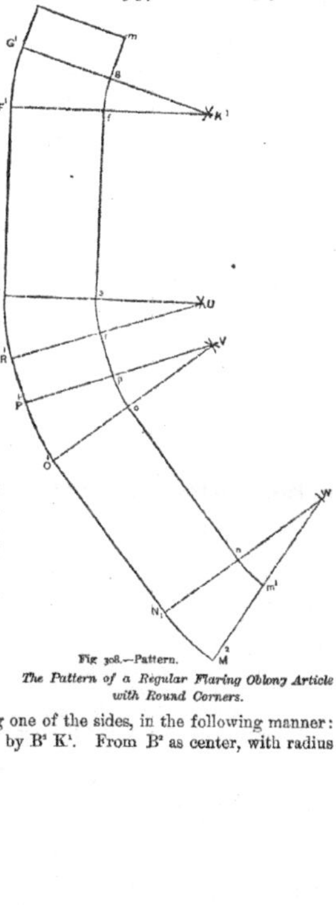
Fig. 308.—Pattern.

*The Pattern of a Regular Flaring Oblong Article with Round Corners.*

as shown by E' B² F' in Fig. 306. To this add the triangle forming one of the sides, in the following manner: From E' as center, with E' B² as radius, describe an arc, as shown by B² K'. From B² as center, with radius

B K of the plan, intersect that arc in the point $K^1$. Draw $B^2$ $K^1$ and $K^1$ $E^1$. To this in turn add the shape of one of the corners, and continue the operation until the entire number of sides are represented, all as shown by $O^1$ $L^1$ $M^1$ $E^1$ $F^1$ $B^3$ $K^1$ $H^1$ $G^1$ $B^2$, which shows the required pattern complete in one piece.

474. *The Pattern of a Regular Flaring Oblong Article with Round Corners.*—In Fig. 307 A C D B is the side elevation of the article and E F G M N O P R the plan. The corners are arcs of circles, being struck by centers H L T S, as shown. Draw the plan in line with the elevation so that the same parts in the different views shall correspond. Through the centers H and L of the plan by which the corners F G and M N are struck, draw F N indefinitely. Prolong the side line of the elevation C D until it cuts F N in the point K, as shown. Then K D is the radius of the inside line of the pattern of the curved part, and K C is the radius of the outside line. Draw the straight line $E^1$ $F^1$ of Fig. 308 in length equal to the straight part of one side of the article, or E F of the plan. Through the points $E^1$ and $F^1$, at right angles to the line $E^1$ $F^1$, draw lines indefinitely, as shown by $E^1$ U and $F^1$ $K^1$. Set the compasses to the radius K C, and putting the pencil point at $F^1$, establish the center $K^1$ in the line $F^1$ $K^1$. Strike the arc $F^1$ $G^1$, which in length make equal to F G of the plan. From $G^1$ draw a line to the center $K^1$, at right angles to which erect $G^1$ $M^1$, in length equal to G M of the plan. In like manner establish the center U in the line $E^1$ U, and from it, with like radius, describe the arc $E^1$ $R^1$. Draw $R^1$ U, at right angles to which erect $R^1$ $P^1$, equal to R P of the plan. At right angles to $R^1$ $P^1$ draw $P^1$ V indefinitely. In the manner above described establish the center V, and from it describe the third arc $P^1$ $O^1$. Draw $O^1$ V. At right angles to it lay off $O^1$ $N^1$, equal to O N of the elevation. Draw $N^1$ W, and draw the arc $N^1$ $M^2$ in the same manner as already described. In the same manner lay off the inner line of the pattern, as shown by *m g f e r p o n m'*. Join the ends $M^1$ *m* and $M^2$ *m'*, thus completing the pattern sought.

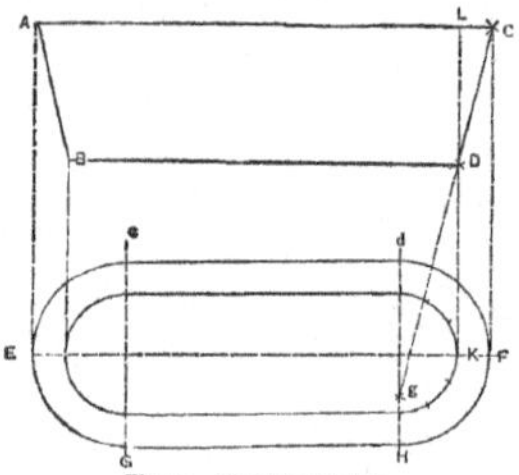


Fig. 309.—Elevation and Plan.

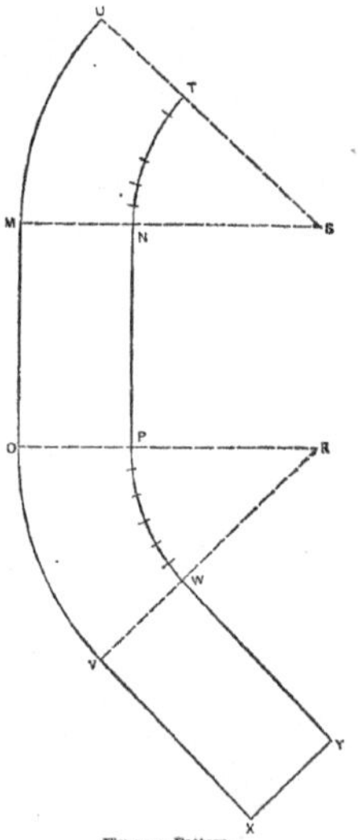


Fig. 310.—Pattern.

*The Pattern of a Regular Flaring Article which in Shape is Oblong with Semicircular Ends.*

475. *The Pattern of a Regular Flaring Article which in Shape is Oblong with Semicircular Ends.*—In Fig. 309, let A B D C be the side elevation of the required article. Below it and in line with it draw a plan, as shown by E *c d* F H G. From D in the elevation erect the perpendicular D L. Then L C represents the flare of the article and C D is the width of the pattern throughout. Across the plan at the point where the curved end joins the straight sides, draw the line *d* H at right angles to the sides of the article. As the plan may be drawn at any distance from the the elevation, this line must be prolonged, if necessary, to meet C D extended. Produce C D until it meets *d* H, as shown by *g*. Then *g* D and *g* C are radii of the curved parts of the pattern. Lay off on a straght line, M O in Fig. 310, the length of the straight part of the article, as shown in the plan by *c d*. At right angles to M O draw M S and O R indefinitely. On these lines set off the spaces M N and O P respectively, both in length equal to C D, the slant hight of the article, which must be the width of the pattern. Set the compasses to the space *g* C for radius, and putting the pencil to M, establish the center S, which must fall somewhere in the line M S. From this center strike the arc M U indefinitely. In like manner, with same radius, describe the arc O V. From the same centers, with radius equal to *g* D, describe the arcs N T and P W. Step off the length of the curved

part of the article upon either the inner or outer line of the plan, and make the corresponding arc of the pattern equal to it, as shown by the spaces in N T and P W. Through the points T and W draw lines from the centers S and R, producing them until they cut the outer arcs at U and V. At right angles to the line S T U or R W V, as the case may be, set off V X Y W, equal to M O P N, which will be the other straight side of the pattern. Then U M O V X Y W P N T will be the complete pattern in one piece. If it were desired to locate the seam midway of one of the straight sections, in adding the last member, as above described, one-half would be placed at each end, instead of all at one end, as we have shown. In like manner changes may be made locating the seam at other points, or for cutting the pattern into several pieces.

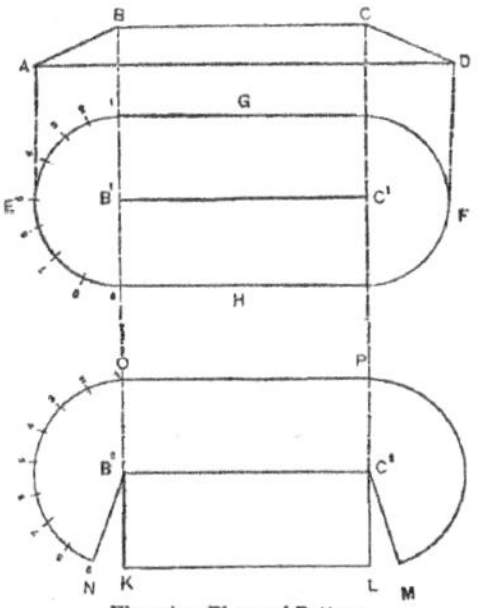


*Fig. 311.—The Pattern of a Raised Cover, Fitting an Oblong Vessel with Round Ends.*

476.—*The Pattern of a Raised Cover, Fitting an Oblong Vessel with Round Ends.*—In Fig. 311, let A B C D represent a side elevation of the cover of which E G F H is the plan or shape of the vessel it is to fit. Various constructions may be employed in making such a cover as this; that is, the joints, at the option of the mechanic, may be placed at other points than shown here; the principle used in obtaining the shape, however, is the same, whatever may be the location of the joints. By inspection of the elevation and plan it will be seen that the shape consists of the two halves of the envelope of a right cone, joined by a straight piece. Therefore, for the pattern we proceed as follows: At any convenient point lay off B² C², in length equal to B¹ C¹ of the plan. From B² and C² as centers, with radius equal to A B or C D of the elevation, describe arcs, as shown by O N and P M. Upon these arcs, measured from O and P respectively, set off the stretchout of the semicircular ends, as shown in plan, thus obtaining the points M and N. From N draw N B², and from M draw M C². From B² and C², at right angles to the line B² C², draw B² K and C² L, in length equal to A B of the elevation, which represents the slant hight of the article. Connect K and L, as shown. Then O N K L M P will be the required pattern.

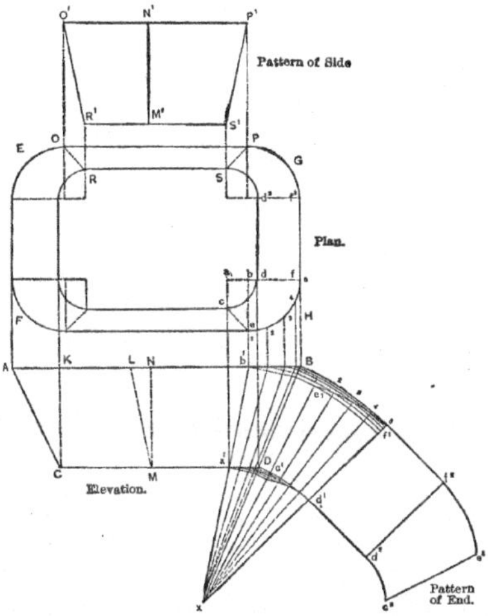


*Fig. 312.—The Patterns of a Flaring Article Oblong in Plan with Rounded Corners, and having Greater Flare at the Ends than at the Sides.*

477. *The Patterns of a Flaring Article Oblong in Plan with Rounded Corners, and having Greater Flare at the Ends than at the Sides.* —In Fig. 312, let A B D C be the elevation of the article and E G H F the plan. A K C of the elevation represents the flare of the ends, while L N M represents the flare of the sides. We will describe the pattern of the sides and ends, the latter including the rounded corners, as cut separately, although the two may be joined in one piece, or the entire rim may be constructed of one piece if required. For the pattern of the side, as indicated by O P S R in the plan, draw R¹ S¹ parallel to the side of the plan, and in length equal to R S. Draw a perpendicular to it, M¹ N¹, in length equal to L M. Through N¹ draw O¹ P¹, also parallel to the side of the plan and equal to O P. The length of O¹ P¹ and R¹ S¹ may be readily determined by using the T-square,

as indicated by the dotted lines. Draw $O^1 R^1$ and $P^1 S^1$. Then $O^1 P^1 S^1 R^1$ will be the pattern of the side. In obtaining the pattern for the corners they must be considered as parts of cones. An elevation of the section of the cone must be constructed with its base on a line parallel to a line drawn through the centers by which the curves in the plan were struck. It is a matter of convenience to draw the elevation of the cone in connection with the elevation of the article, as shown. Through the centers *a b*, by which one of the corners of the plan is struck, draw *a f*. Use the top line A B of the elevation, which is parallel to *a f*, for the base of the cone. From *a* drop a perpendicular to the base of the elevation, thus establishing the point $a^1$. In like manner drop a line from *b* to the upper line of the elevation, establishing the point $b^1$. A corresponding point to *f* is B, and to *d* is D. Draw $b^1 a^1$, producing it indefinitely in the direction of *x*. Also produce B D until it meets $b^1 a^1$ in the point *x*. Then *x* will be the apex of the cone, a section of which constitutes the corner of the article. Divide the plan of the cone *e f* into any convenient number of equal parts, as shown by the small figures. From each of these points drop a perpendicular to the base $b^1$ B of the cone, as shown, and from the points in it thus determined carry lines toward the apex *x*, cutting the line $a^1$ D. Consider $x b^1$ as the axis of the cone. Therefore from the points in $b^1$ B and $a^1$ D, at right angles to $x b^1$, carry points to D B, as shown. From *x* as center, strike an arc corresponding to each of the points in B D, just described, all as shown. From *x* draw any straight line, as $x e^1$, crossing these arcs, which shall be one end of the pattern. From this line, at the point of intersection with the arc corresponding to point 1 in the plan, set off the space of one of the divisions of the plan, stepping to the second arc. From this point set off a corresponding space, stepping to the third arc, and so on for each of the spaces set off in the plan. A line traced through these points, as shown by $e^1 f^1$, will represent one side of the curved part. From each of the points in $e^1 f^1$ draw lines toward the center *x*, cutting the lower set of arcs. Through the points of intersection thus obtained trace the line $c^1 d^1$, which will form the other boundary of one of the curved parts. At right angles to $f^1 d^1$ draw $f^1 f^2$, equal to $f f^2$ of the plan, and $d^1 d^2$, equal to $d d^2$ of the plan. Draw $f^2 d^2$, from which set off a second curved section, as shown by $f^2 e^2 c^2 d^2$, in all respects corresponding to $f^1 e^1 c^1 d^1$, thus completing the pattern of the ends of the article.

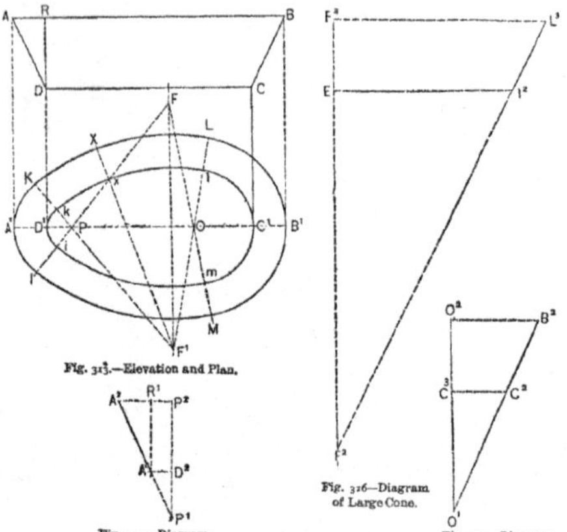

Fig. 313.—Elevation and Plan.

Fig. 314.—Diagram of Small Cone.

Fig. 316—Diagram of Large Cone.

Fig. 315.—Diagram of Middle Cone.

***An Oval or Egg-Shaped Flaring Pan.*** ***(For Pattern see Next Page.)***

478. *The Pattern of an Oval or Egg-Shaped Flaring Pan.*—Let A B C D in Fig. 313 represent the elevation of the article, of which $A^1$ K L $B^1$ M I is the plan. The plan is constructed by means of the centers O, P, F and $F^1$, as indicated. The patterns, therefore, are struck by radii obtained from sections of the several cones of which the article is composed. At any convenient place draw the line $P^2 P^1$, Fig. 314, indefinitely, which let correspond to P of the plan, and upon it construct a section of the article as it would appear if cut on the line $A^1$ P of the plan. Therefore set off, at right angles to it, $A^2 P^2$ equal to $A^1$ P. Make $P^2 D^2$ equal to the straight hight of the article, as shown by R D of the elevation. Make $D^2 A^3$ of the diagram equal to $D^1$ P of the plan. Draw $A^2 A^3$, which will correspond to A D of the elevation. Prolong $A^2 A^3$ until it meets $P^2 P^1$ in the point $P^1$. Then $P^1 A^2$ is the radius of the outside line of the pattern of the portion indicated by K I of the plan, and $P^1 A^3$ is the radius of the line inside of the same part. In like manner draw the line $O^1 O^2$, Fig. 315, corresponding to O of the plan, and construct a section taken on the line O $B^1$, as shown by $O^2 B^2 C^2 C^3$.

Produce $B^2$ $C^2$ until it meets $O^2$ $O^1$ in the point $O^3$. Then $O^3$ $C^2$ and $O^3$ $B^2$ are the radii of the pattern of that portion of the article contained between L and M of the plan. Draw the line $F^3$ $F^3$, Fig. 316, which shall correspond to F or $F^1$ of the plan. Make $F^3$ E equal to the straight hight of the article, and lay off $F^3$ $L^3$ at right angles to it, equal to $F^1$ L of the plan, and E $l^3$ equal to $F^1$ $l^2$ of the plan. Draw $L^3$ $l^3$, which produce until it meets $F^3$ $F^3$ in the point $F^3$. Then $F^3$ $l^3$ and $F^3$ $L^3$, respectively, are the radii of the pattern of those parts shown by K L and I M of the plan. To lay off the pattern after the several radii are obtained, as described above, draw any straight line, in length equal to $F^3$ $F^3$, as shown by $F^4$ $K^1$ in Fig. 317, and from $F^4$ as center, with $F^3$ $l^3$ and $F^3$ $L^3$, Fig. 316, as radii, strike arcs, as shown by $k^1$ $l^1$ and $K^1$ $L^1$, which in length make equal to the corresponding arcs of the plan K L *k l*, as shown. Draw $L^1$ $F^4$. Set the compasses to $O^3$ $C^2$, Fig. 315, and, placing the pencil at $l^1$, find the center $O^5$ in the line $L^1$ $F^4$, from which strike the arc $l^1$ $m^1$, in length equal to *l m* of the plan. In like manner, from the same center, with radius $O^3$ $B^2$ strike the arc $L^1$ $M^1$, equal in length to L M of the plan. Draw $M^1$ $O^5$, which produce indefinitely in the direction of $F^6$. Set the compasses to $F^3$ $l^3$, and, placing the pencil on $m^1$, establish the center $F^6$ in the line $M^1$ $F^6$, and continue the inner line of the pattern, as shown by $m^1$ $i^1$, which in length must equal *m i* of the plan. In like manner, from the same center, with radius $F^3$ $L^3$, describe the arc $M^1$ $I^1$. Draw $I^1$ $F^6$. Set the compasses to $P^1$ $A^3$, and, bringing the pencil point to $i^1$, establish the center $P^3$ somewhere in the line $I^1$ $F^6$. Describe the arc $i^1$ $k^2$, in length equal to *i k* of the plan. In like manner, from the same center, with the radius $P^1$ $A^3$, describe the arc $I^1$ $K^2$, in length equal to I K of the plan. Place the straight-edge against the points $P^3$ and $K^2$ and draw $K^2$ *k*, thus completing the pattern. From inspection it is evident that the pattern might have been commenced at any other point as well as at K *k* of the plan, where we have located the joint. If the joint is desired upon any of the other divisions between the arcs, as L *l*, M *m*, or I *i*, the method of obtaining it will be so nearly the same as above narrated as not to require special description. If the joint is wanted at some point in one of the arcs of the plan, as, for example, at X *x*, draw the line X *x* across the plan, producing it until it meets the center by which that arc of the plan is struck. In laying off the pattern, commence with a line corresponding to X $F^1$, in place of $F^4$ $K^1$, and from it lay off an arc corresponding to the portion of the arc in the plan intercepted by X *x*, as shown by X L *l x*. Proceed in other respects the same as above described until the line *k* $K^2$ is obtained, against which there must be added an arc corresponding to the amount cut from the first part of the plan by X *x*, as above described, or, in other words, equal to *x* X *k* K of the plan.

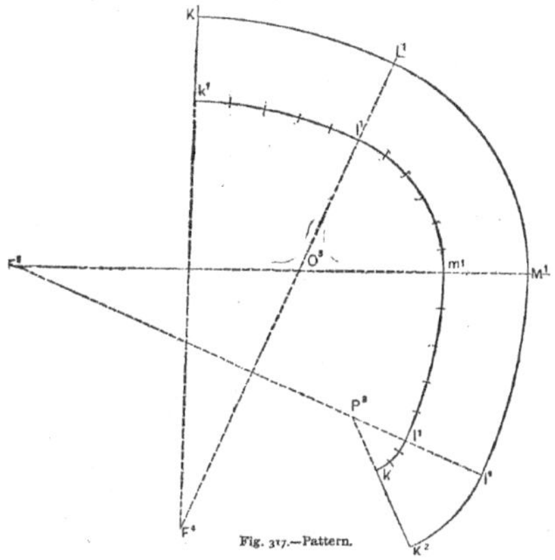


Fig. 317.—Pattern.

*An Oval or Egg-Shaped Flaring Pan.*

479. *The Pattern of a Heart-Shaped Flaring Tray.*—Let E C $G^1$ F G $C^1$ of Fig. 318 be the plan of the article, and I N O K the elevation. By inspection of the plan it will be seen that each half of it consists of two arcs, one being struck from D or $D^1$ as center, and the other from C or $C^1$ as center, the junction between the two arcs being at G and $G^1$ respectively. From $C^1$ draw $C^1$ F, and likewise draw $C^1$ $G^1$. Upon the point $D^1$ erect the perpendicular $D^1$ $C^1$. For the radii of pattern construct a diagram, in which show a profile of the article upon the lines $C^1$ $G^1$ and $D^1$ $C^1$. Draw X P in Fig. 319 in length equal to the straight hight of the article Lay off the perpendiculars X U and P S indefinitely. Upon P S, from P, set off P R equal to $D^1$ $C^1$ of the plan, and on X U, from X, set off X W equal to $D^1$ *c* of the plan. In like manner make P S equal to $C^1$ $G^1$ of the plan, and X U equal to $C^1$ *g* of the plan. Connect U S and W R. Produce P X indefinitely in the direction of Z. Also produce R W until it meets P X in the point Y, and in like manner produce S U until it meets P Z in the point Z. Then Z U and Z S are the radii for that portion of the article contained between

$G^1$ and F of the plan, and Y W and Y R are the radii of that portion shown from $G^1$ to E of the plan. To lay out the pattern after the radii are established, draw any straight line, as $Z^1$ $G^2$ in Fig. 320, in length equal to Z S of the diagram. From $Z^1$ as center, with Z S as radius, describe the arc $G^2$ $F^1$, in length equal to $G^1$ F of the plan. In like manner, with radius Z U, from the same center, describe the arc $g^1$ $f^1$, in length equal to $g$ $f$ of the elevation. Draw $f^1$ $F^1$. Set the compasses to Y R for radius. Place the pencil point at $G^2$, thus establishing the center $Y^1$, which must fall somewhere in the line $Z^1$ $G^2$. From $Y^1$, with radius as named, describe the arc $G^2$ $E^1$, which in length make equal to $G^1$ E of the plan. In like manner, from the same center, with radius Y W, describe the arc $g^1$ $e^1$ equal to the arc $g$ $e$ of the plan. Draw $e^1$ $E^1$, thus completing the required pattern.

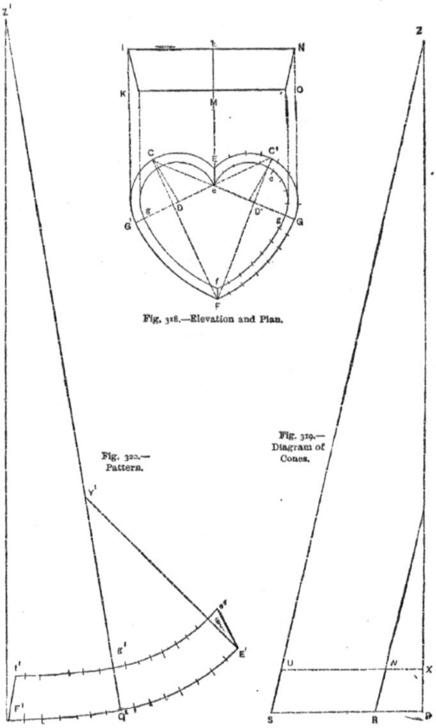


Fig. 318.—Elevation and Plan.

Fig. 319.—Diagram of Cones.

Fig. 320.—Pattern.

*The Pattern of a Heart-Shaped Flaring Tray.*

480. *The Pattern of a Flaring Article, the Top of which is Round and the Bottom of which is Oblong, with Semicircular Ends.*—In Fig. 321, O is the center by which the plan of the top is struck, and P is the center by which one of the semicircular ends is described. The elevation is placed so as to correspond with the plan, as shown by the lines connecting the two, E A, K B, M D and H C. From O erect the perpendicular O $o$, and from P erect the perpendicular P $p$. Prolong the side line C D of the elevation indefinitely in the direction of X. Through the points $p$ and $o$ draw $p$ $o$, which produce until it meets C D prolonged in the point X. Then X is the apex of a cone of which that portion of the article shown by D C $p$ $o$ in the elevation, and by L $p^1$ H $p^2$ N in the plan, is a section. Then X $p$ will represent the axis of the cone. Divide the profile $p^1$ H $p^2$ into any convenient number of equal parts. From each point in it erect a perpendicular to $p$ C, as shown. From the points thus obtained in $p$ C carry lines toward the apex X, cutting $o$ D, as shown. From the points in $p$ C, and also from those in $o$ D, draw lines at right angles to the axis X $p$, cutting the side X C of the cone, as shown. From X as center, strike arcs corresponding to each set of points in X C, as indicated. The arcs from the lower set of points are to receive the stretchout in the following manner: From X draw any straight line, as X $p^3$, meeting the first arc in the point $p^3$, which shall be one end of the stretchout. Set the dividers to the space used in stepping off the plan, and, commencing at $p^3$, step to the second arc, and from that point to the third arc, and so on, as shown in the engraving. A line traced through these points will be the boundary of the lower side of one of the semicircular ends. From each of these points just described draw a line toward the center X, cutting the upper set of arcs, as shown. A line traced through these points of intersection will form the upper edge of the pattern

of the end piece. From the point L', which corresponds to L of the plan, as center, with L' $p^4$ as radius, describe the arc $p^4$ R', and from $p^4$ as center, with radius equal to $p^1$ R of the plan, intersect it at R', as shown. Draw L' R'. Then L' R' $p^4$ is the pattern of one of the sides. To L' R' add a duplicate of the end piece already obtained, all as shown by L' R' E $R^2$ $N^2$, and to $N^2$ $R^2$ add a duplicate of the side just obtained, as shown by $N^2$ $R^2$ $p^5$, thus completing the pattern.

481. *The Pattern of a Flaring Article, the Base of which is a Rectangle and the Top of which is Round, the Center of the Top being toward One End.*—In Fig. 322, let L P N M be the side elevation of the article, of which A D C B is the plan at the base and E G H and K is the plan at the top. Draw two diameters through the plan of the top parallel to the sides of the article, cutting the top in the points E, G, H and K. Draw the lines in the plan A E, A G, D G, D H, etc., and consider the corner pieces E A G, G D H, etc., quarters of inverted scalene cones. The first step will be to obtain a profile or section of each of these quarter cones. From the center F of the plan of the top draw the diagonal lines F A and F D, which shall represent in plan the diagonal sections to be constructed. At any convenient distance outside of the plan draw $A^2$ $O^2$, in length equal to A F, and parallel to it. From the point $O^2$ erect the perpendicular $O^2$ $F^2$, in length equal to the straight hight of the article, as shown by L O of the elevation. From $F^2$, perpendicular to $O^2$ $F^2$, set off $F^2$ $S^1$ equal to F S of the plan. Draw $S^1$ $A^2$ and $A^2$ $F^2$. Then $A^2$ $O^2$ $F^2$ $S^1$ is a section of the article taken diagonally from the center F on the line F A. Divide one-quarter, G E, of the plan of the top, which forms the base of the cone of which the corner is a section, into any convenient number of equal parts, as shown by the small figures 5, 6, 7, etc., and from these points carry lines perpendicular to A F, producing them until they cut $F^2$ $S^1$, and thence at right angles to $F^2$ $A^2$ until they cut $A^2$ $S^1$ prolonged. Then from $A^2$ as center, describe arcs corresponding to these several points, as shown. From $A^2$ draw any straight line, as $A^2$ $E^2$, cutting the first arc in $E^2$. With $E^2$ as a starting point, and with the dividers set to the distance used in spacing the plan E S G, step to the second arc, and thence to the third, and in this manner lay off the stretchout, ending in the point $G^2$. Trace a line through these points in the arcs, as shown, and draw $G^2$ $A^2$. Then $G^2$ $A^2$ $E^2$ will be the pattern of the corner shown by G A E of the plan. In the same general manner construct a diagonal section corresponding to the line F D of the plan,

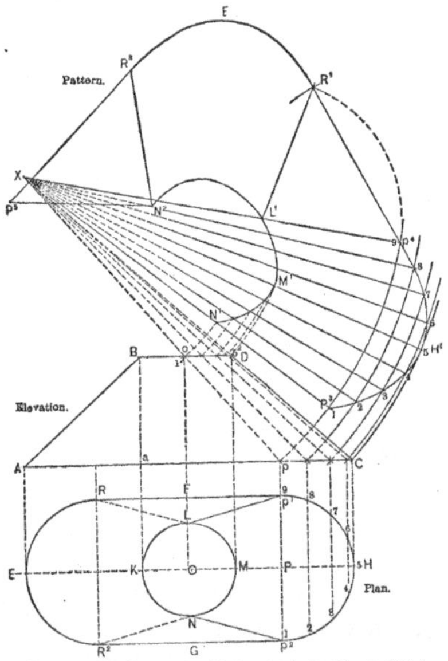


*Fig. 321.—The Pattern of a Flaring Article, the Top of which is Round and the Bottom of which is Oblong, with Semicircular Ends.*

all as shown by $O^1$ $F^1$ $R^1$ $D^1$. Divide the quarter-circle G H of the plan into any convenient number of equal parts, and from the points thus obtained erect lines perpendicular to F D, continuing them until they cut $F^1$ $R^1$, and thence carry them at right angles to $F^1$ $D^1$ until they meet $D^1$ $R^1$ produced. From $D^1$ as center, describe arcs corresponding to the several points in $D^1$ $R^1$ prolonged, upon which, commencing at any point in the first arc, as $G^1$, step off the stretchout of the plan, stepping from arc to arc in the same general manner as explained in connection with the section already constructed. Draw $G^1$ $D^1$ and $H^1$ $D^1$, and trace a line through the points stepped off in the arcs, as shown from $G^1$ to $H^1$. Then $G^1$ $D^1$ $H^1$ will represent the pattern of the corner shown by H D G of the plan. From $H^1$ as center, with $H^1$ $D^1$ as radius, describe an arc, as indicated by the dotted line, and from $D^1$ as center, with radius equal to D C of the plan, intersect it in the point $C^1$. Then $D^1$ $H^1$ $C^1$ represents the pattern of the end D H C of the plan. To this add a duplicate of $H^1$ $D^1$ $G^1$, as shown by $H^1$ $C^1$ $K^1$. From $C^1$ as center, with radius C B of the plan, describe an arc, and from $K^1$ as center, with radius $A^2$ $E^2$ of the pattern for the small corners, describe an arc, cutting it in the point $B^1$. Draw the connecting lines. Then $C^1$ $K^1$ $B^1$ is the pattern of the side shown by C K B of the plan. To $K^1$ $B^1$ add the shape of the pattern of the small corner $E^2$ $A^2$ $G^2$, already obtained, and as shown by $B^1$ $K^1$ $E^1$. From $E^1$ as center, with $E^1$ $B^1$ as radius, describe an arc, as shown by the dotted line, and from $B^1$, with radius equal to B A of the plan, intersect it in the point $A^1$. Draw the connecting lines. Then $E^1$ $B^1$ $A^1$ is the pattern of the end, as shown by E B A of the plan. To this in turn add a duplicate of the pattern of the small corner, and then a duplicate of the pattern of the side, thus completing the required shape.

482. *The Pattern of an Oblong Flaring Article having a Round Top.*—In Fig. 323, A B C D represents the elevation, G H I K the plan of the base, and L M N O the plan at the top. Corresponding points in the two views are connected by dotted lines. In describing the patterns we will consider the corners as quarter cones having oblique bases, and the sides and ends as flat triangles. In other words, the corners are sections of a cone, the elevation of which is shown by A B E, in which A is the apex and B E the oblique base, and the sides are simple triangles, as shown by K O I and I N H of the plan. From G, which represents the apex of one of the cones in plan, draw a line to the center of the top P, as shown by G P, which will represent in the plan a diagonal section of the corner, which must be constructed to obtain the measurements necessary. From any convenient point outside of the plan, as *x*, draw *x p* equal and parallel to G P. From *p* erect the perpendicular *p a*, in length equal to the hight of the article, as shown by F E. Draw *x a*. From $b'$, representing the point of intersection between the line G P and the circle forming the plan of the top of the article, erect the perpendicular $b'$ *b* indefinitely. From *a* draw a perpendicular to *p a*, intersecting $b'$ *b* in the point *b*. Draw *x b*. Then *x b a p* will be the diagonal section of the article taken on the line G P of the plan. Divide the quarter circle L M, being the base of the cone, into any convenient number of equal parts, and from the several points carry lines perpendicular to and cutting *b a*, as shown. From each of the points in *b a* erect lines perpendicular to *x a* and cutting *x b* prolonged, and from each of the points in *x b* prolonged, from *x* as center, describe an arc, as shown. Establish the point 6, corresponding to M in the plan. Draw a line to the center *x*, as shown by *x* $M^1$. Set the dividers to the space used in stepping off the plan L M, and, commencing at the point $M^1$, step to the second arc, and from that point to the next arc, and so on, reaching the last in the point $L^1$. Draw $L^1$ *x*. From $M^1$ as center, with $M^1$ *x* as radius, describe an arc, as shown by *x* $H^1$, and from *x* as center, with radius equal to G H of the plan, intersect that arc in $H^1$, as shown. Draw *x* $H^1$ and $M^1$ $H^1$. To $M^1$ $H^1$ add $M^1$ $H^1$ $N^1$, in all respects a duplicate of $L^1$ *x* $M^1$ reversed. From $N^1$ as center, with $N^1$ $H^1$ as radius, describe the arc $H^1$ $I^1$, and from $H^1$

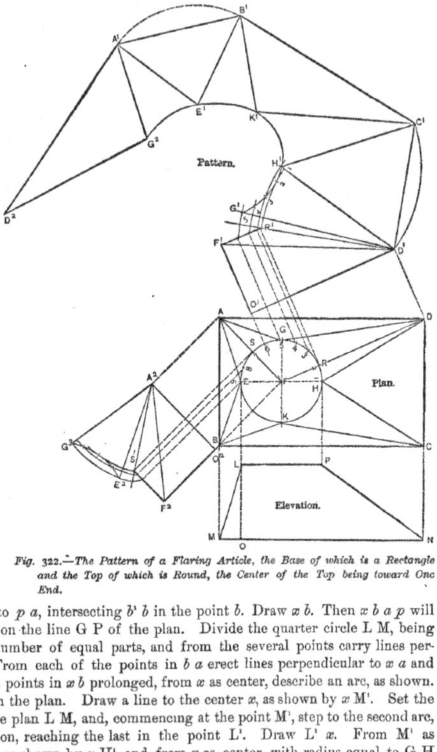


*Fig. 322.—The Pattern of a Flaring Article, the Base of which is a Rectangle and the Top of which is Round, the Center of the Top being toward One End.*

as center, with a radius equal to H I of the plan, intersect this arc in the point I'. Draw H' I' and N' I'. This will complete one-half of the pattern, and the other several sections may be added in the same manner as above described.

483. *The Pattern of an Article having an Elliptical Base and a Round Top.*—Fig. 324 shows the plan and elevation of the article for which the pattern is required. The shape therein possesses some of the general features of a cone, but lines drawn from points 1, 2, 3, etc., in the base, through corresponding points 1', 2', 3', etc., in the top, would reach a center line corresponding to the axis of a cone at different hights, and therefore would never meet. Hence, measurements must be taken upon the top and base direct, instead of being derived from an apex. Divide one-quarter part of the plan of the base into any convenient number of equal spaces, and divide a corresponding part of the plan of the top into the same number of spaces, by lines drawn from the points in the base toward the center of the circle of the top, cutting the arc K L. Also draw the intermediate dotted lines connecting alternate points, as shown in the engraving by 2 1', 3 2', 4 3', etc. Construct a diagram, as shown by A' N' C', Fig. 325, in which the actual distance between corresponding points in base and top shall be shown. Make C' N' equal to the straight hight of the article. At right angles to it set off N' A', in length equal to the distance 1' 1 in plan. From N' set off also spaces corresponding to 2' 2, 3' 3, 4' 4, etc., of the plan, and from each of these points draw a line to C', as shown. Then the lines converging at C' represent the distances which would be obtained by measurements made at corresponding points upon the article itself. Construct a like diagram of the distances represented in the dotted lines in the plan, as shown by C² N² O, Fig. 326. Make C² N² equal to C N of the elevation, and from N² set off at right angles the line N² O. Upon this line make the spaces N² 2, N² 3, N² 4, etc., equal to the length of the dotted lines 1' 2, 2' 3, 3' 4, etc., and from the points thus obtained in N² O draw lines to C². Then these converging lines represent the same distances as would be obtained if measurements were made between corresponding points upon the completed article. For the pattern, commence by drawing any line, P X in Fig. 327, on which set off a distance equal to C' 1 of the first diagram, as shown by 1 1'. Then, with the distance from 1 to 2 of the plan for radius and 1 in pattern as center, describe an arc, which intersect by another arc struck from 1' of the pattern as center and C² 2 of the second diagram as radius, thus establishing the point marked 2 in the pattern. Next, with 1' 2' of the plan as radius, and from 1' of the pattern as center, describe an arc, which intersect by another arc drawn from 2 as center, and with C' 2 as radius of the first diagram, thus locating the point 2' of the pattern. Continue in this manner, locating each of the several points shown from X to Y and from P to R of the pattern, through the several intersections tracing the lines of the pattern, as shown. Then X Y R P will be one-quarter of the

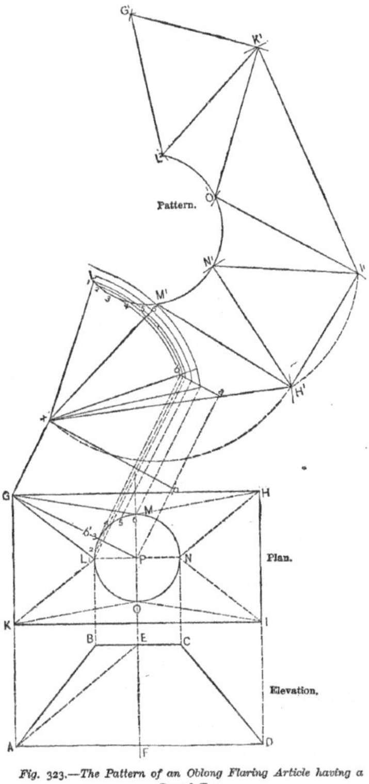


*Fig. 323.—The Pattern of an Oblong Flaring Article having a Round Top.*

required pattern. Repeat this piece three times additional, as shown by V W T U, W X P T and Y Z S R, thus completing the pattern.

484. *The Pattern of a Flaring Article, the Top of which is Round and the Bottom of which is Oblong, with Semicircular Ends, the Center of the Top being Located near One End.*—In Fig. 328, let G K C B be the side elevation of the required article, of which A E R D F P is the plan at the base and Z X I Y the plan at the top. By inspection it will be seen that the article may be resolved into two frustums of cones connected by two flat triangular pieces. Produce B G of the elevation indefinitely in the direction of H, and intersect it in the point H by a line drawn perpendicular to the base B C from the point L, which corresponds to the center of the top of the article, all as shown by L H. Then H is the apex of a cone, of which H B L is a half elevation, and that portion of the article represented by G M L B is a frustum. In like manner produce

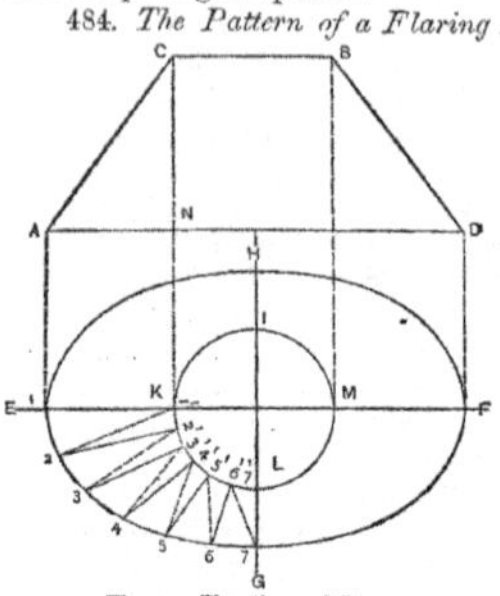


Fig. 324.—Elevation and Plan.

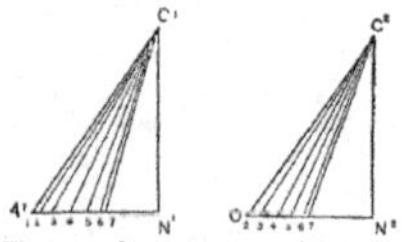


Figs. 325 and 326.—Diagrams of Triangles.

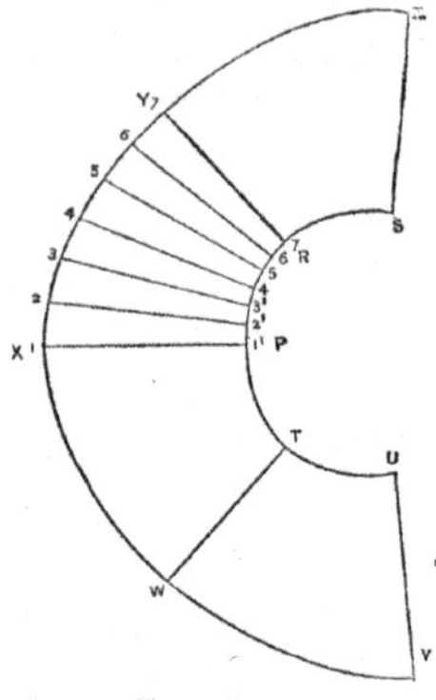


Fig. 327.—Pattern.
*The Pattern of an Article having an Elliptical Base and a Round Top.*

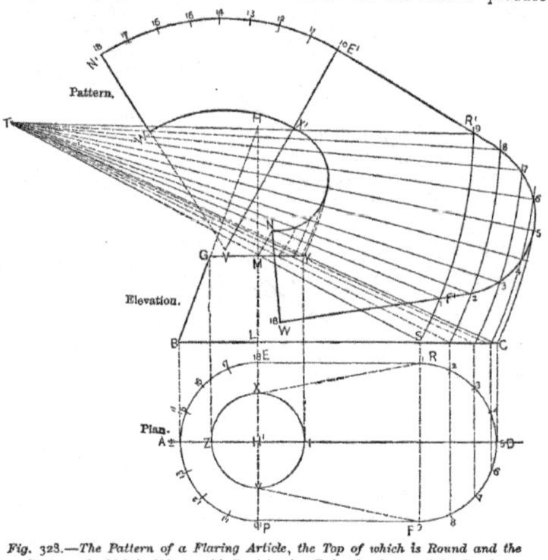


*Fig. 328.—The Pattern of a Flaring Article, the Top of which is Round and the Bottom of which is Oblong, with Semicircular Ends, the Center of the Top being Located near One End.*

C K of the elevation indefinitely in the direction of T. Locate the point S in the base line, corresponding to the junction of the straight side and the semicircular end, as shown by R and F of the plan, and draw a line from it to the point M already obtained, which produce until it meets C K extended in the point T. Then T is the apex of a scalene cone, of which T C S is the half elevation, and of which that part of the article represented in elevation by M K C S is a frustum. The remainder of the envelope of the article is in the shape of two equal triangles, one of which is shown by L M S of the elevation and by P Y F of the plan. For the pattern proceed as follows: Divide the half plan R D F of the scalene cone into any number of equal spaces, and from each of the points erect a line perpendicular to the base B C of the article, and thence carry the lines at right angles to the axis T S of the cone until they cut the side T C. Then from T as center strike arcs corresponding to these several points, all as shown. From T draw a straight line, as T F', intersecting the first arc in the point F'. Set the dividers to the space used in stepping off the plan, and, commencing at the first arc in the point F', step to the second, and from that point to the third, and so on, finally reaching the last in the point R'. Then a line, F' R', traced through these several points will be the pattern of the bottom of the frustum of the scalene cone. From each of the points in S C also draw a line toward T, as shown, cutting M K. From each of the points in M K erect a line perpendicular to the axis T S, cutting the side T C. From T as center draw arcs from these points in T K, as shown. From the points between F' and R' of the pattern draw arcs toward T, intersecting the arcs just described. A line traced through their points of intersection, as shown by N X', will form the pattern of the top of the frustum of the cone. To the two sides N F' and X' R' of the pattern thus far constructed add the two triangles shown, in the following manner: From F' and R' as centers, and with F P of the plan as radius, describe arcs, and from N and X' as centers, with G B of the elevation as radius, describe arcs, cutting the former in the points E' and W. Draw the connecting lines, as shown. Produce E' X' in the direction of V, making E' V equal to the side of the cone, as shown by H B of the elevation. From V as center, with radius equal to H B, describe the arc E' N', in length equal to the stretchout of the plan of the base, shown by E A P, and from its termination, N', draw a line to V, as shown. From the same center, with radius equal to H G, describe an arc from X', and continue it until it intersects N' N in the point W', which will complete the required pattern.

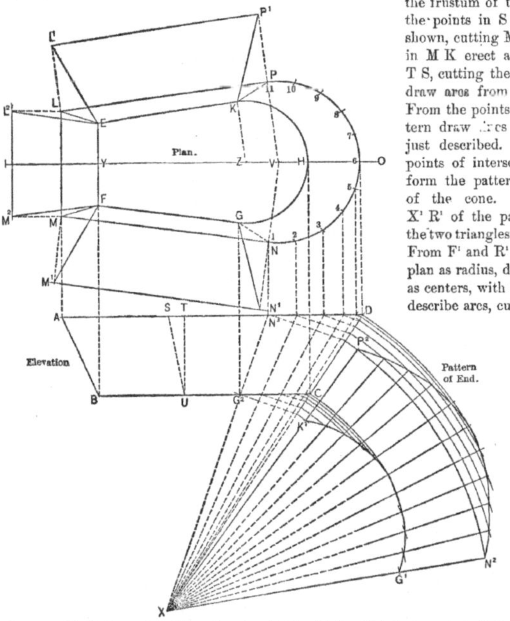


*Fig. 329.—The Patterns of an Oblong Tapering Article, with One End Square and one End Semicircular, having More Flare at the Ends than at the Sides.*

485. *The Patterns of an Oblong Tapering Article, with One End Square and One End Semicircular, having More Flare at the Ends than at the Sides.*—In Fig. 329, let L P O N M be the plan of the article at the top and E K H G F the plan of it at the bottom, and let A D C B be the side elevation. Inasmuch as the article is tapering in plan, the conical part of the pattern will include a little more than shown by a semicircle in plan. The lines showing the junction between the straight sides and the conical part are to be drawn perpendicular to the sides of the article. Therefore lay off in the plan V P and V N, drawn from the center V of the curved part of the plan of the top of the article, perpendicular to the sides L P and M N respectively. And in like manner from Z, the center by which the curved part of the bottom of the article is struck, draw Z G and Z K. Draw L' P' and M' N' parallel to the sides L P and M N, and equal in length to them, as determined by the T-square placed at right angles to them and brought against their ends, and at a distance from them equal to the flaring hight of the side, as shown by S U of the elevation. Connect L' E and P' K, M' F

and $N^1$ G, as shown, which will complete the patterns of the sides. For the pattern of the square end, make Y I equal the slant hight of the end, as shown by A B of the elevation. Through I draw $M^2$ $L^2$, in length equal to M L, determined by the T-square, as already described in connection with the sides. Connect $L^2$ E and $M^2$ F. The rounded end of the article is the section of a cone, of which we must first complete the elevation by producing the side D C of the article in the direction of X indefinitely. From the points G and N in the plan of the bottom and top of the article respectively, drop points on to corresponding lines in the elevation, all as shown by $G^2$ $N^2$, through which draw a line, which produce until it meets D C extended in the point X. Then X is the apex of the cone, of which $G^2$ $N^2$ D C of the elevation is a frustum. Divide the plan N O P of the cone into any convenient number of equal parts, and from them let fall perpendiculars to the base $N^2$ D, from which carry lines at right angles to the axis X $N^2$, producing them until they cut the side X D, as shown. From X as center, strike arcs corresponding to the points thus obtained in X D. From the points in $N^2$ D carry lines toward the apex X, cutting $G^2$ C as shown, from the points in which also carry lines perpendicular to X $N^2$, cutting X D. From X as center strike a similar set of arcs, as shown. From X draw any straight line, as X $P^2$, producing it until it meets the first arc of the set representing the top of the article, as shown in the point $P^2$. From $P^2$ as a starting point, with the dividers set to the same space as used in subdividing the plan, step to the second arc, and thence to the third, and so on according to the number of spaces in the plan, terminating in the point $N^2$. A line traced through these points, as shown by $P^2$ $N^2$, will be one side of the end piece. From these same points draw lines in the direction of X, crossing the second set of arcs. A line traced through the several points of intersection thus formed will be the other side of the pattern.

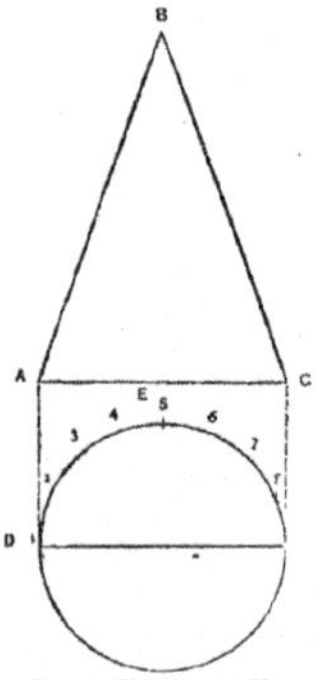

Fig. 330.—Elevation and Plan.

486. *The Envelope of a Right Cone.*—In Fig. 330, let A B C be the elevation of the cone and D E F the plan of the same. Set the compasses to the space B A, or to the slant hight of the cone, for a radius, and from any convenient point as center, as $B^1$ in Fig. 331, strike an arc indefinitely. Connect one end of the arc with the center, as $A^1$ $B^1$. With the dividers step off the circumference of the plan D E F, as shown, and count the spaces until the whole or exactly one-half is completed. Then set off on the arc $A^1$ $C^1$ the same number of steps as is contained in the whole plan, commencing at $A^1$, which point has been connected with the center, as explained, and ending at $C^1$. Draw $B^1$ $C^1$. Then $B^1$ $A^1$ $C^1$ will be the envelope of the cone.

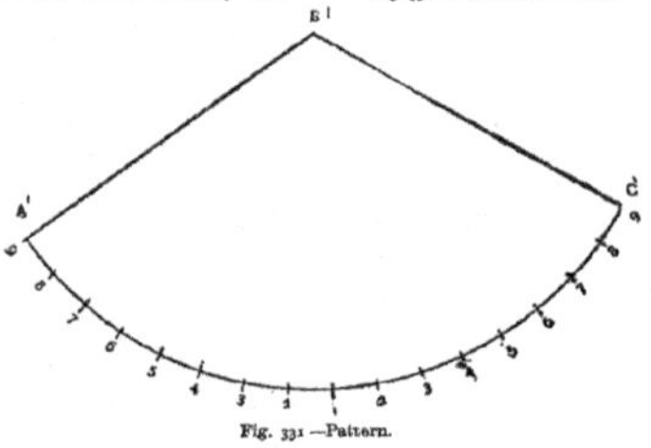
Fig. 331.—Pattern.
*The Envelope of a Right Cone.*

487. *The Envelope of a Right Cone, from which a Section is Cut Parallel to its Axis.*—Let B A F in Fig. 332 be a right cone, from which a section is to be cut, as shown by C D in the elevation. Let $B^2$ L H K be the plan of the cone. Then the line of the cut in plan is shown by $D^2$ $D^1$. For the patterns proceed as follows: Divide that portion of the plan corresponding to the section to be cut off, as shown by $D^1$ $B^2$ $D^2$, into as many spaces as are necessary to give accuracy to the pattern, and divide the remainder of the plan into spaces convenient for laying off the stretchout. From any convenient center, as A, with radius A B, describe an arc, as M N, which make equal to the stretchout of the plan $B^2$ L H K, dividing M N into the same spaces as employed in the plan. From the points in the arc corresponding to that portion of the plan indicated by $D^1$ $B^2$ $D^2$—namely, 8 to 16 inclusive—draw lines to the center A, upon which to set off distances measured from the elevation. From the same points in the plan carry lines vertically, cutting the base of the cone, as shown from B to D, and thence continue them to the apex A, cutting C D as shown. From the points in C D carry lines at right angles, cutting the side of the cone, as shown in the points between C and B. From A as center, with radii corresponding to the points between C and B, cut the corresponding lines drawn

from the same points in the stretchout to A, and through the points of intersection thus obtained trace a line, as shown by D¹ C¹ D². Then the space indicated by D¹ C¹ D² is the shape to be cut from the envelope M A N of the cone to produce the shape, as shown by C D in the elevation.

488. *The Envelope of a Frustum of a Right Cone.*—The principle involved in cutting the pattern for the frustum of a cone, is precisely the same as that for cutting the envelope of the cone itself. The frustum of a right cone is a shape which enters so extensively into articles of tinware, that we have thought it well to illustrate the cutting of a pattern for it by an engraving of somewhat different character from those employed in other and similar cases. In Fig. 333 is shown by plan and elevation an ordinary flaring pan, which is an illustration of the articles employing the shape of a frustum of a cone. For the pattern proceed as follows: Through the elevation draw a center line, K B, indefinitely. Extend one of the sides of the pan, as, for example, D O, until it meets the center line in the point B. Still greater accuracy will be insured by extending the opposite side of the pan also, as shown—the three lines meeting in the point B—which determines the apex of the cone to a certainty. Then B O and B D respectively are the radii of the arcs which contain the pattern. From B or any other convenient point as center, with B O as radius, strike the arc P Q indefinitely, and likewise from the same center, with B D as radius, strike the arc E F indefinitely. From the center B draw a line across these arcs near one end, as P E, which will be an end of the pattern. By inspection and measurement of the plan, determine in how many pieces the plan is to be constructed, and divide the circumference of the plan into a corresponding number of equal parts, in this case three, as shown by K, M and L. With the dividers or spacers step off the length of one of these parts, as shown from M to L, and set off a corresponding distance on the arc E F, as shown. Through the last division draw a line across the arcs toward the center B, as shown by F Q B. Then P Q F E will be the pattern of one of the sections of the pan, as shown in the plan. In the engraving the plan has been placed below and in line with the elevation, in order to better show the correspondence of parts in the two views. In practice this is not necessary; but, since in most cases it can be as well placed there as anywhere else, it is advisable to do so on account of the greater accuracy insured by drawing lines through corresponding points with the T-square, as illustrated by D H, etc. Neither is it necessary to draw more than one-half of the elevation. In this demonstration, as well as in many others in this book, we have not limited ourselves to the smallest number of lines for describing the pattern, but have put in enough others to show the reason for every step taken.

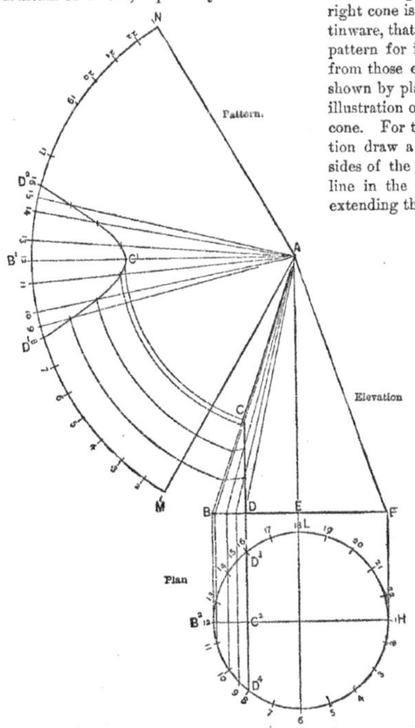


*Fig. 332.—The Envelope of a Right Cone, from which a Section is Cut Parallel to its Axis.*

489. *The Envelope of the Frustum of a Right Cone, the Upper Plane of which is Oblique to its Axis.*—In Fig. 334, let C B D E be the elevation of the required shape. Produce the sides C B and E D until they intersect at A. Then A will be the apex of the cone of which C B D E is a frustum. Draw the axis A G, which produce below the figure, and from a center lying in it draw a half plan of the article, as shown by F G H. Divide this plan into any number of equal parts, and from the points carry lines parallel to the axis until they

cut the base line, and from there extend them in the direction of the apex until they cut the upper plane B D. Place the T-square at right angles to the axis, and, bringing it against the several points in the line B D, cut the side A E, as shown. From A as center, with A E as radius, describe the arc C' E', on which lay off a stretchout of either a half or the whole of the plan, as may be desired, in this case a half, as shown. From the extremities of this stretchout, C' and E', draw lines to the center, as C' A and E' A. Through the several points in the stretchout draw similar lines to the center A, as shown. With the point of the compasses set at A, bring the pencil to the point D in the side A E, and with that radius describe an arc, which produce until it cuts the corresponding line in the stretchout, as shown at D'. In like manner, bringing

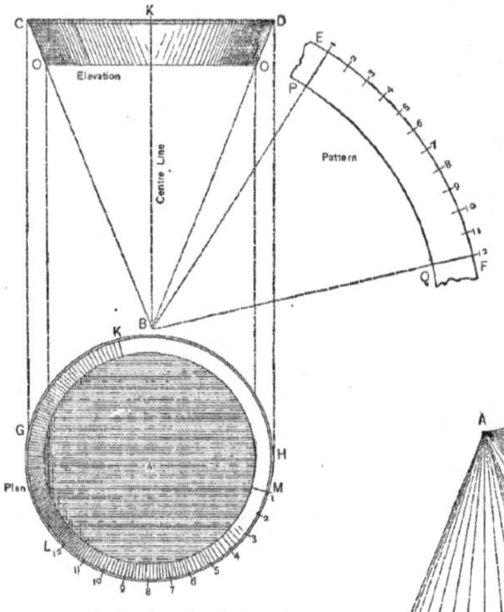


*Fig. 333.—The Envelope of a Frustum of a Right Cone.*

the pencil against the several points between D and E in the elevation, describe arcs cutting the corresponding measuring lines of the stretchout. Then a line traced through these intersections will form the upper line of the pattern, the pattern of the entire half being contained in C' B' D' E'.

490. *The Envelope of a Scalene Cone.*—The difference between a scalene cone and a right cone consists of the base line. In one it is drawn oblique to the axis, while in the other it is at right angles to the axis. In Fig. 335, let G D H be the elevation of a scalene cone, the pattern of which is to be cut. At right angles to the axis D O, and through the point G, draw the line E F. Extend the axis, as shown by D B, and upon it draw a plan of the cone as it would appear when cut upon the

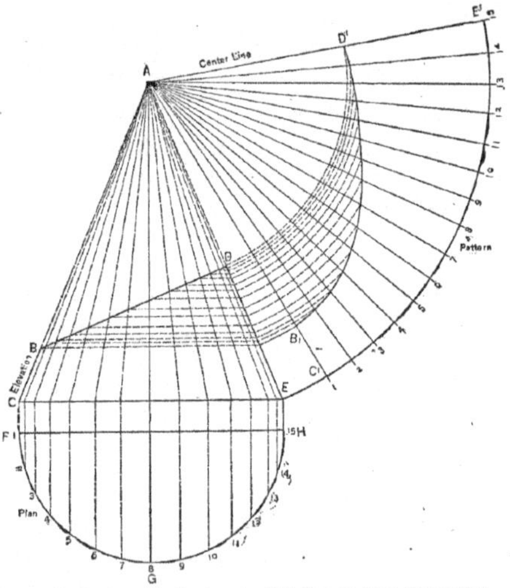


*Fig. 334.—The Envelope of the Frustum of a Right Cone, the Upper Plane of which is Oblique to its Axis.*

line E F, as shown by A B C. Divide the plan into any convenient number of equal parts, and from the points thus obtained drop lines on to E F. From the apex D, through the points in E F, draw lines to the base G H. From D as center, with D G as radius, describe an arc indefinitely, on which lay off a stretchout taken from the plan A B C, all as shown by I M K. From the center D, by which the arc was struck, through the points in the stretchout, draw radial lines indefinitely, as shown. Place the blade of the T-square parallel to the line E F, and, bringing it against the several points in the base line, cut the side D H, as shown from F to H. With one point of the compasses in D, bring the other successively to the points 1, 2, 3, 4, etc., in F H, and describe arcs, which produce until they cut the corresponding lines drawn through the stretchout, as indicated by the dotted lines. Then a line, I L K, traced through these points of intersection, as shown, will complete the required pattern.

491. *The Envelope of a Frustum of a Scalene Cone, or the Envelope of the Section of a Right Cone, contained between Planes Oblique to its Axis.*—In Fig. 336, let F L M K represent the section of the cone the pattern for which is required. Produce the sides F L and K M until they meet in the point N, which is the apex of the cone of which F L M K is a frustum. Through N draw the axis of the cone, which produce in the direction of D indefinitely. From K draw K H at right angles to the axis. At convenient distance from the cone, either above or below it, construct a plan or profile as it would appear when cut on the line K H, letting the center of the profile fall upon the axis produced, all as shown by A D C B. Divide the profile into any number of equal parts, and from the points thus obtained draw lines parallel to the axis, cutting K H. From the apex N, through the points in K H, draw lines cutting the top L M and the base F K. Place the blade of the T-square at right angles to the axis of the cone, and, bringing it successively against the points in L M and F K, cut the side N F, as shown above L, and from H to F. From N as center, with radius N H, strike the arc T S indefinitely, upon which lay off a stretchout from the plan, as shown, and through the points from the center N draw lines indefinitely, as shown. With the point of the compasses still at N, and the pencil brought successively against the points in the side from H to F, describe arcs, which produce until they cut corresponding lines drawn through the stretchout. Then a line traced through these points of intersection, as shown by T U S, will form the lower line of pattern. In like manner draw arcs by radii corresponding to the points in the side at L, which produce also until they intersect corresponding lines drawn through the stretchout. A line traced through these points, as R P O, will be the upper line of the pattern sought.

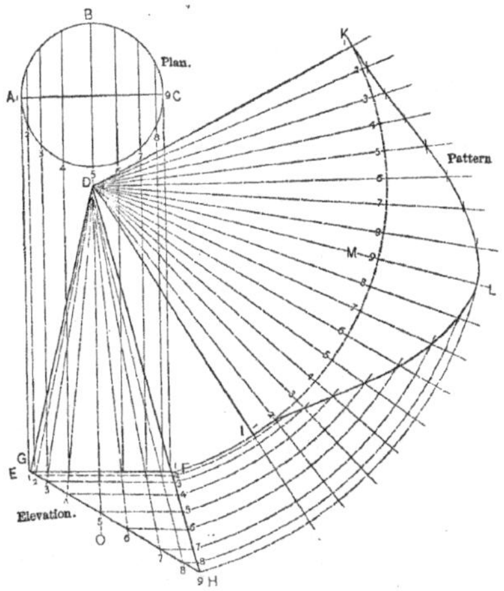


*Fig.* 335.—*The Envelope of a Scalene Cone.*

492. *The Envelope of the Frustum of a Cone, the Base of which is an Elliptical Figure.*—This shape is very frequently used in pans and plates, and therefore we have employed the representation of an ordinary oval or elliptical pan in our engraving by way of illustration. (See Fig. 337.) Draw an elevation of either a side or end of the article, and corresponding to and in line with it lay off the plan, as shown, employing for this purpose any rule for constructing the ellipse which employs centers. In Fig. 337, let that part of the plan lying between H and L be an arc whose center is at U, and let those portions between V and H and L and W be arcs whose centers are respectively R and S. In drawing the plan, let it be composed of two lines, one of which shall represent the plan of the vessel at the top and the other the plan of the vessel at the bottom.

A C D B represents an elevation of the vessel, and is so connected with the plan as to show the relationship of corresponding points. After having drawn the plan, the next step is to construct the diagram shown in Fig. 338. Draw the horizontal line H U indefinitely, and at right angles to it draw H A, indefinitely also. Make H U, Fig. 338, equal to H U of the plan, Fig. 337. Make H C of Fig. 338 equal to the vertical hight of the vessel, as shown in the elevation by D X. Draw the line C G parallel to H U, making C G in length equal to U N of the the plan, Fig. 337. Through the points U and G thus established draw the line U G, which continue until it meets H A in the point A. Then A U will be the radius by which to describe that portion of the pattern which is included between the points H and L of the plan. With A U as radius, and from any convenient point as center—as, for example, A, Fig. 339—draw the arc H L, which in length make equal to H L of the plan, Fig. 337, as shown by the points 1, 2, 3, etc. From the same center, and with the radius A G of Fig. 338, describe the parallel arc N O. From the points H and L of the arc first drawn, draw lines to A, thus intercepting the arc N O and determining its length. In the diagram, Fig. 338, set off from H, on the line H U, the distance H R, making it equal to R H of the plan, Fig. 337. Also, upon the line C G, from the point C, set off C I, equal to R N of the plan, Fig. 337. Then, through the points R and I thus established, draw the line R B, which produce until it intersects A H. Then R B will be the radius for those portions of the pattern lying between V and H and L and W of the plan, Fig. 337. From the point H, on the line H A, Fig. 339, set off the distance H B, equal to R B of Fig. 338. Then, with B as center, describe the arc E H, and from corresponding center C, at the opposite end on pattern, describe the arc L K. From the same centers, with B I as radius, describe the arcs N M and O P, all as shown. Make H E and L K in length equal to H E and L K of the plan, Fig. 337. From E and K, respectively, draw lines to the centers B and C, intercepting the arcs N M and O P in the points M and P. Then E K P M will be one half of the complete patterns of the vessel.

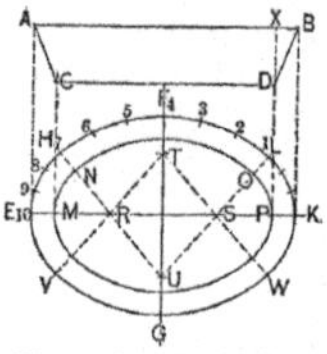


Fig. 337.—Elevation and Plan.

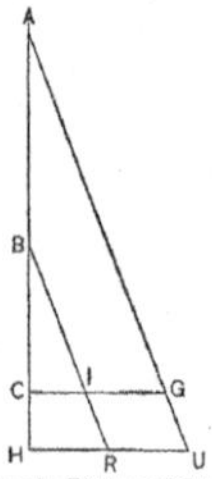


Fig. 338.—Diagram of Radii.

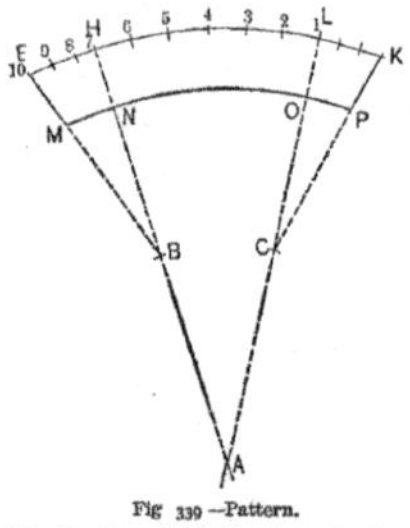


Fig 339 —Pattern.
*The Envelope of the Frustum of a Cone, the Base of which is an Elliptical Figure.*

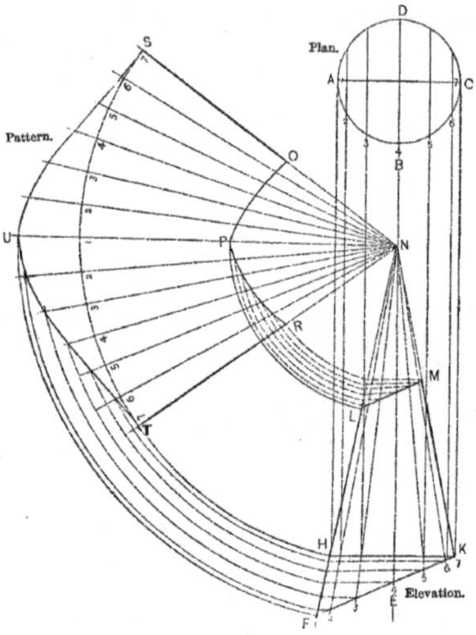


*Fig. 336.—The Envelope of a Frustum of a Scalene Cone.*

493. *The Pattern of a Flaring Article which Corresponds to the Frustum of a Cone whose Base is a True Ellipse.*—In Fig. 340, let G H F E be the elevation of one side of the article, L M U R the elevation of an end, E' R' F' U' the plan of the article at the base, and T V S P the plan

at the top. Produce E G and F H of the side elevation until they meet in the point I. At any convenient place draw the straight line D A of Fig. 341, in length equal to I $E^2$. Make D B equal to I $G^1$. From A and B draw perpendiculars to D A indefinitely, as shown by A O and B N. Divide one-quarter of the plan $E^1$ $R^1$ into any convenient number of equal parts, as indicated by the small figures. From the points thus determined draw lines to the center C, and also carry lines perpendicular to the base of the article E F, as shown, from which line continue them toward the apex I, cutting the top G H, as shown. Take the distances C 5, C 4, C 3, etc., of the plan and set off corresponding distances from A on A O, as shown by A 5, A 4, A 3, etc. From these points in A O draw lines to D, cutting B N. From D as center, describe arcs corresponding to the several points in A O, as shown. From any convenient point in the first arc draw a straight line to D, as shown by W D. This will form one side of the pattern. From W as a starting point, lay off the stretchout of the plan $E^1$, $R^1$, $F^1$, etc., using the same length of spaces as employed in dividing it, stepping from one arc to the next each time, as shown. A line traced through these points will be the outline of the plan, one-half of the entire envelope being shown in the pattern from W to Z. From these points also draw lines to the center D, and from D intersect them by arcs drawn in the same manner as before described, corresponding to the several points in B N. A line traced through the points of intersection between these arcs and the radial lines from D will form the upper line of the pattern, as shown. Then W X Y Z will constitute the pattern of one-half of the envelope, to which add a duplicate of itself for the complete pattern.

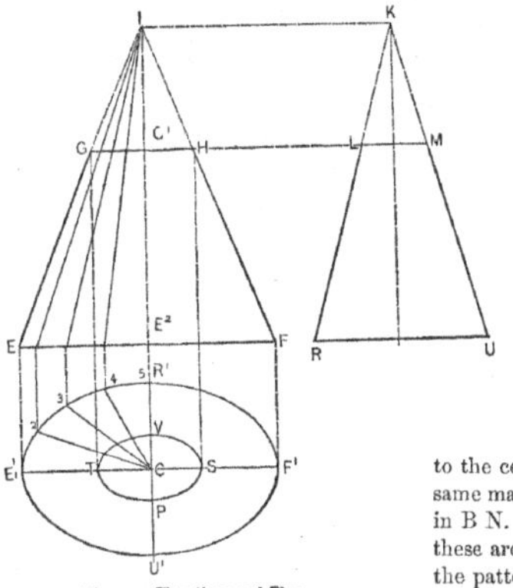


Fig. 340.—Elevations and Plan.

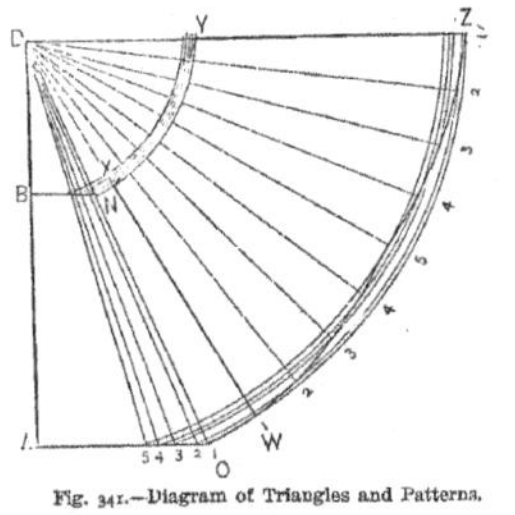


Fig. 341.—Diagram of Triangles and Patterns.

*The Pattern of a Flaring Article which corresponds to the Frustum of a Cone whose Base is a True Ellipse.*

494. *Patterns of a Tapering Article with Equal Flare throughout, which Corresponds to the Frustum of a Cone the Base of which is Elliptical (Struck from Centers), the Upper Plane of the Frustum being Oblique to the Axis.*—In Fig. 342, let H F G A be the shape of the article as seen in side elevation. The plan is shown by I L N O. In order to indicate the principle involved in the development of this shape, we have introduced lines which show the construction of the figure. It may be remarked at the outset that a conical figure having an elliptical base, or, in other words, whose base corresponds to a figure struck from centers, exhibits throughout its extent the peculiar properties of its base. In other words, a conical solid, the base of which is an elliptical figure struck from various centers, resolves itself into sections of cones, the several bases of which correspond to the circles, arcs of which compose the elliptical base. Thus, by inspection of the engraving, it will be seen that the shape H F G A is made up of sections of cones corresponding to the arcs of which the plan I L N O is composed. Those parts of the figure shown in plan by K U T M and R U T P may be considered as segments cut from a right cone, the radius of the base of which is either O K or L R, and the apex E of which is to be ascertained by producing the lines $K^1$ D and $M^1$ B, through the points D and B, until they meet in E. By further examination it will be seen that the points $K^1$ and $M^1$ are established by vertical lines carried to the base from the points K and M of the arc struck from the center O, and the points D and B, representing the points of intersection above referred to, fall somewhere in vertical lines corresponding to U T of the plan. These points of intersection between the lines drawn from $K^1$ and $M^1$, and U and T, are determined by producing the sides H F and A G of the given figure until they meet the vertical lines drawn from U and T. The parts shown in the plan by K U R and M T P may be con-

sidered as segments cut from a right cone, the radius of the base of which is either U I or T N. The apex of this cone is to be found by means of an end elevation, in which are drawn lines corresponding to the points R and K of the plan, to $B^1$, and which in hight correspond to the points B and D of the side elevation already determined, all as shown in $R^2$ $B^1$ and $K^3$ $B^1$. The conditions of this figure, as stated, are that it shall have equal flare throughout; therefore, the pitch of the sides $O^2$ $C^1$ and $L^2$ $C^1$ in the end elevation must be equal to those of the side elevation, determined by the established flare of the article. By this means the point $C^1$ in the end elevation is determined. From $C^1$ the point C in the side elevation is derived, as indicated by the

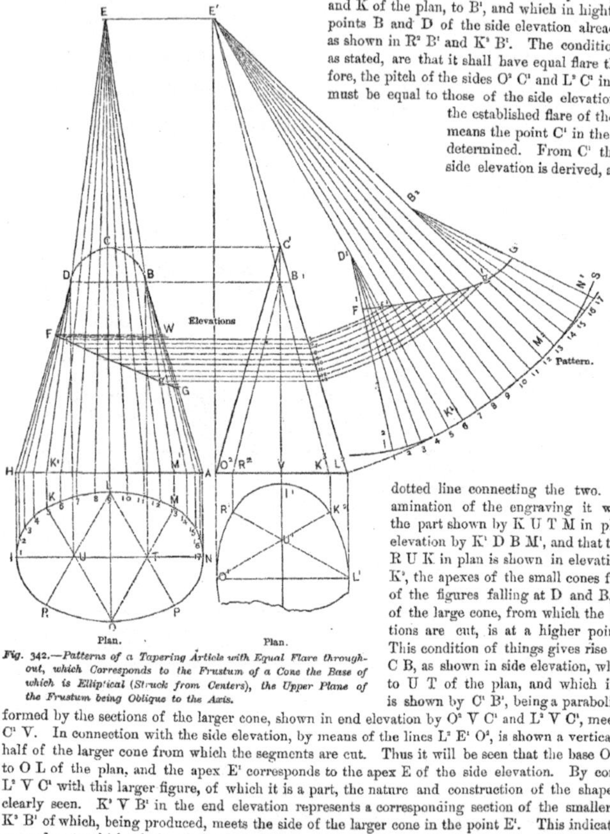


*Fig. 342.—Patterns of a Tapering Article with Equal Flare throughout, which Corresponds to the Frustum of a Cone the Base of which is Elliptical (Struck from Centers), the Upper Plane of the Frustum being Oblique to the Axis.*

dotted line connecting the two. By further examination of the engraving it will be seen that the part shown by K U T M in plan is shown in elevation by $K^1$ D B $M^1$, and that the part shown by R U K in plan is shown in elevation also by $R^2$ $B^1$ $K^3$, the apexes of the small cones forming the ends of the figures falling at D and B, while the apex of the large cone, from which the two middle sections are cut, is at a higher point, shown at E. This condition of things gives rise to the shape D C B, as shown in side elevation, which corresponds to U T of the plan, and which in end elevation is shown by $C^1$ $B^1$, being a parabolical curve. It is formed by the sections of the larger cone, shown in end elevation by $O^2$ V $C^1$ and $L^2$ V $C^1$, meeting on the line $C^1$ V. In connection with the side elevation, by means of the lines $L^2$ $E^1$ $O^2$, is shown a vertical section of one-half of the larger cone from which the segments are cut. Thus it will be seen that the base $O^2$ $L^2$ corresponds to O L of the plan, and the apex $E^1$ corresponds to the apex E of the side elevation. By comparing section $L^2$ V $C^1$ with this larger figure, of which it is a part, the nature and construction of the shape will be more clearly seen. $K^3$ V $B^1$ in the end elevation represents a corresponding section of the smaller cone, the side $K^3$ $B^1$ of which, being produced, meets the side of the larger cone in the point $E^1$. This indicates a correspondence of parts which admits of the figure being constructed in the way we have specified. Having thus described the nature of the figure, the manner of drawing the two elevations, both of which are necessary in developing the patterns, becomes evident without further explanation. For the patterns we proceed as follows:

Divide one-half of the plan into any convenient number of equal parts, as shown by the small figures, and from the points thus established carry lines vertically, cutting the base line H A, and thence carry them toward the apexes of the various cones from the bases of which they are derived. That is, from the arc K M draw lines toward the apex E, and from the points derived from the arc I K carry lines toward the apex D, and in like manner from the points derived from the arc M N carry lines in the direction of the apex B, all of which produce until they cut the top line F G of the article. From the points in F G thus established carry lines to the right, cutting the slant lines of the cones to which they correspond. Thus, from the points occurring between F and *f*, draw lines cutting B A, being the slant of the small cone, as shown by the points immediately below W. In like manner, from the points between *g* and G, carry lines cutting the same line, as shown by G. The slant line of the large cone is shown only in end elevation, and therefore the lines corresponding to the points between *f* and *g* must be carried across until they meet the line $B^1$ $L^2$. Commence the pattern by taking any convenient point, as $E^1$, for center, and $E^1$ $L^2$ as radius, and strike the arc $L^2$ S indefinitely. Upon this arc, commencing at any convenient point, as $K^4$, set off that part of the stretchout of the plan corresponding to the base of the larger cone, as shown by the points 5 to 13 in the plan, and as indicated by corresponding points from $K^4$ to $M^2$ in the arc. From the points thus established draw lines indefinitely in the direction of the center $E^1$, as shown. From $E^1$ as center, with radii corresponding to the points 5 to 13 inclusive, established in the line $B^1$ $L^2$ already described, cut corresponding radial lines just drawn, and through the points of intersection thus established draw a line, all as shown by $f^1$ $g^1$. Next take A B of the side elevation as radius, and setting one foot of the compasses in the point $K^4$ of the arc, establish the point $D^1$ in the line $K^4$ $E^1$, and in like manner, from $M^2$, with the same radius, establish the point $B^2$ in the line $M^2$ $E^1$, which will be the centers from which to describe those parts of the patterns derived from the smaller cone. From $D^1$ and $B^2$ as centers, with radius B A, strike arcs from $K^4$ and $M^2$ respectively, as shown by $K^4$ $I^2$ and $M^2$ $N^1$, upon which set off those parts of the stretchout corresponding to the smaller cones, as shown by the arcs K I and M N of the plan. From the points thus established, being 5 to 1 and 13 to 17 inclusive, draw radial lines to the centers $D^1$ and $B^2$, as shown. For that part of the pattern shown from $F^1$ to $f^1$, set the dividers to radii, measuring from B, corresponding to the several points immediately below W of the side elevation, and from $D^1$ as center cut the corresponding radial lines drawn from the arc. In like manner, for that part of the pattern shown from $G^1$ to $g^1$, set the dividers to radii measured from B, corresponding to the points in the line B A at G, with which, from $B^2$ as center, strike arcs cutting the corresponding measuring lines, as shown. Then $F^1$ $G^1$ $N^1$ $I^2$ will be one-half of the pattern sought—in other words, corresponding to I K L M N of the plan. The whole pattern may be completed by adding to it a duplicate of itself.

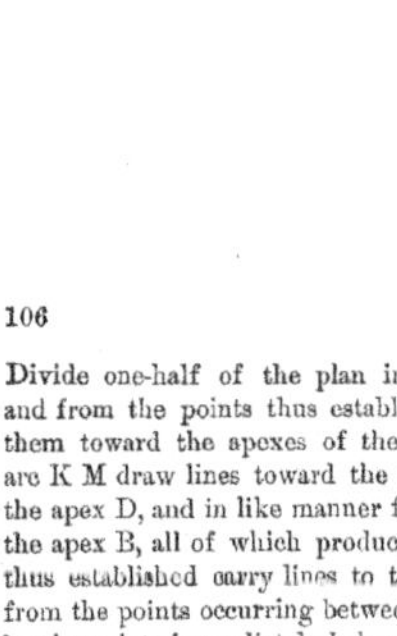


Fig. 343.—Elevation and Plan.
*An Irregular Flaring Article, both Top and Bottom of which are Round, the Top being Smaller than the Bottom, and Tangent at One Point in Plan.*

495. *The Pattern of an Irregular Flaring Article, both Top and Bottom of which are Round, the Top being Smaller than the Bottom, and the two being Tangent at One Point in Plan.*—In Fig. 343, let B D E C be the side elevation of the article, one-half of the plan of the bottom being shown in F H G, and one-half of the plan of the top by F K L. For the pattern proceed as follows: Produce the side E D indefinitely in the direction of A. Produce the side C B until it meets the other in the point A. Having the plan drawn directly in line with the elevation, so that like points in each correspond, all as shown in the engraving, divide the plan of the base and the plan of the top into the same number of equal spaces, as shown by $a^1$, $b^1$, $c^1$, $d^1$, etc., and *a*, *b*, *c*, *d*, etc., respectively. This may be done by dividing the base and cutting the circle of the top by lines drawn from these points to the point F. From F as center, with F $a^1$, F $b^1$ as radii, describe arcs, as shown, cutting F G. From F G continue them at right angles until they cut the base C E, whence carry them toward the apex A, cutting the top B D. From any convenient point, as $A^1$ in Fig. 344, as center, with radius A E of Fig. 343, describe an arc, as shown by $G^3$ $G^3$. In like manner, with radii A $i^4$, A $h^4$, A $g^4$, etc., describe arcs indicated by $i^3$ $i^3$, $h^3$ $h^3$, $g^3$ $g^3$, etc., in the pattern. From the same center $A^1$, with corresponding radii taken from A of the elevation, to the intersections made by the radial lines with the top

B D, describe arcs, as shown in the pattern by $i^2$ $i^2$, $g^2$ $g^2$, $h^2$ $h^2$, etc. Draw any straight line from A' to the first arc corresponding to the points in the base, as shown by A' C', which will represent one side of the required pattern. Set the dividers to the space used in stepping off the plan of the base, and, starting with C', lay off the stretchout, stepping from arc to arc, as shown. Trace a line, C' E' C' through these points, which will be the bottom of the required pattern. From these same points draw lines to the center A', cutting the set of smaller arcs. Trace a line, B' D' B', through the intersections of these lines with arcs of corresponding numbers, which will be the top line of the pattern. From the last points in the line C' E' C' draw a line toward the center A', as shown, reaching B', which will complete the pattern.

496. *The Pattern for an Irregular Flaring Article which is Elliptical at the Base, Round at the Top, the Top being so Situated with Respect to the Base as to be Tangent to One End of it when Viewed in Plan.*—In Fig. 345, let D G F E be the side elevation of the article and K N M one-half of the plan of the base. The half plan of the top is shown by K W L, the base and top being tangent in plan at the point K. The pattern for this shape is to be obtained by cutting the surface up into triangles so small that there is no apparent curve between points. To do this, proceed as follows: Divide the plans of the top and base into the same number of equal parts, as shown by 1, 2, 3, etc., in the base and 1', 2', 3', etc., in the top, and connect similar points in the two by lines, as shown by 6 6', 5 5', etc. Also connect each point in the plan of the top with the next lower number in the plan of the base, as shown by the diagonal dotted lines in the engraving, as 6 7', 5 6', etc. At any convenient point draw A U, in length equal to D E of the elevation, and lay off U T at right angles to it. Let A represent all points in the circle which is the plan of the top of the article. Lay off from U the distance from each of the several points in the circle to the corresponding point in the ellipse. Thus make U 7 equal to 7' 7 of the plan, U 6 equal to 6' 6, etc. Draw the radial lines $A^2$, $A^3$, $A^4$, etc. In like manner construct a corresponding section, as shown by C B V, using for the spaces in B V the length of the diagonal or dotted lines between the circle and the ellipse in the plan. Draw $C^2$, $C^3$, $C^4$, etc. By means of these two sets of lines, converging at A and C respectively, we have the actual dimensions of the triangles into which we have imagined the surface of the article to be divided, and which in plan are shown by 7 7' 6, 7' 6 6', 6 6' 5, etc. These are to be used in describing the pattern as follows: At any convenient place draw the straight line P R in Fig. 346, in length equal to G F of the elevation, or, what is the same, equal to A 7 of the first diagram. As we have shown but half of the plan, the pattern will also appear as one-half of the whole shape, and therefore P R will form its central line. From P as center, with radius C 6 of the second diagram, describe an arc, which intersect by a second arc struck from R as center, with radius 7 6 of plan, thus establishing the point 6 of the pattern. Then with radius A 6 of the first diagram, from 6 of the pattern as center, describe an arc, which cut with another arc struck from 7' of the pattern as center, and 7' 6' of the plan as radius, thus locating the point 6' of the pattern. Continue this process, locating in turn 5 5', 4 4', etc., until points corresponding to all the points laid off in the plan are established. Draw lines through these points. Then O P R S will be one-half of the required pattern.

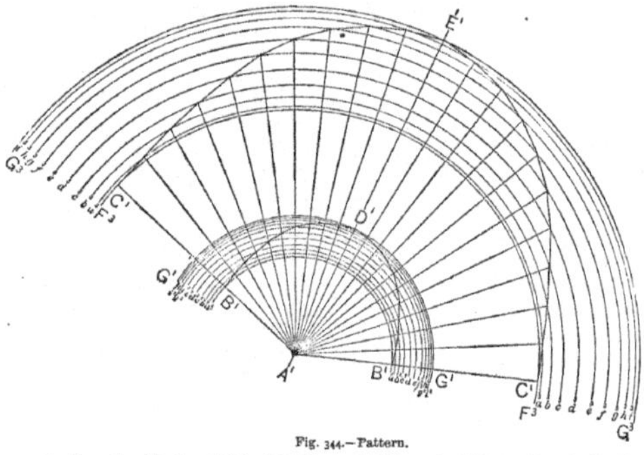

Fig. 344.—Pattern.

*An Irregular Flaring Article, both Top and Bottom of which are Round, the Top being Smaller than the Bottom, and the two being Tangent at One Point in Plan.*

497. *Pattern for a Scale Scoop.*—In Fig. 347, let A B C D represent the side elevation of a scale scoop, being a style in quite general use, and E F H G a section of the same as it would appear cut

upon the line B D, or, what is the same, so far as concerns the development of the patterns, an end elevation of the scoop. The following rule also applies to other forms. The curved line A B C, representing the top of the article, may be drawn at will, being, in this case, a free-hand curve. For the patterns proceed as follows: From the center K, by which the profile of the section or end elevation is drawn, draw a horizontal line, which produce until it meets the center line of the scoop in the point O. Produce the line of the side D C until it meets the line just drawn in the point X. Then X is the apex and X O the axis of a cone, a section of the envelope of which each half of the scoop may be supposed to be. Divide one-half of the profile, as shown in end elevation by E G, into any convenient number of spaces, and from the points thus obtained carry lines horizontally, cutting the line B D, as shown, and thence carry lines to the points X, cutting the top B C, as shown. With X D as radius, and from X as center, describe an arc, as shown by L N, upon which lay off the stretchout of the scoop, as shown in end elevation. From the points in L N thus obtained, draw lines to the center X, as shown. From the points in B C, formed by the lines drawn from B D to the point X, drop lines cutting the side D C, as shown. With X as center, and radii corresponding to each of the several points between D and C, describe arcs, which produce until they cut radial lines drawn from the arc L N to the center X of corresponding numbers. Then a line traced through the points thus obtained, as shown by L M N, will be the profile of the pattern of one-half of the required article.

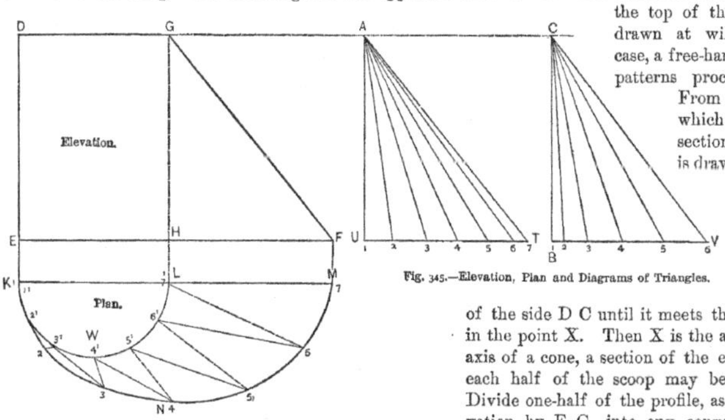


Fig. 345.—Elevation, Plan and Diagrams of Triangles.

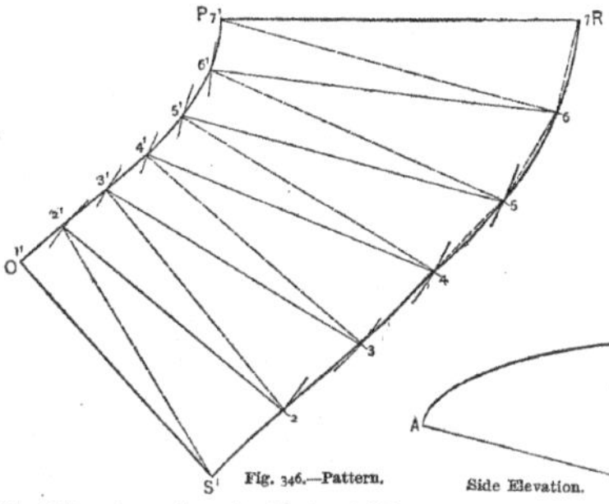

Fig. 346.—Pattern.

*The Pattern for an Irregular Flaring Article which is Elliptical at the Base, Round at the Top, the Top being so Situated with Respect to the Base as to be Tangent to One End of it when Viewed in Plan.*

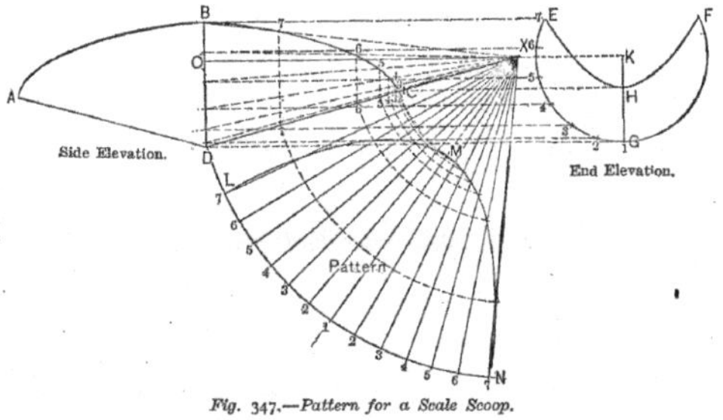


*Fig. 347.—Pattern for a Scale Scoop.*

498. *An Irregular Section through an Elliptical Cone.*—In Fig. 348 is shown an irregular section cut from a cone, the base of which is elliptical. Forms somewhat similar to this are in use for various purposes.

Without naming a list of the articles in which the principles here explained are used, we will present a single demonstration. Application of the same principles may be made in constructing similar shapes to the one here illustrated, for whatever use required. In the engraving B $E^2$ C D is the plan of the cone from which the irregular section, shown by *r o x s*, is cut. An end elevation of the cone and also of the article required is shown in $E^4$ $A^3$ $D^1$. A E of the side elevation is the center of the shape. By inspection of the several views it will be seen that three patterns are required: the top, or cover, the bottom and the rim. The stretchout of the bottom is obtained by stepping off the length of the line *s x* in the side elevation and laying the same down on $B^2$ $C^2$, as shown in the pattern. Through the points in the line $B^2$ $C^2$ thus obtained draw measuring lines in the usual manner. Since B $E^2$ C D of the plan represents a straight section through the cone on the line F G, and as the shape of the article we are seeking is a curved or warped surface cutting through the cone above the base, we cannot use the plan B $E^2$ C D in laying off the width of the bottom, but must obtain a line in it corresponding to the continuous point of contact made between the edge of the bottom face and the rim. To do this we proceed as follows: Divide the quarter of the plan, as shown by $E^2$ C, into any convenient number of equal parts, as shown by the small letters, *a*, *b*, *c*, etc. From these points carry lines vertically to the base line E G of the cone, and thence continue them toward the apex A, crossing the rim, as shown in the side elevation. From the center I of the plan draw lines to those same points, *a*, *b*, *c*, etc. From the points formed in the line *s x* of the article drop points back on to the plan, cutting the radial lines of corresponding numbers. Through the points of intersection thus obtained trace a line, as shown by the second line in the plan, which will represent the line of contact between the envelope of the cone and the bottom of the rim, as seen in plan. Upon the several measuring lines drawn through the stretchout $B^2$ $C^2$, set off on either side the length upon lines of corresponding numbers, drawn from the second line in the plan to the center line I C. Through the points thus obtained trace a line, as shown by $B^2$ $E^6$ $C^2$ $D^3$, which will be the pattern of the bottom of the article. The pattern of the top is obtained in the same way. Drop the points back into the plan from the line R O, thus obtaining the inner line in the plan, which represents a continuous point of contact between the envelope of the cone and the top of the article. Lay off a stretchout of *r o*, as shown by $B^1$ $C^1$, through which draw measuring lines in the usual manner, and upon these lines set off distances, measured on corresponding lines in the plan, from the center line I C to the inner line just obtained, by dropping points from R O, as shown. Then $B^1$ $E^5$ $C^1$ $D^2$ will be the pattern of the upper piece. For the pattern of the rim we proceed as follows: Produce the base line F G of the cone indefinitely, as shown from $E^3$, upon which erect a perpendicular, $E^3$ $A^3$, at any convenient place, in length equal to A E of the side elevation. Drop all the points in the line *r o* and *s x* on to $A^3$ $E^3$, by lines carried at right angles to the axis A E, and from these points in $A^3$ $E^3$ produce lines indefinitely, as shown. Upon the base line F G prolonged, measuring from $E^3$, set off the length of the radial lines in the plan, measuring from I to the

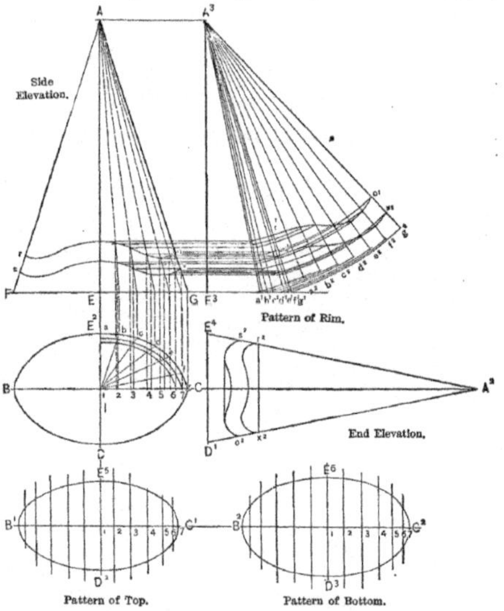

*Fig.* 348.—*An Irregular Section through an Elliptical Cone.*

outer line of the plan, as shown, thus making $E^3\,A^1$ equal to I $a$, and $E^2\,b^1$ equal to I $b$, etc. From the points $a^1$, $b^1$, $c^1$, $d^1$, etc., thus obtained, draw lines to $A^2$, cutting both sides of horizontal lines last described. From $A^2$ as center draw arcs corresponding to the points $a^1$, $b^1$, $c^1$, $d^1$, etc., as shown. Set the dividers to the space used in stepping off the plan $E^2$ C, and lay off its stretchout, stepping from any convenient point in the arc corresponding to $a^1$ to the arc corresponding to $b^1$, and from this to the arc corresponding to $c^1$. From the points thus obtained, indicated by $a^2$, $b^2$, $c^2$, $d^2$, etc., draw radial lines to $A^2$ from the various points of intersection between the horizontal lines drawn from $A^2\,E^2$. With the radial lines drawn from the points $a^1$, $b^1$, $c^1$, $d^1$, etc., describe arcs, which produce until they cut the lines of corresponding number drawn from the points $a^2$, $b^2$, $c^2$, etc. Through the points of intersection thus obtained trace lines, as shown by $r^1\,o^1$ and $x^1\,s^1$. Then $r^1\,o^1\,s^1\,x^1$ is the pattern of one-quarter of the rim.

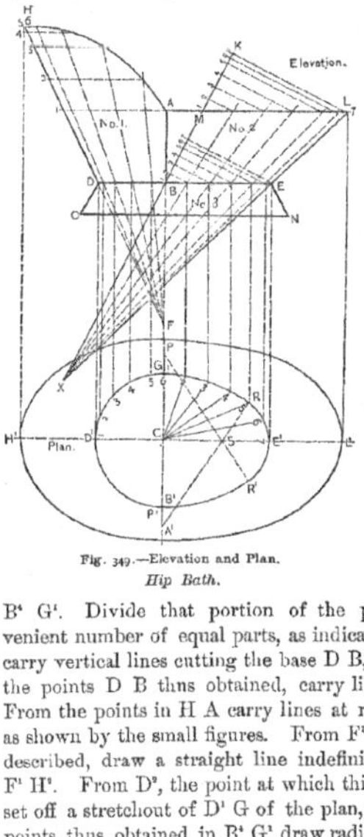

Fig. 349.—Elevation and Plan. *Hip Bath.*

499. *Patterns for a Hip Bath.*—In Fig. 349, let H A L N O be the elevation of the bath, of which $D^1$ G $E^1$ $B^1$ is a plan on the line D E. Let the section $A^5$ $M^1$ $B^6$, Fig. 350, represent the flare which the bath is required to have through its sides on a line indicated by A B in elevation. By inspection of the elevation it will be seen that three patterns are required, which, for the sake of convenience, we have numbered in the various representations 1, 2 and 3. In the following demonstration the various parts are treated as sections of cones, that of No. 1 being an irregular frustum of a right cone, that of No. 2 the frustum of a scalene cone, while the foot, or No. 3, is a section of a cone whose base is oval or egg-shaped. For the patterns, commencing with No. 1, we proceed as follows: Produce the line H D indefinitely, and also A B, likewise indefinitely, until they meet in the point F. Then F is the apex of the cone of which No. 1 is a section. From any convenient point, as $F^1$, Fig. 351, for center, with F D as radius, describe an arc, as shown by $B^4$ $G^1$. Divide that portion of the plan corresponding to piece No. 1, as shown by $D^1$ G, into any convenient number of equal parts, as indicated by the small figures 1, 2, 3, etc. From the points thus obtained carry vertical lines cutting the base D B, as shown. From the apex F of the cone, through the points D B thus obtained, carry lines cutting the upper boundary H A of the piece. From the points in H A carry lines at right angles to the axis A F, cutting the side D H, as shown by the small figures. From $F^1$ in Fig. 351, the center by which the arc $B^4$ $G^1$ was described, draw a straight line indefinitely, cutting $B^4$ $G^1$ near the center, as is shown by $F^1$ $H^2$. From $D^2$, the point at which this straight line crosses the arc, measuring both ways, set off a stretchout of $D^1$ G of the plan, all as indicated by the small figures. Through the points thus obtained in $B^4$ $G^1$ draw radial lines indefinitely. From the center $F^1$ upon the straight line $F^1$ $H^2$, which it will be seen corresponds to F H of the elevation, set off points corresponding to the points in F H, all as indicated by the small figures, 1, 2, 3, etc. From F as center, with radii corresponding to these points, strike arcs, which produce both to the right and the left until they cut radial lines of corresponding numbers. Then a line traced through these points, as shown by $A^4$ $H^2$ $A^5$, will be the boundary of the pattern upon its upper side, and the whole pattern will be contained by $B^4$ $G^1$ $A^4$ $H^2$ $A^5$. Since the small diagram $A^5$ $M^1$ $B^6$, Fig. 350, represents a section of the article upon the line A B in elevation, set off the angle A M B in elevation equal to $M^1$ $A^5$ $B^6$ of the diagram, and through the points M B draw a line indefinitely. Produce the line L E until it meets the line drawn through M B in the point X. Then X is the apex and X K is the axis of the scalene cone of which M L E B of the elevation is a section. Divide that portion of the plan corresponding to this piece, No. 2, into any convenient number of equal parts, as indicated by the small figures, and from the points thus obtained carry lines vertically, cutting the base E B, as shown. From the apex X, through the points thus obtained, carry lines cutting the top of the piece L M, as shown. From the points in E B, and also from the corresponding points in L M, draw lines at right angles to the axis X K, cutting it as shown in the two sets of points marked with the small figures 1, 2, 3, 4, etc. Lay off

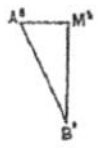
Fig. 350.—Flare at Sides. *Hip Bath.*

$X^1$ $K^1$, Fig. 352, at any convenient point, equal to X K of the elevation, in which set off the points corresponding to the points just obtained in X K, all as indicated by corresponding figures. From each set of points in $X^1$ $K^1$ erect lines indefinitely, perpendicular to $X^1$ $K^1$, all as shown. In the lines drawn from the points, commencing at $B^1$, set off lengths corresponding to lengths measured from C of the plan, on radial lines drawn to the points stepped off in G $E^1$, and through the points thus obtained draw radial lines from $F^1$, as shown, producing them until they cut corresponding lines drawn from the points commencing at $M^1$. From $X^1$ as center, with radii corresponding to the intersections between the radial lines and the perpendiculars drawn from the points at $B^1$ and $M^1$, describe arcs indefinitely, as shown. From $X^1$ draw any straight line, as $X^1$ $M^2$, as shown, crossing the arcs just described, which will form a basis of measurement for one side of the pattern. From the point $B^2$, where the line $X^1$ $M^2$ crosses the first arc corresponding to the set of points, commencing at $B^1$, step off the stretchout of the plan G $E^1$, using the same spaces as first employed, stepping from arc to arc, as shown. Then a line traced through the points thus obtained, as shown by $B^2$ $E^2$ $B^2$ will be the edge of the pattern corresponding to B E of the elevation. Through the points in $B^2$ $E^2$ $B^2$, from $X^1$, draw radial lines, which produce until they cut arcs of corresponding numbers drawn from the points in the lines at $M^1$. Then a line traced through these points, as shown by $M^2$ $L^2$ $M^2$, will be so much of the line of pattern corresponding to the top of the article in elevation as shown from M to L.

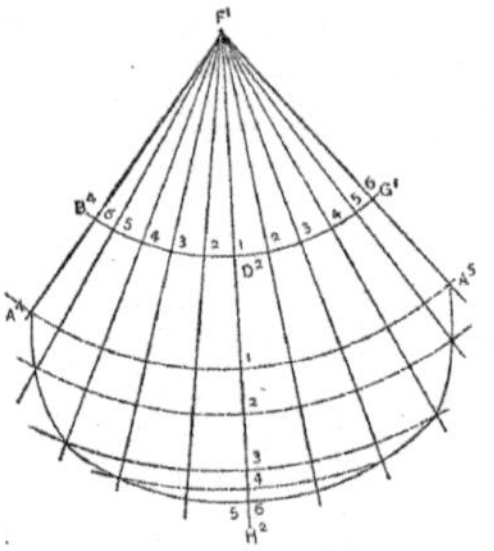

Fig. 351.—Pattern of Back (No. 1).
*Hip Bath.*

From $B^2$ and $B^2$ respectively as centers, with radius equal to $A^5$ $B^4$ of the small section to the left of the elevation, describe an arc, and from $M^2$ and $M^2$ as centers, with radius equal to $M^1$ $A^5$ of the small section, describe arcs intersecting those first drawn in the points $A^2$ and $A^2$. Continue the line of the outside of the pattern from $M^2$ and $M^2$, respectively, to $A^2$ and $A^2$. Then $A^2$ $L^2$ $A^2$ $B^2$ $B^2$ will be the pattern of No. 2. For pattern of No. 3, first construct a section of radii, as shown in Fig. 353 of the engravings. The plan corresponding to the line D E of No. 3 in elevation, as shown by $D^1$ G $E^1$ $B^1$, is struck from several centers. From C a semicircle, $B^1$ $D^1$ G, is struck. Arcs G R and $B^1$ $R^1$ are struck from centers P and $P^1$, and the arc R $P^1$ is struck from the center S. Therefore, to obtain the radii by which the pattern may be described, we must construct sections of the several cones to which the arcs composing the plan may be supposed to belong. Draw any straight line, as $O^1$ T, indefinitely, at right angles to which set off $O^1$ $N^1$ indefinitely. From $O^1$, measuring on the line $O^1$ T, set off $O^1$ $C^1$ equal to the vertical hight of piece No. 3 measured in elevation, and from $C^1$ draw $C^1$ $E^4$ perpendicular to $O^1$ T. Since the plan $D^1$ G $E^1$ $B^1$ corresponds to D E of the elevation, or the upper edge of the piece—the pattern for which we are about to describe—measurements must be made upon the corresponding line in the section, which is $C^1$ $E^4$. On $C^1$ $E^4$ set off the length of the radii by which the several sections of the plan were struck. Make $C^1$ $E^3$ equal to S $E^1$ of the plan, and $C^1$ $D^3$ equal to C $D^1$ of the plan, and $C^1$ $E^4$ equal to P $R^1$ of the plan. Since the flare of the base, or No. 3, is to be equal throughout its extent, the several radii as seen in the section will be parallel. Therefore, from the points

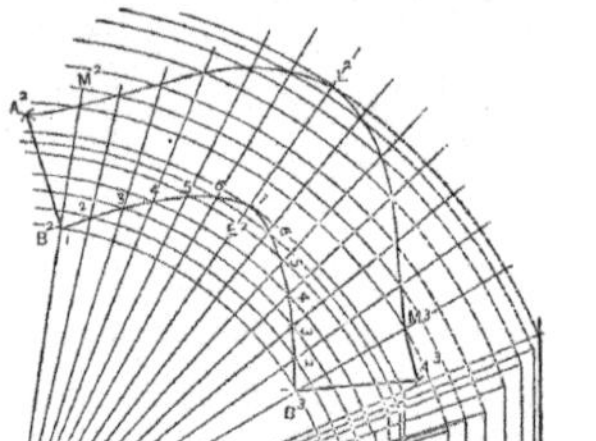

Fig. 352.—Pattern of Front Part (No. 2).
*Hip Bath.*

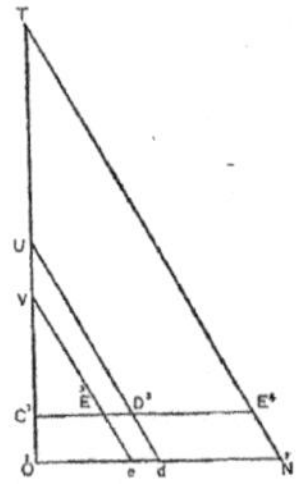

Fig. 353.—Diagram of Cones for Radii of Pattern of Foot.
*Hip Bath.*

$E^3$, $D^3$, $E^4$ draw lines cutting the line $O^1$ $N^1$ at angles corresponding to the flare of the piece, as indicated by E N or D O of the elevation. Produce these lines in the opposite direction until they meet the vertical line $O^1$ T in the points V, U and T respectively. Then V $E^3$ is a radius of that part of the pattern the plan of which is shown by R $E^1$ $R^4$, and U $D^3$ is the radius of that part of the pattern shown in plan by G $D^1$ $B^1$, and T $E^4$ is the radius of those portions of the pattern shown in plan by G R and $B^1$ $R^1$. In describing the pattern it is immaterial with which point we commence, since in the plan but one-half has been divided. We will start at the point D. Therefore, from any convenient center, as $U^1$ in Fig. 354, with radius equal to U $D^3$, describe the arc $D^2$ $G^2$, upon which set off the stretchout of $D^1$ G in the plan. In like manner, from the same center, with the same radius, describe the arc $D^2$ $B^8$, which make of corresponding length. From $G^2$, through the center $U^1$, draw the line $G^2$ $T^2$ indefinitely. Set the dividers to the radius T $E^4$ of the diagram, and measure from $G^2$ along the line $G^2$ $T^2$. Establish the center $T^2$, from which strike the arc $G^2$ $R^2$, which in length make equal to the stretchout of G R of the plan. In like manner, from $B^8$, also through the center $U^1$, draw the line $B^8$ $T^1$, and with the dividers set as just described, measuring from $B^8$, establish the center $T^1$, from which describe the arc $B^8$ $R^4$. From R draw a line to the center $T^2$, as shown. Set the dividers to V $E^3$ of the section for radius, and, measuring from $R^2$, establish the center V, from which describe the arc $R^2$ $R^3$, in length equal to the stretchout of the plan from R to $R^1$. In like manner, using the centers thus established and using lengthened radii of the diagram, or, in other words, setting the dividers from the points V, U and T respectively, to the lower line of the diagram, Fig. 353, describe the outer line of the pattern, determining the length of the several arcs in it by the lines drawn from the several centers produced, all as shown. Then $R^3$ $r^3$ $r$ $R^4$ will be the pattern of the foot of the article, or No. 3 of the elevation.

500. *Patterns of a Coal Hod.*—In Fig. 355 we show by elevation and section an ordinary funnel coal hod, to be constructed in four pieces. The pieces composing the front of the funnel are to be seamed together on the line S R, the back and the front are to be joined on the line R Q, and there is to be a seam between the base or rim and the body of the hod. The inner line of the plan *a* 2 8 *x* represents the bottom of the hod or the top of the foot, as indicated by $a^1$ $8^1$ in the elevation. The outer line of the plan A $B^1$ C D E shows the shape of the hod at a point through the upper portion corresponding to T X in the elevation. A $B^1$ *s n* shows the shape of the spout, while $O^1$ 1 *y* represents an imaginary section taken through piece No. 3, and is introduced to better show correspondence between parts. By tracing the course of the dotted lines connecting the several portions the reader will understand the relationship between the representations in plan and the corresponding points in elevation. For the patterns we will commence with piece No. 1 in the elevation, or the back of the hod. Subdivide the inner line of the plan corresponding to the bottom of this piece, as shown by the small figures 1, 2, 3, 4, etc., and from these points carry lines vertically, cutting the base of the hod in the points $1^1$, $2^1$, $3^1$, etc. Produce the side X $8^1$ of the piece No. 1 in elevation until it meets the center line in the point H. Then H is the apex of a cone, of the envelope of which the required piece is a part. Through the points already obtained in the bottom line of the piece, as above described, draw lines from the apex H, producing them until they cut the upper line R X of the piece, all as indicated by the dotted lines and by the points $1^2$, $2^2$, $3^2$, etc. The axis of the cone, of the envelope of which piece No. 1 is a part, is represented by I H. Place the T-square at right angles to the axis, and, bringing it against the points in the curved line R X, cut the axis, as shown by the points $3^3$, $4^3$, $5^3$ etc. From the measurements obtained by these several steps the pattern for the piece No. 1 may be described. To avoid confusion of lines, H I, with all the points above described, is transferred to Fig. 356, as shown by $H^1$ $I^1$. At right angles to $H^1$ $I^1$, from the point $2^2$, draw a straight line, as indicated, in which set off points equal to the distances L 1, L 2, L 3, L 4, etc., in the plan, and through these points, from $H^1$, draw lines, as shown, which produce indefinitely. With the T-square at right angles to $H^1$ $I^1$, and brought successively against the several points at the top of the line, as indicated by the small figures $1^3$, $2^3$, $3^3$, etc., draw lines, producing them until they meet the radial lines just mentioned of corresponding numbers. From $H^1$ as center, with radii corre-

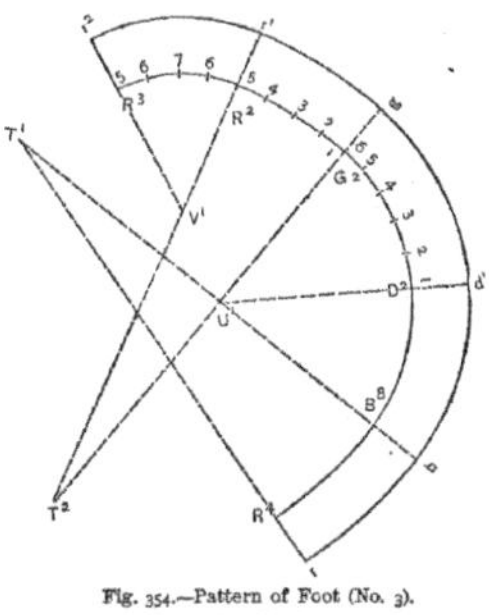


Fig. 354.—Pattern of Foot (No. 3). *Hip Bath.*

sponding to the points of intersection thus obtained, describe the arcs, as shown. In like manner, from the same center, with radii corresponding to the intersections with the line 2², describe similar arcs. The stretchout of the piece, the pattern of which we are describing, is obtained from the plan, as indicated by the portion 1 8 *y*. Take the space in the dividers used in stepping off the plan and, starting with the arc drawn from the intersection of the line 2² with the line corresponding to 1 of the plan, step to the next arc, and thence

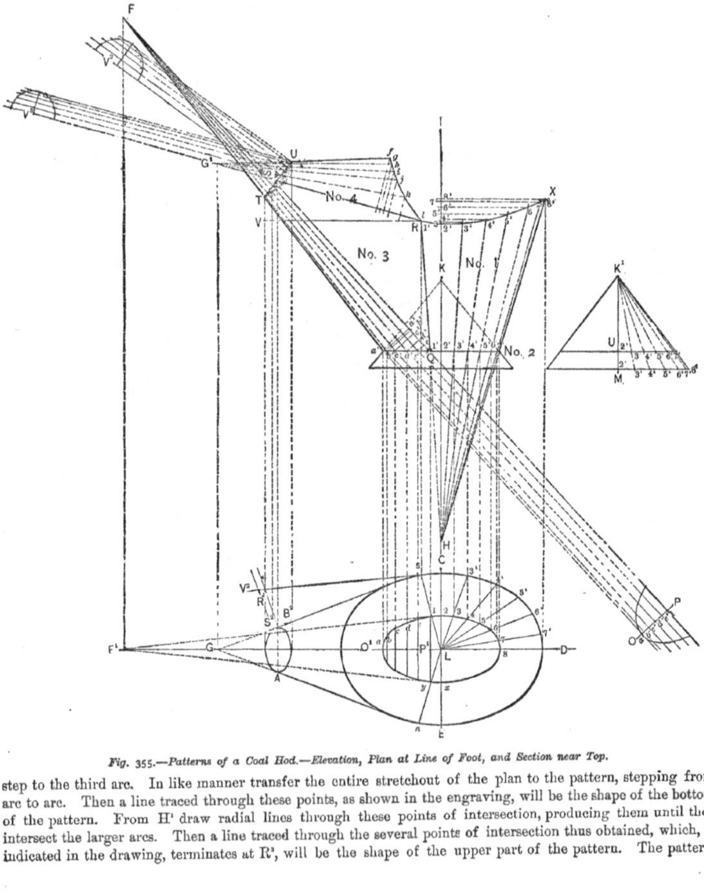

*Fig. 355.—Patterns of a Coal Hod.—Elevation, Plan at Line of Foot, and Section near Top.*

step to the third arc. In like manner transfer the entire stretchout of the plan to the pattern, stepping from arc to arc. Then a line traced through these points, as shown in the engraving, will be the shape of the bottom of the pattern. From H¹ draw radial lines through these points of intersection, producing them until they intersect the larger arcs. Then a line traced through the several points of intersection thus obtained, which, as indicated in the drawing, terminates at R³, will be the shape of the upper part of the pattern. The patterns of the remaining pieces are obtained, in the main, in the same general manner, and the steps for each are clearly indicated in the engravings. It is not necessary, therefore, to give a detailed demonstration of each. There are a few points differing, however, from the pattern just explained to which we will call attention. In the foot, or flange (No. 2 in the elevation), the same points are used as described in connection with No. 1. The elevation of No. 2 has been drawn to one side from the elevation of the complete hod, in order to avoid confusion of lines. The various parts or points appear in it as though it occupied its normal position. The pattern is given in Fig. 357. $K^2$ $M^1$ corresponds in all respects with $K^1$ M of the elevation. Inasmuch as both boundary lines in the elevation are straight lines, this pattern is simpler than that of No. 1, of which but one line was straight. Therefore all the points of distance from the axis in each appear in one line, as they did in the lower line of piece No. 1. Thus the points in the line U of the elevation appear upon $U^1$ of the diagram, and all the points on the line M in the elevation appear upon the line $M^1$ of the diagram. Distances taken from the plan L 1, L 2, L 3, etc., are set off on $U^1$, as shown, and lines drawn from $K^2$ through these points cut the line $M^1$, giving the requisite points in it. All other steps belonging to this pattern are identical with those of No. 1. In piece No. 3 a new condition arises. By inspection of it in the elevation, it will be seen that the profile or section presented in the plan is not taken at right angles to its axis, F Q; therefore the first requirement is to ascertain the shape of the profile which will fit it when placed at right angles to the axis, as, for instance, on the line $a^1$ $a^2$. To obtain this we proceed as follows: At any point in convenient proximity to the base of the cone, draw a line at right angles to the axis, as O P, upon which to construct the new profile. Subdivide the plan in the usual manner, as indicated by the points *a*, *b*, *c*, etc. From these points draw vertical lines, cutting the base in the elevation, as shown by $a^1$, $b^1$, $c^1$, etc. From these points, parallel to the axis of the cone F Q, draw lines cutting O P, as indicated by $a^2$, $b^2$, $c^2$, etc. By this means we have the subdivisions in O P corresponding to the divisions in $O^1$ $P^1$ of the plan. Therefore, to complete the profile, set off on each of the lines drawn through O P, measuring to each side of it, the distance from corresponding points in $O^1$ $P^1$ in the plan to the circumference, and trace a line through these points. Having thus obtained a section of the cone at right angles to its axis, the remaining steps connected with describing piece No. 3, so far as concerns its conical part, are identical with those described in connection with piece No. 1. From the points $a^1$, $b^1$, $c^1$, $d^1$, etc., in the bottom line of the piece draw lines at right angles to the axis, obtaining the points indicated by $a^2$, $b^2$, $c^2$, etc. From the same points in the bottom line of the piece carry lines toward the apex F, cutting the upper line of the piece T U, and from the points thus obtained in T U draw lines in like manner at right angles to the axis. The line F Q then is transferred with all its points to Fig. 358, as indicated by $F^2$ $Q^2$, and is there used in exactly the same manner as the corresponding line in the pattern of the piece No. 1. For the pattern of the portion of No. 3 which is flat, and which is outside of the shape derived from the cone, we proceed as follows: By describing that portion of the pattern derived from the cone, we have obtained the point indicated in the

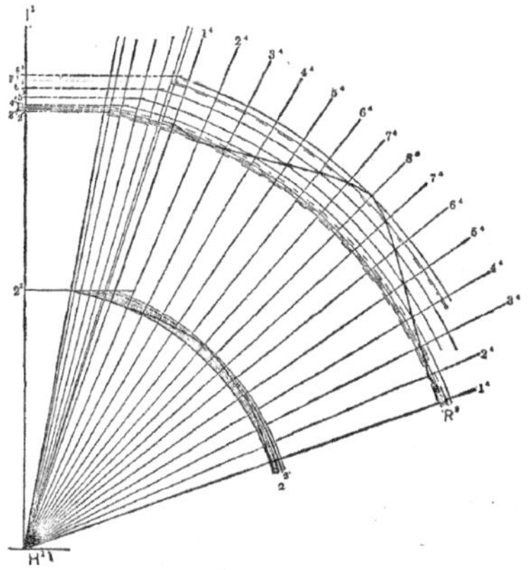

Fig. 356.—Pattern of Back Part of Body (No. 1).
*Coal Hod.*

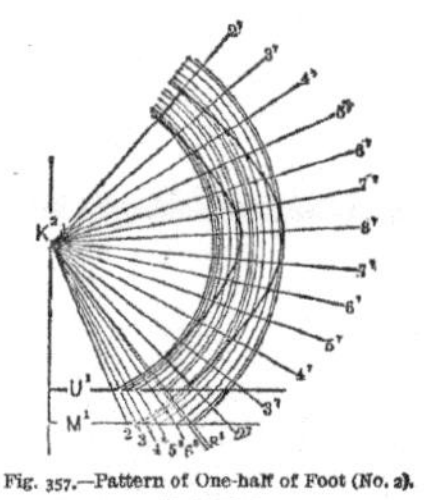

Fig. 357.—Pattern of One-half of Foot (No. 2).
*Coal Hod.*

engraving by $2^1$. By inspection of the elevation it will be seen that ending at the point $2^1$ will be one boundary line of the inner surface. This line in length will be equal to $2^2$ $R^2$ of piece No. 1. Therefore, with the dividers set to this distance as a radius, and with one leg at the point $2^1$ as center, describe the arc *u v*. It then remains to find the point in this arc at which the remaining boundary line of the pattern will intersect. By inspection of the elevation it will be noticed that the division line between the conical part of piece No. 3 and the plane surface, or, in other words the line F Q, does not intersect with $G^1$ R within the boundaries of the elevation. Therefore it will be noticed that a small portion of the boundary line of the pattern lying between the end of the division line between pieces Nos. 3 and 4 and the cone part of piece No. 3, must be established. S in the elevation represents the intersection of the lines F Q and the line $G^1$ R. Upon the line $F^2$ $Q^2$ in the pattern establish a corresponding point, as indicated by $S^1$. From this point draw a line at right angles to $F^2$ $Q^2$ until it meets the corresponding distance line already drawn. Then from $F^2$, with radius corresponding to this point of intersection, describe an arc in a similar manner to the arcs already drawn. The point $S^2$, at which this arc intersects the stretchout line $F^2$ $1^2$, will be the point in the pattern corresponding to S in the elevation, and therefore serves as a center from which to measure in order to obtain the point sought in the arc *u v*, to which a line from it is to be drawn. The next step, therefore, is to obtain the radius by which to strike the intersecting arc. By inspection of the elevation it will be seen that this radius will be equal in length to the actual distance from S to R. From the fact that the line S R lies in a plane which is neither parallel to the general plane of the elevation nor parallel to the horizontal line drawn through the elevation, it follows not only that we cannot use the distance from R to S in the elevation, but that a special operation must be performed in order to obtain the actual distance indicated. This operation consists of dropping on to the line of the plan G *s* a point corresponding to the point S in the elevation, as shown by $S^2$ in the plan. From R in the elevation draw the horizontal line R V. From the intersection of this line with S $S^2$ take the distance to S, and set it off from $S^2$ at right angles to the line of the plan G *s*, all as shown by $S^2$ $V^2$. Through the points thus obtained draw the lines *s* $V^2$. Then *s* $V^2$ is the actual length of the line S R in the elevation, and is the distance to be used in Fig. 358 as radius by which to describe an arc intersecting the arc *u v* in the pattern. Therefore, with $S^2$ as center, and with radius S $V^2$ of the plan, describe the arc *w z*, cutting the arc *u v* in the point $R^1$. Connect the points $S^2$ and $2^1$ with this point of intersection, $R^1$, by straight lines, as shown. Inasmuch as the point S in the elevation, from which we have worked to obtain the measurement just used, lies outside of the piece No. 3, the boundary line of the pattern, representing that portion of the line T U of the elevation between the line bounding the conical part and the division line between it and No. 4, will fall somewhat below $S^2$ in the pattern, and is to be traced along points corresponding to the line T U in the elevation, the means for doing which, in full-size work, will be easy to perceive, but to indicate which becomes somewhat difficult in a diagram of a scale so small as the accompanying engrav-

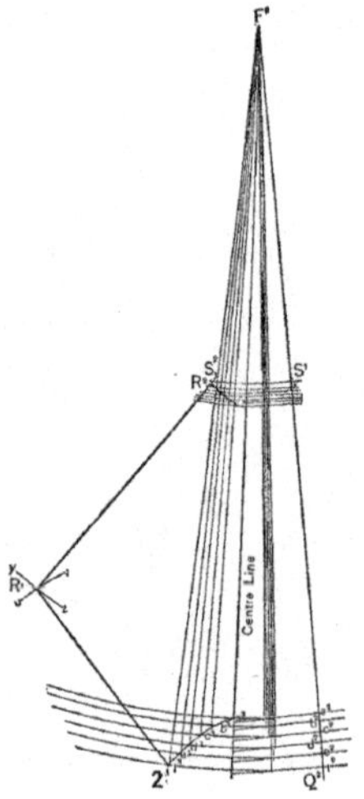


Fig. 358.—Pattern of One-half of Front (No. 3). *Coal Hod.*

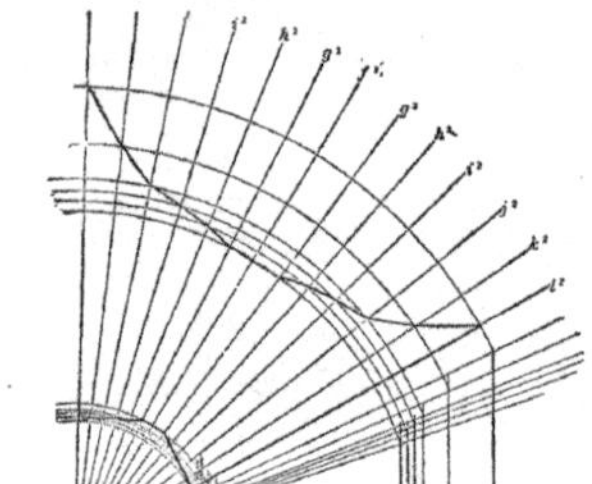

Fig. 359.—Pattern of Hood (No. 4) *Coal Hod.*

ings. In the development of the pattern for piece No. 4 there is nothing different from the steps explained in connection with one or the other of the preceding. The profile corresponding to the end S U is assumed to be a semicircle, all as indicated by the dotted lines. Connect S U with the section. From this sectional view a profile of the piece, as it would appear if cut through the point U at right angles to the axis of the cone, is to be constructed, all as shown in the second profile, connected with S U in like manner by dotted lines. Through the points in S U, from the apex, lines are drawn, which are produced until they cut the opposite end of the pattern, as shown by the points *f*, *g*, *h*, etc. From these points lines are drawn at right angles to the axis. In developing the pattern the line $G^1$ R with all its points is transferred to Fig. 359, as shown by $G^2$ $R^3$. These points are used in describing the pattern in the same manner as explained in connection with the preceding pieces. Respecting the engravings here presented, it is to be remarked that working to so small a scale as is necessary in a book of this character, some inaccuracies in proportions, etc., are altogether unavoidable. In the plan the inner and outer lines, in actual work, would not be parallel. The distance between them, measured upon the line 2 C, would be less than measured upon the line 8 D, because the point $2^3$ is nearer the apex than $8^3$, and also because, the plan being elliptical, the flare is not so great laterally as longitudinally. The difference is so slight, however, as to be somewhat difficult to indicate in an engraving of small scale. The process employed and described in connection with the patterns, however, will be found accurate, and we think it has been presented in such a manner as to enable any one to lay out full-size work without doubt or difficulty.

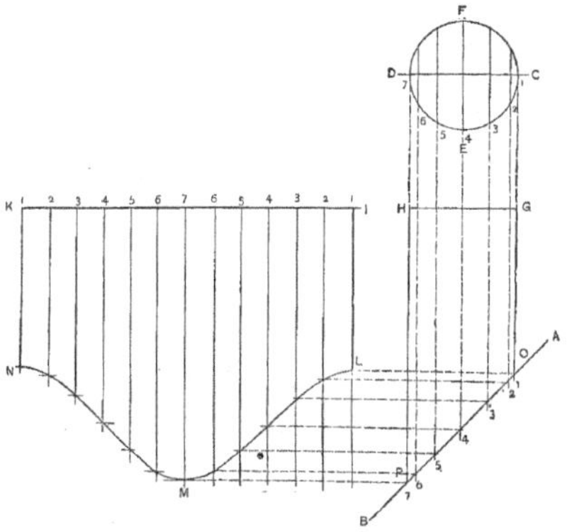

*Fig. 360.—The Pattern of a Round Pipe to Fit Against a Roof of One Inclination.*

501. *The Pattern of a Round Pipe to Fit Against a Roof of One Inclination.*—In Fig. 360, let A B be the pitch of the roof and C F D E the profile of the pipe which is to miter against it. Let G O P H be the elevation of the pipe as it is required to be. Draw the profile in line with the elevation, as shown by C F D E, and divide it into any convenient number of equal parts. Place the T-square parallel to the sides of the pipe, and, bringing it successively against the divisions of the profile, cut the pitch line, as shown by A B. Lay off a stretchout in the usual manner, at right angles to and opposite the end of the pipe, as shown by I K, and draw the measuring lines. Reverse the T-square, placing it at right angles to the pipe, and, bringing it successively against the points in A B, cut the corresponding measuring lines. A line traced through the points thus obtained, as shown by L M N, will finish the pattern.

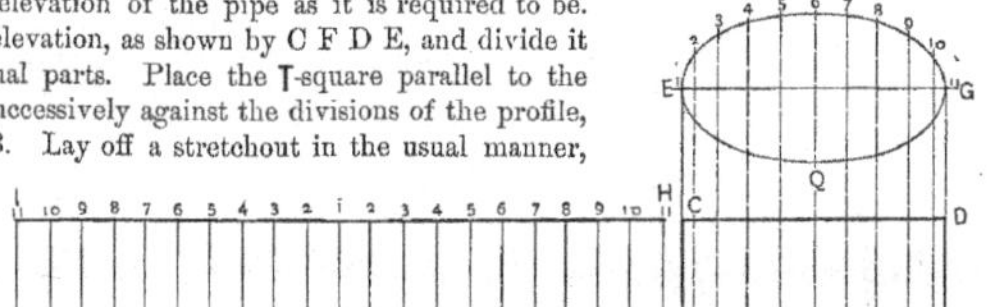

*Fig. 361.—The Pattern of an Elliptical Pipe to Fit Against a Roof of One Inclination.*

502. *The Pattern of an Elliptical Pipe to Fit Against a Roof of One Inclination.*—In Fig. 361, let N O D O be the elevation of an elliptical pipe fitting against a roof, represented by A B. Let E F G be the section or profile of the pipe. Draw the profile in convenient proximity to the elevation, as shown, and divide it

into any convenient number of equal parts. Place the T-square parallel to the sides of the pipe, and, bringing it against the points in the profile, drop lines cutting the roof line A B, as shown. Opposite to the end of the pipe, and at right angles to it, lay off a stretchout, as shown by H I, and through the points in it draw measuring lines in the usual manner. Reverse the T-square, placing it at right angles to the pipe, and, bringing it successively against the points in A B, cut the corresponding measuring lines, as indicated. A line traced through these points, as shown by K L M, will be the required pattern. In the illustration the long diameter of the ellipse, or E G, is shown as crossing the roof. The same rule applies if the pipe is placed in the opposite position—that is, with Q F crossing the roof—the only change required being in the position of the profile, which, of course, would require to be turned around, and drawing the elevation to correspond

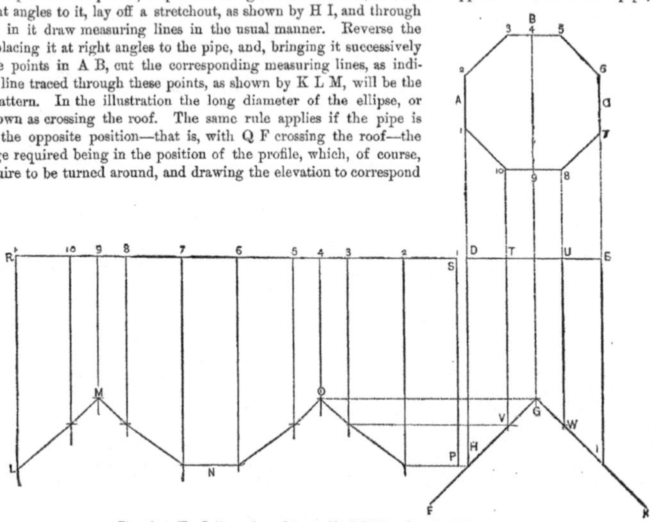


*Fig. 362.—The Pattern of an Octagon Shaft Fitting Over the Ridge of a Roof.*

with it. Otherwise proceed in all respects as above. From this it is evident that a pattern for the pipe, when its section lies diagonally, may be described by the same rule.

503. *The Pattern of an Octagon Shaft Fitting Over the Ridge of a Roof.*—In Fig. 362, let A B C be the section and D H G I E the elevation of an octagon shaft mitering against a roof, represented by the lines F G and G K. Draw the section in line with the elevation, as shown, and from the angles drop lines, giving T V and U W of the elevation. Drop the point G back on to the section, thus locating the points 9 and 4. Opposite the end of the shaft, and at right angles to it, draw a stretchout line, as shown by S R, and through the points in it draw measuring lines in the usual manner. Place the T-square at right angles to the shaft, and, bringing it successively against the points in the roof line formed by the intersection with it of the angle lines in the elevation, and also against the point G, representing the ridge of the roof, cut the corresponding measuring lines. Then a line traced through the points thus obtained, all as shown by P O N M L in the engraving, will be the pattern required.

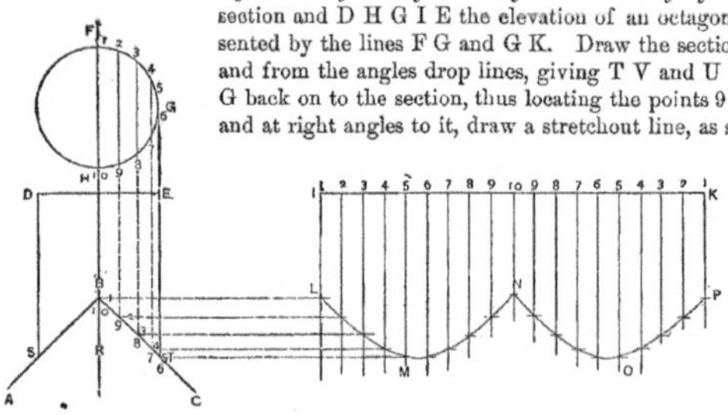


*Fig. 363.—The Pattern of a Round Pipe to Fit Over the Ridge of a Roof.*

504. *The Pattern of a Round Pipe to Fit Over the Ridge of a Roof.*—Let A B C in Fig. 363 be a sec-

tion of the roof and D S B T E an elevation of the pipe. Draw a profile of the pipe in line, as shown by F G H. Since both inclinations of the roof are to the same angle, both halves of the pattern will be the same. Therefore space off but one-half of the profile for dropping the points on to the roof line. Lay off a stretchout, however, equal to the whole profile, numbering the points in both halves correspondingly. Draw measuring lines through these points in the usual manner. Place the T-square parallel to the sides of the pipe, and, bringing it against the points in the profile, cut the roof line, as shown from B to T. Reverse the T-square, placing it at right angles to the lines of the pipe, and, bringing it successively against the points dropped upon the roof line, cut the corresponding measuring lines. A line traced through the points, as shown by L M N O P, will form the required pattern.

505. *The Patterns of a Cylinder Mitering over the Peak of a Gable Coping having a Double Wash.*—Let A B C in Fig. 364 be the elevation of a coping to surmount a gable, the profile of which is D E F E' D', which, as will be seen, shows a double wash, E F and F E'. Let M O P N be the elevation of a pipe or shaft which is required to miter over this double wash at the peak of the gable. For the pattern proceed as follows: In line with the pipe or shaft construct a profile of the same, as shown by G' L' K' H', which divide into any convenient number of equal parts, and from the points thus obtained drop lines vertically on to the elevation.

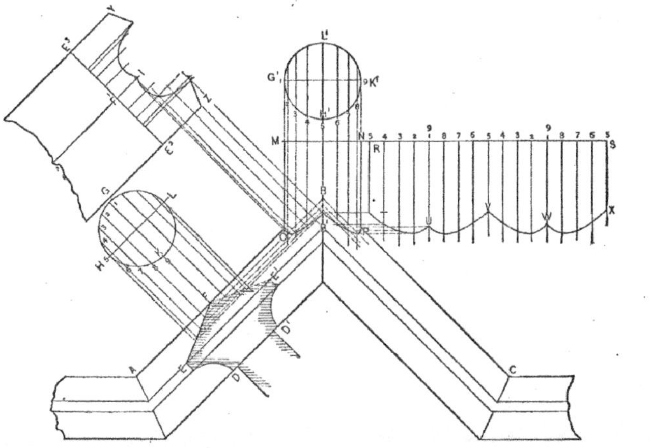


*Fig. 364.—The Patterns of a Cylinder Mitering over the Peak of a Gable Coping Having a Double Wash.*

Draw a corresponding profile, as shown by H G L K, directly over the profile of the coping, all as shown, which divide into the same number of equal parts, beginning at a corresponding place in the profile, and from the points in it drop lines on to the profile of the coping, cutting the washes E F and F E', and thence carry the lines parallel to the lines of the coping, producing them until they intersect the lines dropped from the profile G' L' K' H'. Through the points of intersection thus obtained trace a line, as shown from O to P, then O B' P will be the miter line in elevation. In line with the end M N of the shaft, and at right angles to it, lay off a stretchout of the profile G' H' K' L', as shown by R S, in the the usual manner, through the points in which draw measuring lines. Commence numbering these measuring lines with the figure corresponding to the point at which the seam is desired to be, in this case 5. Place the T-square at right angles to the shaft, and, bringing it against the points in the miter line O B' P, cut the corresponding measuring lines. Then a line traced through these points of intersection, as shown by T U V W X, will be the pattern required. In case it should be desired to miter the coping against the base of the shaft, the pattern for it may be obtained from the same lines in the following manner: At right angles to the lines of one side of the coping, as A B, lay off a stretchout of the wash of the coping, E F E', all as shown by E² F' E². In this stretchout line set off points corresponding to the points in E F E', obtained by the lines previously dropped from the profile G H K L. Place the T-square at right angles to A B, and, bringing it against the points in the miter line O B', cut lines of corresponding numbers drawn through the stretchout E² E², all as indicated by the dotted lines. Then a line traced through these points of intersection, as shown by Z I Y, will be the pattern of the wash required to miter against the base of the shaft. In case the shaft is octagonal in shape, the same general rules apply. Less divisions, however, will be required in the profile, it only being necessary to drop points from the angles, being, in this respect, identical with Section 503.

506. *The Pattern of a Flange to Fit Around a Pipe and Against a Roof of One Inclination.*—Let L M, Fig. 365, be the inclination of the roof and P R T S an elevation of the pipe passing through it. N O then represents the length of the opening which is to be cut in the flange, the width of which will be the same as the diameter of the pipe. Let A B D C be the size of the flange desired, as it would appear if viewed from a point directly above the pipe. Immediately in line with the pipe draw the profile G H I K, putting it in the center of the plan of the flange A B D C, or otherwise, as required. Divide one-half of the profile in the usual manner, and carry lines vertically to the line L M, representing the pitch of the roof, and thence, at right angles to it, indefinitely. Carry points in the same manner from A and B. Draw C' D' parallel to L M. Make C' A' equal to A C, or the width of the required flange, and draw A' B' parallel to C' D'. Then C' A' B' D' will be the pattern of the required flange. Draw E' F' through it at a point corresponding to E F of the plan, crossing the lines drawn from the profile. From E' F' set off on each side, on each of the measuring lines crossing it, the width on corresponding lines, measuring from E F in the plan to the profile. Through the points thus obtained draw a line, which will give the shape of the opening to be cut, all as shown by G' H' I' K'.

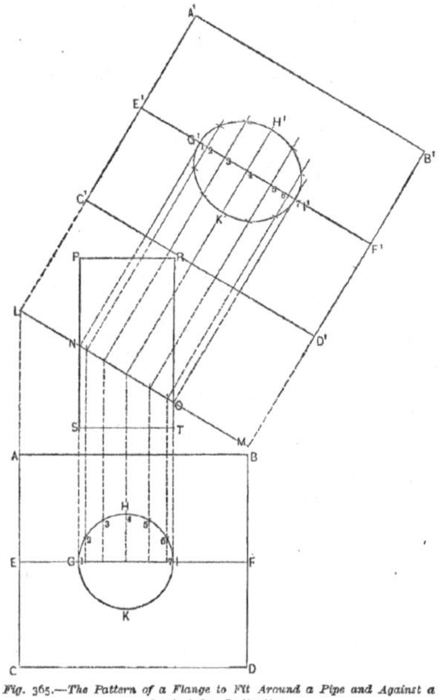


*Fig. 365.—The Pattern of a Flange to Fit Around a Pipe and Against a Roof of One Inclination.*

507. *A Conical Flange to Fit Around a Pipe and Against a Roof of One Inclination.*—In Fig. 366, is shown, by means of elevation and plan, the general requirements of the problem. A B represents the pitch of the roof, G H K I represents the pipe passing through it, and C D F E the required flange fitting around the pipe at the line C D and against the roof at the line E F. The flange, as we have drawn it, becomes a section of the envelope of a right cone. By prolonging E C and F D until they intersect at W, the apex is found,

and by continuing these same lines in the opposite direction, to L and M respectively, and drawing the line L M, a section of the cone is described, from the envelope of which the flange is cut. In connection with the elevation just described, we have shown a plan of the several parts, or a representation of them as they would appear if viewed from above. S T represents the pipe and N O the the flange. While the pipe is made to pass through the center of the cone, as may be seen by examining the base line L M in the elevation, and also P R of the plan, it does not pass through the center of the oblique cut E F in the elevation, or, what is the same, N O of the plan. For the pattern of the flange proceed as shown in Fig. 367, which in the lettering of its parts is made to correspond with Fig. 366, just described. Divide the plan P X R into any convenient number of parts—in this case twelve—and from each of the points thus established erect perpendiculars to the base of the cone, obtaining the points 1', 2', 3', etc. From these points draw lines to the apex of the cone W, cutting the oblique line E F and the top of the flange C D, as shown. Inasmuch as C D cuts the cone at right angles to its axis, the line in the pattern corresponding to it will be an arc of a circle; but with E F, which cuts the cone obliquely to its axis, the case is different. A measurement in the pattern is required at each point, corresponding to the divisions given in the plan. Accordingly, the several points in E F, obtained by the lines from the plan drawn to the apex W, must be transferred to one of the sides of the cone. From the points 0², 1², 2², 3², in E F, draw lines at right angles to the axis of the cone W X, cutting the side W M, as shown. We now have all the points necessary to use in describing the pattern. With W as center, and with W M as radius, strike the arc P' R' indefinitely, and, with the same center and with W D as radius, strike the arc C' D' indefinitely, which will form the boundary of the pattern at the top. At any convenient distance from W M draw W P', a portion of the length of which will form the boundary of one end of the pattern. On P' R', commencing with P', set off spaces equal in length and the same in number as the divisions in the plan P X R, all as shown by 0², 1², 2², 3², etc. From these points draw lines to the center W, as shown. With one point of the dividers set at W and the other brought successively to the points cut in W M by the horizontal lines drawn from E F, cut the corresponding lines in the stretchout of the pattern, as indicated by the curved dotted lines. A line traced through these points, as E' F', will represent the lower side of the pattern. As we used but one-half of the plan in laying out the stretchout, the pattern C' E' F' D' thus obtained is but one-half of the piece required. In use it is to be doubled. The seam can be made to come through the short side at C E, or through the long side at D F, at pleasure.

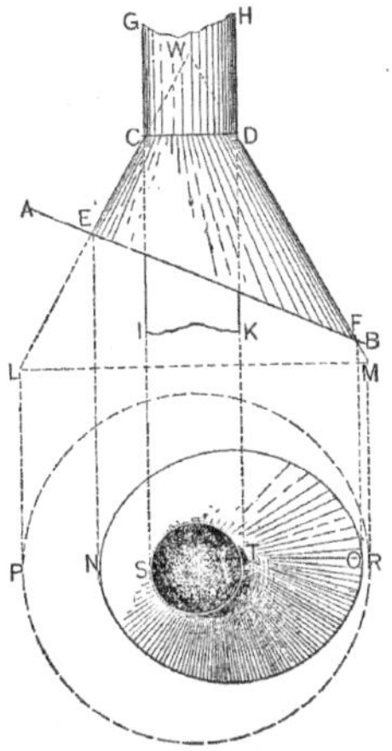


Fig. 366.—Elevation and Plan.
*A Conical Flange to Fit Around a Pipe and Against a Roof of One Inclination.*

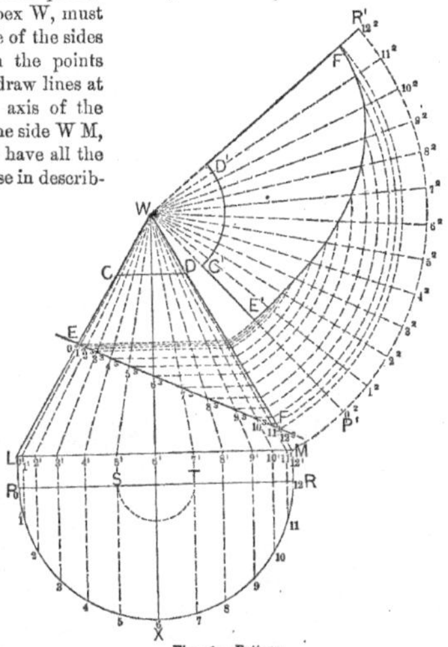


Fig. 367.—Pattern.
*A Conical Flange to Fit Around a Pipe and Against a Roof of One Inclination.*

508. *The Pattern of a Flange to Fit Around a Pipe and Over the Ridge of a Roof.*—In Fig. 368, let A B B C be the section of the roof against which the flange is to fit, and let O P S R be the elevation of the

pipe required to pass through the flange. Let the flange in size be required to extend from A to C over the ridge B. By inspection it will be seen that the process of describing the pattern is identical with that in Section 506. Produce C B, as shown by B A', making B A' equal to B A. Proceed as in the manner described in the problem just referred to. Divide the profile D E F G into any number of equal parts in the usual manner, and from the points so obtained carry lines vertically to the line A' C, and thence, at right angles to it, indefinitely. Also carry lines in a similar manner from the points A' and C. Draw H L. Make H I the width of the required flange, and draw I K parallel to H L. Connect K L. Through that part of the flange in which the center of the required opening is desired to be, draw the line A² C', crossing the lines drawn from the profile. From each side of this line, on the several measuring lines, set off the same distance as shown upon the corresponding lines between D F of the profile and the circumference. A line traced through the points thus obtained, as shown by D' E' F' G', will be the required opening to fit the pipe. Through the center, across the flange, draw the line N M, which represents the line of bend corresponding to the ridge B of the section of the roof.

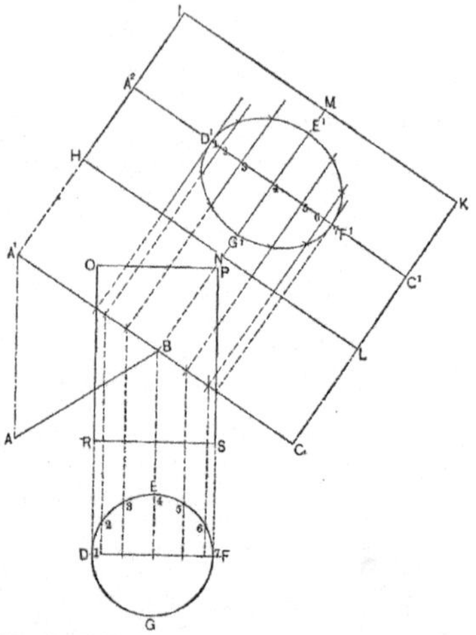

*Fig. 368.—The Pattern of a Flange to Fit Around a Pipe and Over the Ridge of a Roof.*

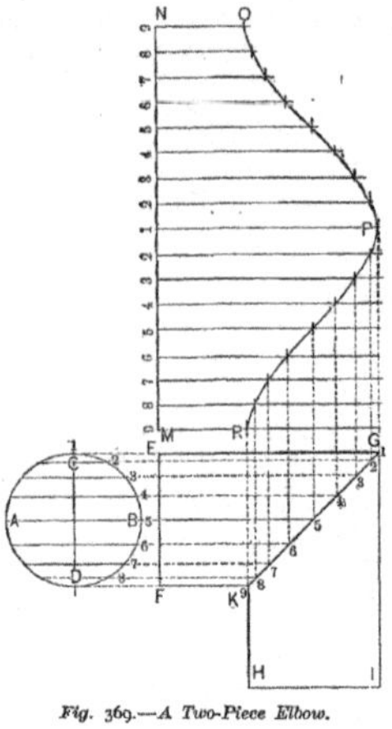

*Fig. 369.—A Two-Piece Elbow.*

509. *A Two-Piece Elbow.*—In Fig. 369, let A C B D be the profile of the pipe in which the elbow is to be made. Draw an elevation of the elbow as it is required to be, as shown by E G I H K F. Draw the diagonal line G K, which represents the joint to be made. Draw the profile of the pipe in line with one arm of the elbow, as shown. Divide the profile into any convenient number of equal parts. Place the T-square parallel to the lines of the arm of the elbow, opposite the end of which the profile has been drawn, and, bringing the blade successively against the several points in the profile, drop corresponding points on the miter or joint line K G, as shown by the dotted lines. Opposite the end of the same arm, and at right angles to it, lay off a stretchout line, M N, divided in the usual manner, and through the divisions draw measuring lines, as shown. Place the blade of the T-square at right angles to the same arm of the elbow, or, what is the same, parallel to the stretchout line, and, bringing it successively against the points in K G, cut the corresponding measuring lines, as shown. A line traced through these points, as indicated by R P O, will form the required pattern.

510. *A Three-Piece Elbow.*—In Fig. 370, let E M L I H K N F be the elevation of a three-piece elbow. Draw the profile A B C in line with one arm, as shown, and divide it into any convenient number of equal parts. Draw the joint or miter lines M N and L K. Place the blade of the T-square parallel to the arm of the elbow opposite the end of which the profile has been drawn, and, bringing it against the points in the profile,

drop corresponding points upon the miter line M N. Shift the movable head of the T-square, so that the blade lies parallel to the second section of the elbow, or the same thing may be accomplished by using the 45-degree set-square, and, bringing it against the points in M N, drop like divisions upon L K. At right angles to the second section, lay off a stretchout of the profile A B C, as shown by P O, through the points in which draw measuring lines in the usual manner. Placing the T-square so that the blade shall come at right angles to this section, or, what is the same, parallel to the stretchout line, bring it successively against the several points in the miter lines M N and L K, and cut the corresponding measuring lines. Then lines traced through these points, as shown by D X Y and G W Z, will be the pattern of the middle section. For the end sections of the elbow proceed as follows: Opposite the end of and at right angles to the arm draw a stretchout, as R S, through the divisions in which draw measuring lines in the usual manner. Placing the T-square at right angles to the arm, and bringing it successively against the points in K L, cut the corresponding measuring lines, as shown. Then the line T U V, traced through the points thus obtained, forms the pattern of an end section.

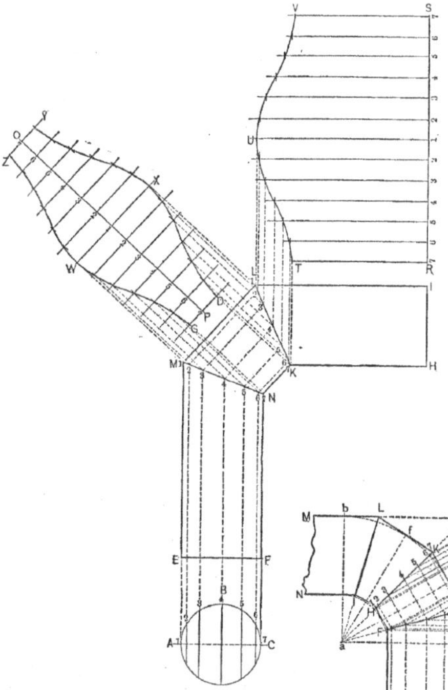

*Fig.* 370.—*A Three-Piece Elbow.*

511. *A Four-Piece Elbow.*—To draw the elevation of a four-piece elbow proceed as follows: Lay off the arms E G F D and N I L M in Fig. 371 at right angles to each other, and draw the diagonal line *a d*, upon which they would intersect if produced indefinitely. Establish the point *a* on this diagonal line at convenience, and from it draw the lines *a b* and *a c* at right angles to

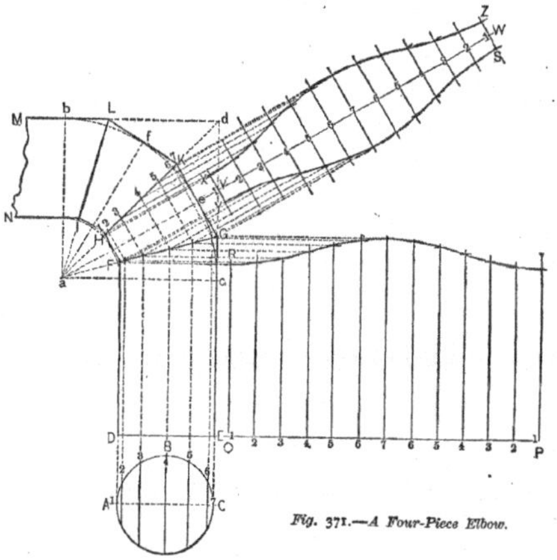

*Fig.* 371.—*A Four-Piece Elbow.*

the two arms of the elbow respectively. Draw *c d* and *b d*, thus completing the square *a b d c*. From *a* as center, and with *a b* as radius, describe the arc *b f e c*, as shown, which divide into three equal parts, thus obtaining the points *f* and *e*. Through *f* and *e*, to the center *a*, draw the lines *f a* and *e a*, which will represent the centers of the middle sections of the elbow, at right angles to which the sides of the same are to be drawn. Through *f*, and at right angles to *f a*, draw L K, meeting M L in the point L, and stopping on the line *a d* at the point K. Through *e*, and at right angles to *e a*, draw a line, commencing in the point K and terminating in G where it meets the line E G. In like manner draw the lines of the inner side of the elbow, as shown by F H and H I. Draw the miter or joint lines F G, H K and L I, as shown. For the patterns proceed as follows: In line with one arm of the elbow draw a profile, as shown by A B C, which divide into any convenient number of equal parts. Place the T-square parallel to this arm of the elbow, and, bringing the blade against the points in the profile, drop corresponding points upon the miter line F G. Change the T-square so that its blade shall be parallel to the lines of the second section of the elbow, and, bringing it against the points in F G, cut corresponding points on H K. Opposite the end of and at right angles to the lower arm of the elbow, lay off the stretchout line O P, as shown, through the divisions in which draw the usual measuring lines. Place the T-square at right angles to the arm of the elbow, and, bringing it successively against the points in the miter line F G, cut the corresponding measuring lines. Then a line traced through the points thus obtained, as shown from R to T, will be the pattern of one of the arms. Produce *a e*, representing the middle of the second section in the elbow, as shown by V W, upon which lay off a stretchout, and through the points in the same draw measuring lines. Placing the T-square parallel to *a e*, or, what is the same, at right angles to the section in the elbow, bring it against the several points in the miter lines H K and F G, and cut the corresponding measuring lines. Then lines traced through the points thus obtained, as shown from X to Z and Y to S, will give the pattern.

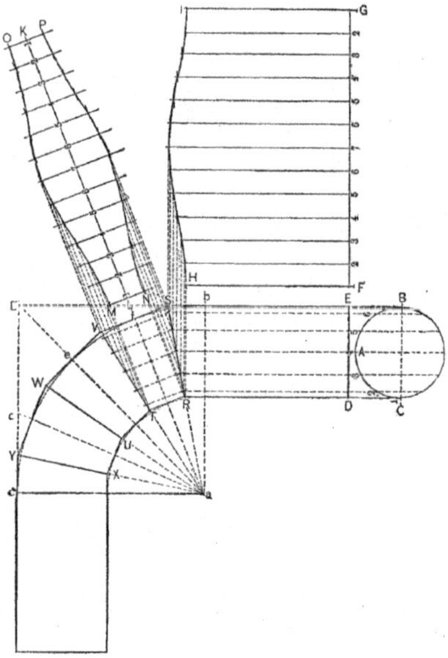

*Fig. 372.—A Five-Piece Elbow.*

512. *A Five-Piece Elbow.*—The elevation of a five-piece elbow may be drawn as follows: Lay off the two arms (Fig. 372) at right angles to each other. Draw the line *g a* indefinitely, upon which they would meet if sufficiently prolonged. Establish the point *a* in this diagonal line with reference to the curve which it is desired the elbow shall have, and from it, at right angles to the two arms of the elbow respectively, draw *a b* and *a c*. From *a* as center, with *a b* as radius, describe the arc *b f e c*, which divide into four equal parts, thus obtaining the points *d*, *e* and *f*, from which draw lines to *a*, all as shown by *d a*, *e a* and *f a*. Then these lines represent center lines of the several sections of which the elbow is composed, and at right angles to which the sides are to be drawn. Through *f*, and at right angles to *f a*, draw V S, joining the side of the arm E S in the point S, and a corresponding line drawn through *e* in the point V. In like manner draw the line T R, representing the inner side of the same section. The remaining sections are to be obtained in the same way. As but one section is necessary for use in cutting the patterns, the others may or may not be drawn, all at the option of the pattern cutter. Draw the miter or joint lines S R and V T. Opposite one arm

draw a profile, as shown by B A C, which subdivide in the usual manner. Place the T-square parallel to the lines of the arm, and, bringing the blade against the several points in the profile, drop corresponding points upon the miter line S R. Shift the T-square so that the blade shall be parallel to the part V S R T, and transfer the points in S R to V T, as shown. For the pattern of the arm, at right angles to it and opposite the end lay off a stretchout, as shown by F G, through the points in which draw the usual measuring lines. Place the T-square at right angles to the arm, and, bringing it against the points in R S, cut the corresponding measuring lines, as shown. Then a line traced through these points, as shown from H to I, will be the pattern. For the pattern of the sections prolong the line *a f*, as shown by L K, upon which lay off a stretchout, through the points in which draw the measuring lines in the usual manner. Placing the T-square at right angles to the section, or, what is the same, parallel to the stretchout line, bring it against the several points in the lines R S and T V, and cut the corresponding measuring lines. Then lines traced through the points thus obtained, all as shown by N P and M O, will be the pattern sought.

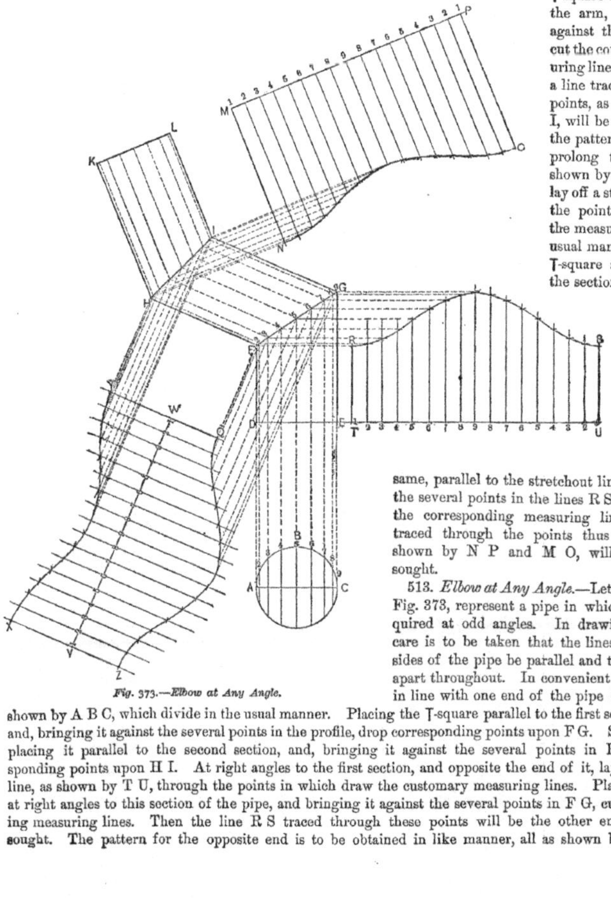


*Fig. 373.—Elbow at Any Angle.*

513. *Elbow at Any Angle.*—Let D F H K L I G E, Fig. 373, represent a pipe in which elbows are required at odd angles. In drawing the elevation care is to be taken that the lines representing the sides of the pipe be parallel and the same distance apart throughout. In convenient proximity to and in line with one end of the pipe draw a profile, as shown by A B C, which divide in the usual manner. Placing the T-square parallel to the first section of the pipe, and, bringing it against the several points in the profile, drop corresponding points upon F G. Shift the T-square, placing it parallel to the second section, and, bringing it against the several points in F G, drop corresponding points upon H I. At right angles to the first section, and opposite the end of it, lay off a stretchout line, as shown by T U, through the points in which draw the customary measuring lines. Placing the T-square at right angles to this section of the pipe, and bringing it against the several points in F G, cut the corresponding measuring lines. Then the line R S traced through these points will be the other end of the pattern sought. The pattern for the opposite end is to be obtained in like manner, all as shown by M N O P, and

therefore need not be described in detail. For the pattern of the middle section proceed as follows: At right angles to it lay off a stretchout, W V, with the customary measuring lines. Placing the T-square at right angles to the section, bring it successively against the points in G F and I H, and cut the corresponding measuring lines, as shown. Then lines traced through these points, as shown by Y X and Q Z, will be the pattern sought. The positions of the longitudinal joints in the several sections of this elbow, as well as those of all others, are determined by the order in which the measuring lines drawn through the stretchout are numbered. In the present instance we have allowed the joints to come on the back of the pipe, or, in other words, upon D F H K, which corresponds to the point 1 in the profile. Hence, in numbering the measuring lines in the several stretchouts, we have placed 1 at the commencement and ending, while if we had desired the joint to come on the opposite side, or at the point corresponding to 9 of the profile, we would have commenced and ended with that figure in numbering the measuring lines, the figure 1 in that case in regular order coming where 9 now occurs.

514. *A Pipe Carried Around a Semicircle by means of Cross Joints.*—In Fig. 374, let F E D be the semicircle around which a pipe, of which A C B is a section, is to be carried by means of any suitable number of cross joints, in this instance ten. Divide the semicircle F E D into the same number of equal parts as

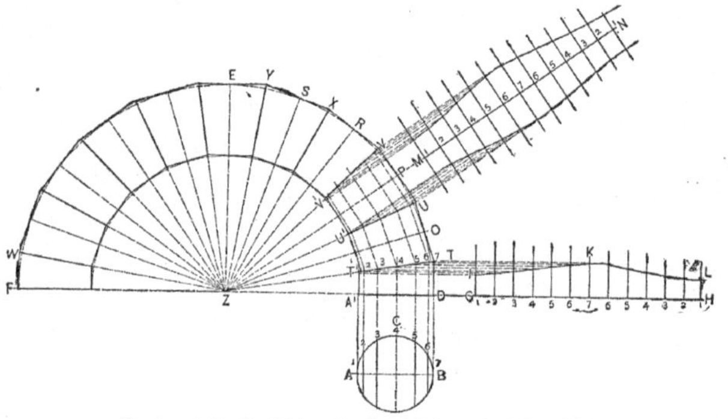


*Fig. 374.—A Pipe Carried Around a Semicircle by means of Cross Joints.*

there are to be joints, which, as just stated, in the present case is ten, all as shown by D, O, P, R, S, E, etc. From each of these points, D, O, P, R, etc., draw lines to the center Z, as shown. Obtain points intermediate between E S, S R, R P, etc., as shown in the engraving by Y, X, V, etc., through which draw lines from the center Z indefinitely. Connect the points E S, S R, R P, etc., by drawing lines at right angles to Y Z, X Z, V Z, etc. From D draw D Z. Set off a space, D A', equal to the diameter of the pipe. Draw a semicircle, as indicated by the dotted line, and obtain the inner line of the pipe in the same manner as just described for the outer line. Draw the profile of section A B C directly below and in line with one end of the pipe, all as shown in the engraving. As may be seen by inspection of the diagram, two patterns are required, one corresponding to the half section occurring at the end, and the other corresponding to the full sections composing the body of the pipe. The pattern for the latter may be obtained as shown in the engraving, or, if preferred, it may be obtained by making a duplicate of one-half of the larger piece. For the pattern of the end section proceed as follows: Divide the profile A B C in the usual manner into any convenient number of equal parts, and from the points thus obtained carry lines upward at right angles to Z D, cutting T' T. Prolong the line Z D, and upon it place a stretchout from the profile A C B, perpendicular to which draw measuring lines in the usual manner. With the T-square placed parallel to Z D, and brought successively against the points in T' T, cut the measuring lines of corresponding numbers. Then a line traced through the points of intersection thus obtained will be the shape of the pattern sought, all as shown by I K L.

Then G I K L H will be the complete pattern for one of the end sections. For the pattern of the large sections lay off a stretchout opposite the center of any one of them, and upon a line radial from the center, as shown by M N, and through the points in it draw measuring lines in the usual manner. Place the T-square parallel to the stretchout line, and, bringing it against the several points in the miter lines U' U and V' V, which are obtained by carrying the points from T' T by lines drawn parallel to each section through which they pass, cut the corresponding measuring lines, all as shown, thus completing the pattern.

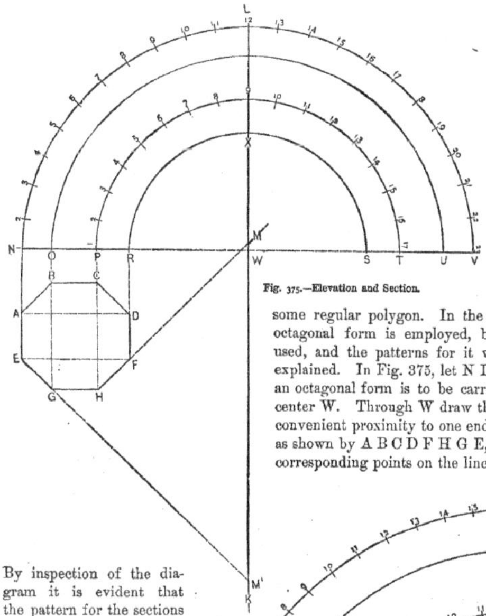


Fig. 375.—Elevation and Section.

515. *To Form a Semicircle in a Pipe by means of Longitudinal Seams.*—By the nature of the problem the pipe resolves itself, with respect to its section or profile, into some regular polygon. In the illustration presented in Fig. 375 an octagonal form is employed, but any other regular shape may be used, and the patterns for it will be cut by the same rule as here explained. In Fig. 375, let N L V be some semicircle around which an octagonal form is to be carried. Draw N V, passing through the center W. Through W draw the perpendicular L K indefinitely. In convenient proximity to one end of the semicircle construct a profile, as shown by A B C D F H G E, letting points in it fall directly below corresponding points on the line N V, all as shown in the engraving.

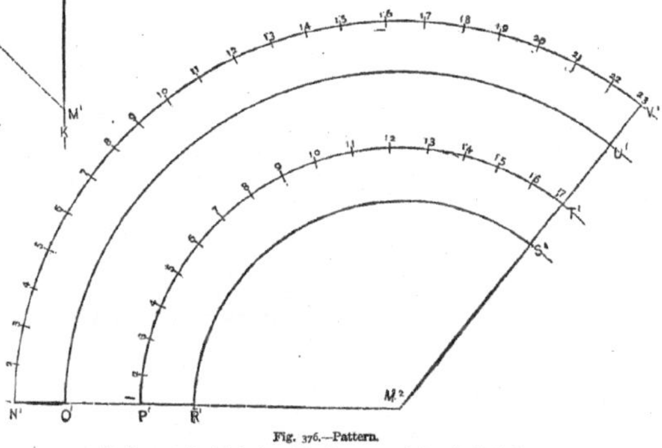


Fig. 376.—Pattern.
*To Form a Semicircle in a Pipe by means of Longitudinal Seams.*

By inspection of the diagram it is evident that the pattern for the sections corresponding to O U T P in the elevation, may be pricked directly from the drawing as it is now constructed, and that the patterns for the sections represented by E A and D F of the profile, will be plain straight strips of the width of one side of the figure, as shown by either E A or D F, and in length corresponding to the length of the sweep of the elevation on the lines N L V and R X S respectively. But for the two sets of pieces, represented by N V U O and P T S R in the elevation, additional steps are to be taken. Prolong the side H F of the profile until it cuts the center line L K of the elevation in the point M. Then M F and M H are the radii of the pieces corresponding to P T S R of the elevation. Prolong the side E G of the elevation until it cuts the center line in the point M'. Then M' G and M' E are the radii of the pieces corresponding to N V U O of the elevation. These radii are to be used as shown in Fig. 376. From M' in Fig. 376 as center, using each of the several radii in turn, strike arcs indefinitely, as shown by N' V', O' U', P' T' and R' S'. Step off the length N V in the elevation, Fig. 375, and make N' V' of Fig. 376 equal to it. Draw N' O' and V' U' radial to M'. In like manner establish the length of P' T', and draw P' R' and T' S', also radial to the center, as shown. It is evident by inspection of Fig. 376 that if the patterns for the two pieces are struck from a common center, as we have shown, it is only necessary to step off the length upon one member. By drawing radial lines, as shown, the other arcs will be intercepted at the proper points. This rule may be employed for carrying any polygonal shape around any curve which is the segment of a circle. The essential points to be observed are the placing of the profile in correct relationship to the elevation and to the central line L K. Then prolong the

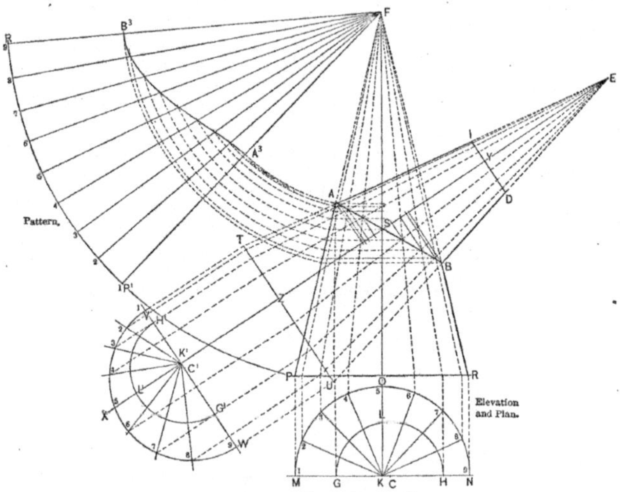


*Fig. 377.—Two-Piece Elbow in Tapering Pipe.*

oblique sides until they cut the central line, thus establishing the radii by which they may be struck. In the case of elliptical curves, by resolving them into segments of circles and applying this rule to each section, as though it were to be constructed alone and distinct from the others, no difficulty will be met in describing patterns by the principles here set forth. The several sections may be united so as to produce a pattern in one piece by joining them upon their radial lines. This principle is further explained in the problem of the patterns for the curved molding in an elliptical window cap. See Section 569.

516. *Two-Piece Elbow in Tapering Pipe.*—In Fig. 377 is represented an elbow constructed in two pieces occurring in taper pipe. The several steps required for the development of the pattern are as follows: Produce the sides A I and B D of the upper piece of the elbow until they meet in the point E. Then E is the apex and E B and E A the sides of a cone of which I D B A is a section. Produce the axis E S to any convenient point, as Z, through which draw T U at right angles to the axis. Produce the sides I A and D B until they meet T U in the points T and U, as shown. The next step is to construct a section of the cone as it would appear when cut on the line T U. Through any convenient point below the lower section, and at

right angles to the axis of the lower section, draw a straight line, as M N. From the points A and B of the miter line between the two sections, drop lines parallel to the axis of the lower section, cutting M N in the points G and H. From the point C, midway between G and H, as center, with C G or C H as radius, describe the semicircle G L H. Then G L H may be regarded as a plan of the miter A B upon a horizontal plane. From the extremities P and R of the bottom line of the lower section drop points parallel to its axis, as shown at M and N. From the point K, midway between M and N, and also in line with the axis S O produced, with radius K M or K N, describe the semicircle M O N. Then M O N is the plan of the lower section at the base P R. Divide M O N into any convenient number of equal parts in the usual manner, and from the points thus obtained carry lines vertically to the base P R; also from the same points draw lines to the center K, cutting G L H. Produce the sides P A and R B until they meet in the point F. From the points in P R carry lines toward the apex F, cutting A B. Produce the axis of the upper section beyond Z to any convenient point, as K'. Through K', at right angles to the axis Y Z produced, draw V W. Upon V W set off C', the same distance from K' that C is from K in the line M N. From C', with radius equal to that used in describing the profile G L H, describe the semicircle G' L' H', in which set off points corresponding to the divisions in G L H. From K', through these points in H' L' G', draw radial lines indefinitely, as shown by K' 2, K' 3, etc. From the apex E, through the points in the miter line A B, obtained from the plan of the lower section, as already described, draw lines cutting T U, as shown by the dotted lines. From the points T U draw lines parallel to the axis, cutting the radial lines drawn from K' through the points in the profile H' L' G'. Through the points of intersection between lines of corresponding numbers thus obtained trace a line, as shown by V X W. Then V X W is the plan of the cone, of which I D B A is a section as it would appear when cut on the line T U. From the same points in A B, as already described, draw lines at right angles to the axis, cutting it as shown by the points above and below S. From any convenient point, as E' in Fig. 378, draw the straight line E' Z', which make equal to E Z of the elevation. Set off E' Y' equal to E Y of the elevation; also set off in E' Z' points corresponding to the points in E Z of the elevation, obtained from the miter line A B above described. From Z', at right angles to E' Z', set off Z' U', upon which, measuring from Z', set off distances corresponding to each of the radial lines, measuring from K' to the profile V X W, all as indicated by the small figures. From the points thus obtained in Z' U', draw lines to E'. Intersect each of these lines by a line drawn from the corresponding point in E' Z', all as shown in the diagram. From E' as center, with radii E'1, E'2, E'3, etc., describe arcs, as shown between $B^2$ and $A^2$, upon which set off the stretchout of the profile V X W, by stepping from arc to arc. Through these points draw lines to the center E'. From E' as center, with radii corresponding to the several intersections between the lines drawn from the points in E' Z', and the lines drawn from E' to the points in Z' U', describe arcs which shall intersect lines of corresponding numbers drawn from the arc $B^2$ $A^2$ to the center E'. Inasmuch as the point Y', corresponding to Y of the elevation, represents all these points, each of these lines will be intersected by an arc corresponding to the intersection of the line Y' and the line drawn from E' to one of the points in Z' U', all as shown in the diagram. Then lines traced through the intersections thus obtained, as shown from D' to I', and also from B' to A', will be the top and bottom lines of the required pattern. Connect D' B' and I' A'. Then D' I' A' B' is the half pattern of the upper section. For the pattern of the lower section, from any center, as F, Fig. 377, with F P as radius, describe the arc P' R', upon which lay off a stretchout of the plan M O N, making points in P' R' to correspond with the points in the plan. Through these points draw lines to the center F, as shown. From points in the miter line A B draw lines at right angles to the axis F O, cutting the side F P, as shown in the points below A. From F as center, with radii

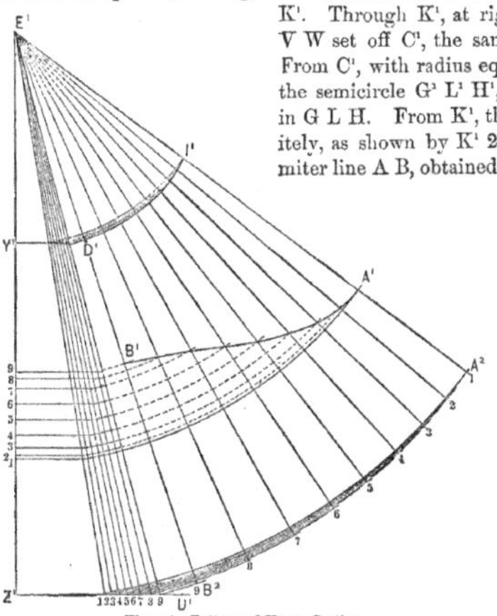

Fig. 378.—Pattern of Upper Section.
*Two-Piece Elbow in Tapering Pipe.*

corresponding to the several points below A just mentioned, describe arcs, as shown by the dotted lines, each of which produce until it cuts the line bearing a corresponding number drawn from the arc $P^1$ $R^1$ to the center F. A line traced through these points, as shown by $A^2$ $B^2$, will be the upper line of the required pattern of the lower section, and $A^2$ $P^1$ $R^1$ $B^2$ will be the half pattern of the lower section.

517. *Three-Piece Elbow in Tapering Pipe.*—In Fig. 379 is shown a three-piece elbow occurring in taper pipe, in which the flare is uniform throughout the three sections. The usual method of constructing the patterns for such an elbow would be the same as have been described for the two-piece elbow in the last demonstration. A short method, however, is available, both in three-piece and in two-piece elbows. Having described the ordinary method as applied to a two-piece elbow, the short method may be described in connection with the three-piece elbow as follows: The sections of which the elbow MORFEPNL is composed are cut from the right cone, shown by E G F. As drawn, the lower section of the elbow P R F E corresponds with the lower section of the cone, E F being the base common to both. The second section of the elbow O R P N corresponds with $O^1$ R P $N^1$, and the third section of the elbow M O N L corresponds with $M^1$ $O^1$ $N^1$ $L^1$ of the cone. The principle upon which the patterns are cut is that by which the envelope of any section of the cone is described. The essential point requiring attention, therefore, is the means by which the lines P R and $N^1$ $O^1$, which divide the cone into sections, shall be located so that the several sections of the cone shall, when joined together, constitute the elbow that is required. To find the angle of the miter line, or the line of cut through the cone, lay off the angle of the elbow, as A B C. Bisect this angle by the line D B. Then D B represents the direction across the cone at which the cut must be made. Having thus obtained the direction of the line, at any required hight draw P R parallel to D B. In like manner, for the second section, at any convenient point against the axis lay off the angle S U I, corresponding to the angle desired between the second and third sections. Bisect this angle, as shown by U T. From U as center, with any convenient radius, describe the arc $O^2$ T $N^2$. Upon either side of the cone, according to convenience, locate a point representing the length of one of the sides of the second section, as, for example, $O^1$. Set the dividers to $O^2$ T, and from

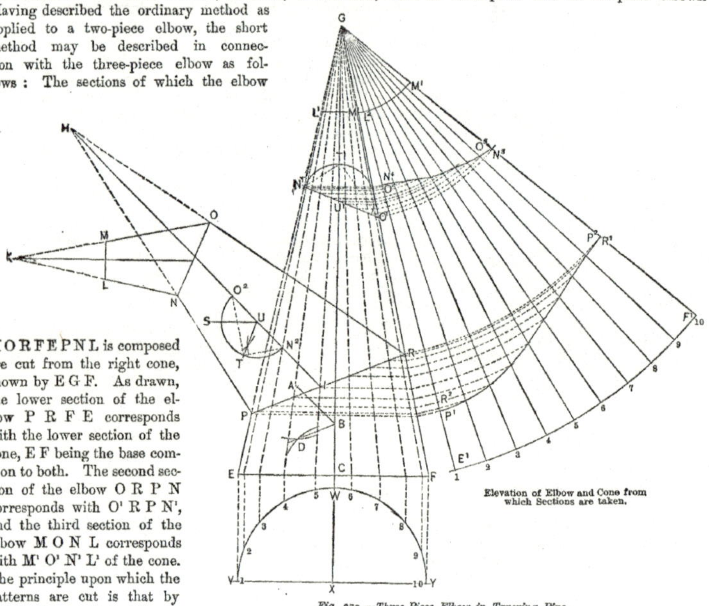


*Fig.* 379.—*Three-Piece Elbow in Tapering Pipe.*

O' as center, with this radius, cut the axis of the cone in the point T'. From T' as center, with radius T N', cut the side of the cone in N'. Draw N' O', which will be the line of cut dividing the second and third sections. A simpler method of obtaining these lines is as follows : The lower section of the elbow R F E T corresponds with the cone already. At the required hight draw P R. Upon the sides of the cone set off R O' equal to P N. Set off P N' equal to R O. Then draw N' O' as before. Having thus obtained the lines of cut through the cone, the patterns may be described as follows : Draw the plan V W Y, its center X falling upon the axis of the cone produced, which divide in the usual manner into any convenient number of equal parts. Through the points thus obtained erect perpendiculars to the base E F, and thence carry them toward the apex G, cutting the miter lines P R and N' O'. With the T-square at right angles to the axis G C, and brought successively against the points in N' O' and P R, cut the side G F of the cone, as shown by the points above O' and below R. From G as center, with radius G F, describe the arc E' F', upon which lay off the stretchout of the plan V W Y, as shown by the small figures 1, 2, 3, etc., and from these points draw lines to the center G. From G as center, describe arcs corresponding to the several points established in G F from the miter lines already described, which produce until they intersect lines of corresponding numbers drawn from the center G to the arc E' F'. Through these points of intersection trace lines, as shown by O² N² M² and R² P². From G as center, with radius G M', describe the arc L² M'. Then L² M' O³ N⁴ is the pattern of the upper section, and O² N³ P³ R² is the pattern of the third section.

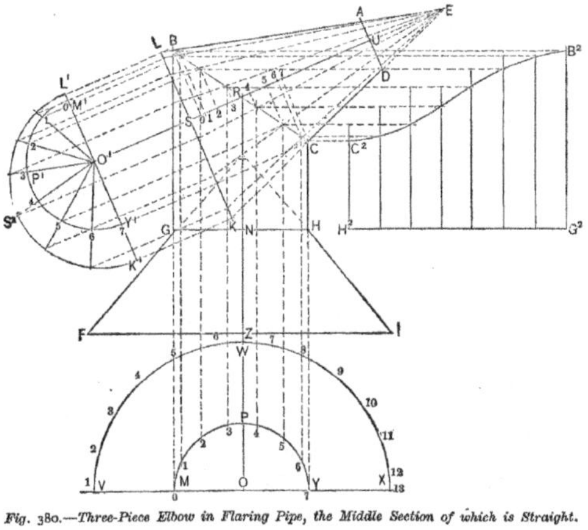


*Fig. 380.—Three-Piece Elbow in Flaring Pipe, the Middle Section of which is Straight.*

518. *Three-Piece Elbow in Flaring Pipe, the Middle Section of which is Straight.*—In Fig. 380, let F G B A D C H I be an elevation of the elbow, the lower section F G H I and the upper section B A D C of which both flare, while the middle section B C H G is straight. For the patterns we proceed as follows : Produce the sides B A and C D of the upper section until they meet in the point E. Then E is the apex and E B and E C sides of a cone, of which A D C B is a section. At any point outside of the section, at right angles to the axis U R, draw L K, and produce the sides E B and E C until they meet it, as shown by E L and E K. In line with the middle section draw the profile M P Y, as shown. Divide M P Y in the usual manner into any convenient number of equal parts, as shown by the small figures. Through the points thus obtained carry lines vertically, cutting the miter line B C. Construct a section of the cone as it would appear if cut on the line L K as follows: Produce the axis U R to any convenient point, as O'. From O' as center, with radius equal to O M, as shown by O' M', describe the semicircle M' P' Y'. Divide this semicircle into the same number of equal parts, as M P Y already described, and through the points from O' draw radial lines indefinitely, as shown by O' 1, O' 2, O' 3, etc. Through the points in B C already obtained draw lines from the apex E, cutting L K, and thence, parallel to the axis, drop points intersecting the lines drawn from O'. Through the points thus obtained trace a line, as shown by L' S² K', which will be the profile of the cone when cut on the line L K. From the points in B C also, at right angles to the axis U S, draw lines cutting U S, as shown by the small figures, 0, 1, 2, 3, etc. At any convenient point draw E' S', Fig. 381, equal to E S of the elevation. Set off E' U' equal to E U of the elevation. Likewise set off points, as shown by the small figures, corresponding to the points in U S of the elevation. From S', at right angles to E' S', draw S' K², equal to S K of the eleva-

tion. On this line, measuring from $S^1$, set off distances corresponding to the length of the several radial lines between the center $O^1$ and the profile $L^1$ $S^2$ $K^1$, all as shown by the small figures, 0, 1, 2, 3, etc. From these points draw lines to $E^1$. At right angles to $E^1$ $S^1$ draw lines corresponding to $U^1$ and the points indicated by the small figures, which produce until they intersect the corresponding radial lines drawn from $E^1$. From E as center, with radii corresponding to the several points in $S^1$ $K^1$, describe arcs indefinitely. From E draw the straight line $E^1$ 7, which will form the boundary of one side of the pattern. Commencing at 7, which is in the outer arc, step off the stretchout of the plan $L^1$ $S^2$ $K^1$, stepping from 7 to the second arc, as shown at 6, and from there to the third arc, as shown at 5, and so on, each time stepping to the next arc. From the points in the arcs thus obtained draw measuring lines to $E^1$. From $E^1$ as center, with radii corresponding to the intersection between the lines drawn perpendicular to U S and the radial lines drawn from $E^1$ to the points in $S^1$ $K^1$, describe arcs intersecting measuring lines already drawn. Then a line traced through these points, as shown by $C^1$ $B^1$, will form one boundary of the required pattern. From $E^1$ as center, with radius equal to E D of the elevation, or, what is the same, with radius corresponding to the intersection of the line drawn from U, with a radial line corresponding to $K^2$, describe the arc $D^1$ $A^1$. Then $D^1$ $A^1$ $B^1$ $C^1$ is half of the required pattern. For the pattern of the middle section, at right angles to its straight end, G H, lay off a stretchout taken from the plan M P Y, as shown by $H^2$ $G^2$, Fig. 380, through the points in which draw measuring lines in the usual manner. Place the T-square at right angles to this section of the pipe, or, what is the same, parallel to the stretchout line, and, bringing it successively against the points in B C, cut measuring lines of corresponding numbers. Then a line traced through the points thus obtained, as shown by $B^2$ $C^2$, will be the shape of the pattern corresponding to the line B C in the elevation. For the lower section of the elbow produce the sides F G and I H until they meet in the point T. Then T is the apex, and T F and T I sides of a right cone, of which G H I F is a frustum. From the center O of the plan, which is in line with the axis of the cone, with radius equal to Z I of the elevation, describe the semicircle V W X, which will be the plan of the cone at the base. Divide V W X into any number of equal parts for use in laying off the stretchout. From any convenient center, as $T^1$ in Fig. 382, with radius equal to T F of the elevation, describe the arc $F^1$ $I^1$ indefinitely, upon which set off a stretchout of the plan V W X in the usual manner. From $T^1$ as center, with radius equal to T G of the elevation, describe the arc $G^1$ $H^1$. From the last point in $F^1$ $I^1$ (13) draw a line toward the center $T^1$, cutting the smaller arc in the point $H^1$. Connect $G^1$ $F^1$. Then $G^1$ $F^1$ $I^1$ $H^1$ is half the pattern of the lower section of the elbow.

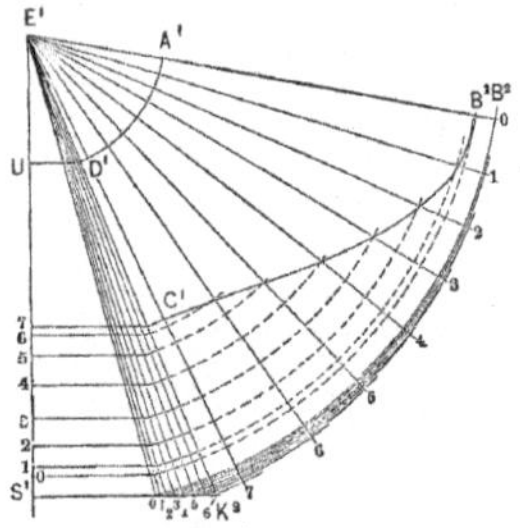


**Fig. 381.—Pattern of Upper Section.**
*Three-Piece Elbow in Flaring Pipe, the Middle Section of which is Straight.*

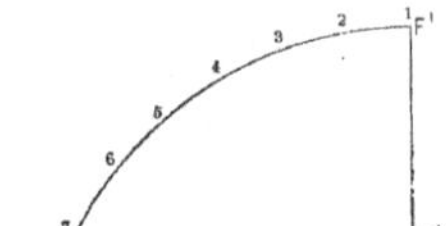


**Fig. 382.—Pattern of Lower Section.**
*Three-Piece Elbow in Flaring Pipe, the Middle Section of which is Straight.*

519. *Three-Piece Elbow, the Middle Section of which Tapers.*—In Fig. 383, let D E G L N M K F be an elbow, the middle section (F G L K) of which tapers, the upper and lower sections being straight. For the patterns proceed as follows: Opposite and in line with the upper straight section draw the half profile A B C, which divide in the usual manner into any convenient number of equal parts, as indicated by the small figures, 1, 2, 3, etc. Draw the miter line F G between the sections, and from the points in the profile A B C drop lines, cutting F G as shown. Opposite the end D E, and at right angles to that section of the pipe, lay off the stretchout $D^1$ $E^1$ of the plan A B C, through the points in which draw the usual measuring lines. With the T-square placed parallel to this stretchout line, or, what is the same, at right angles to the lines of the section, and brought successively against the points in the miter line F G, cut corresponding measuring lines, as shown. Through the points thus obtained trace a line, as shown from $F^1$ to $G^1$. Then $D^1$ $E^1$ $G^1$ $F^1$ will be the half of the required pattern of the upper section. For the pattern of the lower

section proceed in the same general manner. Draw the half profile P O R in line with it, which divide into any convenient number of equal spaces, from the points in which carry lines vertically, cutting the miter line K L bounding the section. Opposite the straight end of this section, and at right angles to it, draw the stretchout line M' N', in length equal to the half section P O R. Through the points in M' N' draw the usual measuring lines. Place the T-square at right angles to the lines of the section, and, bringing it successively against the points in K L, cut measuring lines of corresponding numbers, as shown. Then a line traced through these points, as shown by K' L', will be the shape of the pattern of the lower pieces. For the pattern of the middle section proceed as follows: Produce the sides L G and K F until they meet in the point H. Then H is the apex, and H L and H K are sides of a cone of which F G L K is a section. Through the point V, which represents the intersection of the axis of the upper section of the pipe with the miter line F G, draw H V, which produce indefinitely in the direction of U. At right angles to H U, and at any convenient point outside the section F G L K, draw T L. Produce the sides F K and G L until they meet this line. The next step is to construct a section of the cone as it would appear if cut on the line T L. Produce H U, as shown by H S'. At right angles to H S', through S', draw the line T' L' indefinitely. From S' as center, with radius equal to S O of the plan of the lower section, describe the arc P' O' R, which divide into the same number of equal parts as the profile P O R. From S', through these points, draw radial lines indefinitely. Through the points in the miter line K L, obtained from the profile P O R already described, draw lines from the apex H, cutting T L, and from this line carry them parallel to the axis H U, until they intersect radial lines drawn from S'. Through these points of intersection trace a line, as shown by T' U² L'. Then this line is the profile of the cone as it would appear if cut on the line T L. From the points in the miter line K L draw lines at right angles to the axis H U, cutting H U, as shown in the points 0 4 and 4 8. In like manner cut H U by lines drawn at right angles to it from the points in F G, also shown by the points between 0 4 and 4 8. From any convenient point, as H' in Fig. 384, draw the line H' U', in length equal to H U of the elevation. At right angles to H' U' set off U' T', equal to U T of the elevation. In H' U' set off points corresponding to the points in H U in the elevation. With the dividers take the distance upon each of the several lines radiating from S' in the profile to the line T' U² L', and set off like distances from U' on U' T', all as shown by the small figures from 8 to 0. From these points draw lines to H', intersecting them by lines drawn at right angles to H' U', from the points of like numbers in that line already described. Having thus obtained measurements of the middle section at the several points required they are spread, and the pattern itself is described as follows: From H' as center, with radius corresponding to the several points in U' T', describe arcs upon which to lay off the stretchout of the profile T' U² L'. Draw

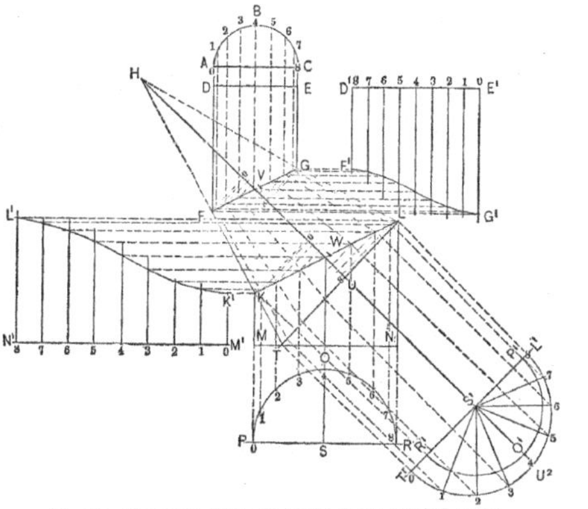


*Fig. 383.—Three-Piece Elbow, the Middle Section of which Tapers.*

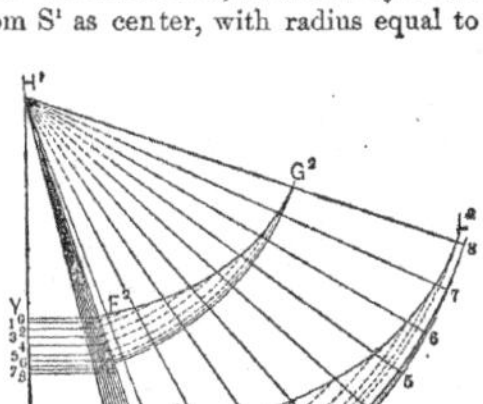


**Fig. 384.—Pattern of Middle Section.**
*Three-Piece Elbow, the Middle Section of which Tapers.*

any straight line from $H^1$, as shown by $H^1$ 0. Set the dividers to the space 0 1 in the section $T^1$ $U^2$ $L^1$, and, commencing at 0 in Fig. 384, step to the second arc, and from the point last set off step to the third arc, and thus continue until the stretchout $F^2$ $U^1$ $L^1$ has been laid off, stepping from arc to arc, as described. From the points in the stretchout thus obtained draw measuring lines to the center $H^1$, all as shown. From $H^1$ as center, with radii corresponding to the intersections of the lines drawn perpendicular to $H^1$ $U^1$ with the radial lines drawn from $H^1$ to $U^1$ $T^1$, describe arcs, which produce until they intersect measuring lines of corresponding numbers, all as indicated in the engraving. Then lines traced through the points of intersection thus obtained, as shown by $F^2$ $G^2$ $L^2$ $K^2$, will be the pattern sought.

520. *A Two-Piece Elbow in Elliptical Pipe.*—The only difference to be observed in cutting the patterns for elbows in elliptical pipes, as compared with the same operations in connection with round pipes, lies with the profile or section. The section is to be placed in the same position as shown in the rules for cutting elbows in round pipe, but it is to be turned broad or narrow side to the view, as the requirements of the case may be. In round pipe there is, of course, no such distinction possible. In Fig. 385 is shown a right angled two-piece elbow in an elliptical pipe, which shows the flat side to the front. The same rule would apply in cutting the patterns if the elbow occurred in a pipe showing the narrow side to the view, the only change being in the placing of the section. The demonstration which follows, together with the reference given above to the rules for cutting elbows in round pipe, will be sufficient to enable the mechanic to cut the patterns of any required elbow in elliptical pipe. Let A C E F D B be the elevation of the elbow at the required angle. Draw C D, which forms the miter line. In line with one arm of the elbow draw a section, as shown by G H I K, which divide in the usual manner, and by means of the T-square placed parallel to the arm, drop points upon the miter line, as shown. Opposite the end of the arm lay off a stretchout, and through the points in it draw the usual measuring lines. Reversing the T-square, placing it at right angles to the arm, and bringing it in contact with the several points in the miter line, cut the corresponding measuring lines. A line traced through these points, as shown by L P O, will constitute the required pattern.

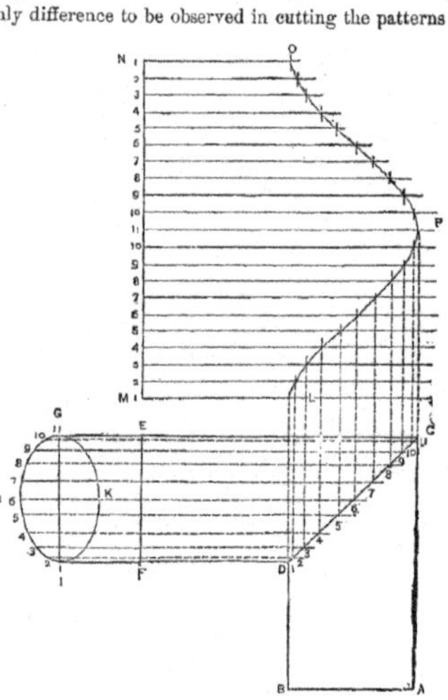

*Fig. 385.—A Two-Piece Elbow in Elliptical Pipe.*

521. *A T-Joint between Pipes of the Same Diameters.*—Let D F G H M I K E in Fig. 386 represent a junction between two pipes of the same size at right angles, of which A B C and $A^1$ $B^1$ $C^1$ are sections. As the two pipes have like sections, the miter lines F L and K L appear straight in elevation. Space both sections into the same number of equal parts, as shown, and drop points on to the miter lines. Lay off two stretchouts, N O at right angles to the upper pipe and R T at right angles to the lower pipe. Set the T-square at right angles to the upper pipe, and, bringing the blade against the several points on the miter lines, cut the corresponding measuring lines drawn through the stretchout, as indicated by the dotted lines. Then N $F^1$ U V W O will be the pattern for the upper piece. By inspection of the elevation and sections it will be seen that only a portion of the measuring lines are required to be drawn through the stretchout R T. It will be noticed that 7 comes at the middle of the required opening, while 4 represents the position of the edges. Therefore, draw the lines 4, 5, 6, 7, 6, 5, 4, as shown. Place the blade of the T-square at right angles to the lower section of pipe, and, bringing it against the several points in the miter lines, cut the corresponding measuring lines, as shown by the dotted lines. A line, X Y Q O, traced through these points will bound the opening to be cut in the pattern for the lower pipe. For the pattern of the pipe, from the points 1 in the stretchout draw the lines R P and

T S, in length equal to the length of the pipe. Connect P S. Then P R T S will be the required pattern. The seam in the pipe may be located as shown in the engraving, or at some other point, at pleasure.

522. *A T-Joint between Pipes of Different Diameters.*—In Fig. 387 it is required to make a joint at right angles between the smaller pipe D F G E and the larger pipe H K L I. For this purpose both a side elevation and an end view are necessary. At a convenient distance from the end of the smaller pipe in each view draw a section of it. Space these sections into any suitable number of equal parts, commencing at corresponding points in each, and setting off the same number of spaces, all as shown by A B C and A' B' C'. From the points in A B C draw lines downward through the body of the large pipe indefinitely. From the points in

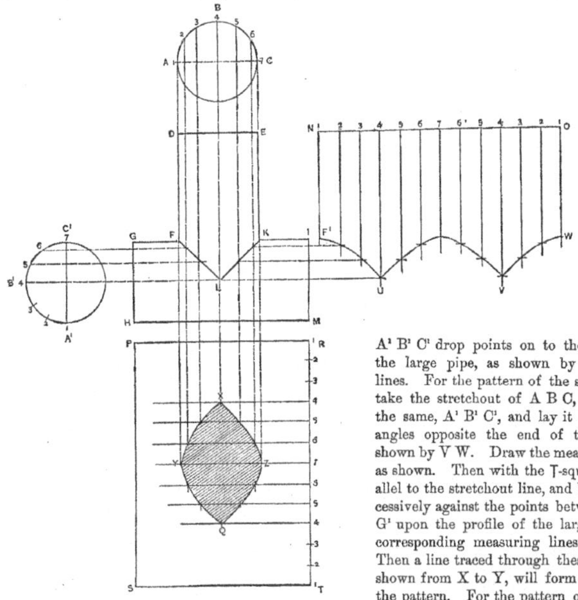


*Fig. 386.—A T-Joint between Pipes of the Same Diameter.*

A' B' C' drop points on to the profile of the large pipe, as shown by the dotted lines. For the pattern of the smaller pipe take the stretchout of A B C, or, what is the same, A' B' C', and lay it off at right angles opposite the end of the pipe, as shown by V W. Draw the measuring lines, as shown. Then with the T-square set parallel to the stretchout line, and brought successively against the points between F' and G' upon the profile of the large pipe, cut corresponding measuring lines, as shown. Then a line traced through these points, as shown from X to Y, will form the end of the pattern. For the pattern of the larger pipe the stretchout is taken from the profile view F' G' L', and laid off at right angles to the pipe opposite one end, as shown by N P. A corresponding line, M O, is drawn opposite the other end, and the connecting lines M N and O P are drawn, thus completing the boundary of the pattern. For the shape of the opening to be cut in the pattern, in spacing the profile of the large pipe F' G' L', the points 1, 2, 3 and 4 are made to correspond to the points dropped from the section of the small pipe, the other divisions of the profile being taken at will simply for the purpose of obtaining a correct stretchout. From these points (1 2 3 4) in the stretchout, therefore, measuring lines are drawn, intersecting those previously dropped from corresponding points in the profile A B C, giving points through which the line R S T U is traced, which forms the shape of the opening. If for any reason it be desired to show a correct elevation of the junction between the two pipes, the miter line F G is obtained by intersecting the lines dropped from A B C with lines of corresponding numbers from F' G' in the profile of the large pipe.

523. *A* T*-Joint between Pipes of Different Diameters, the Smaller Pipe Setting to One Side of the Larger.*—In Fig. 388, let A B C be the size of the small pipe and F' H' M' be the size of the large pipe, between which a right-angled joint is to be made, the smaller pipe being set to one side of the axis of the large pipe, as indicated in the profile. Draw an elevation, as shown by D F I L M K G E. Also draw a section, as shown by D' F' M' H' E'. Place a profile of the small pipe above each, as shown by A B C and A' B' C', both of which divide into the same number of equal parts, commencing at the same point in each. Placing the T-square parallel to the small pipe, and, bringing it successively against the points in the profile A' B' C', drop lines cutting the profile of the large pipe, as shown from F' to H'; and in like manner drop lines from the points in the profile A B C, continuing them through the elevation of the larger pipe indefinitely. For the pattern of the small pipe set off a stretchout line, V W, at right angles to and opposite the end of the pipe, and draw the measuring lines, as shown. These measuring lines are to be numbered to correspond to the spaces in the profile, but the place of beginning determines the position of the seam in the pipe. In the illustration given we have located the seam at the shortest part of the pipe, or, in other words, at the line corresponding to the point 10 in the section. Therefore we commence numbering the stretchout lines with 10. Place the T-square at right angles to the small pipe, and, bringing the blade successively against the points in the profile of the large pipe from F' to H', cut the corresponding measuring lines, as shown. A line traced through the points thus obtained, as shown by X Y Z, will form the end of the required pattern. For the pattern of the large pipe lay off a stretchout of the end view, locating the seam where desired, as above described in connection with the small pipe. In this instance we have located the seam on a line corresponding to point 13 in the profile. Therefore, in laying off the stretchout, as shown on O R, we commence with this number. After laying off the stretchout opposite one end of the pipe, draw a corresponding line opposite the other, as shown by N P, and connect N O and P R, thus completing the outline. In spacing the profile of the large pipe, the spaces in that portion against which the small pipe fits are made to correspond to the points obtained by dropping lines from the profile of the small pipe upon it, as shown by 1 to 7 inclusive. This is done in order to furnish points in the stretchout corresponding to the lines dropped from the profile A B C, as shown. No other measuring lines than those which represent the portion of the pipe which the small pipe fits against, are required in the stretchout. Accordingly the lines 1 to 7 inclusive are drawn from O R, as shown, and are cut by corresponding lines dropped from A B C. A line traced through the several points of intersection gives the shape S T U, which is the opening in the large pipe. If it be necessary for any purpose to show a correct elevation of the junction between two pipes, the miter line F H G is obtained

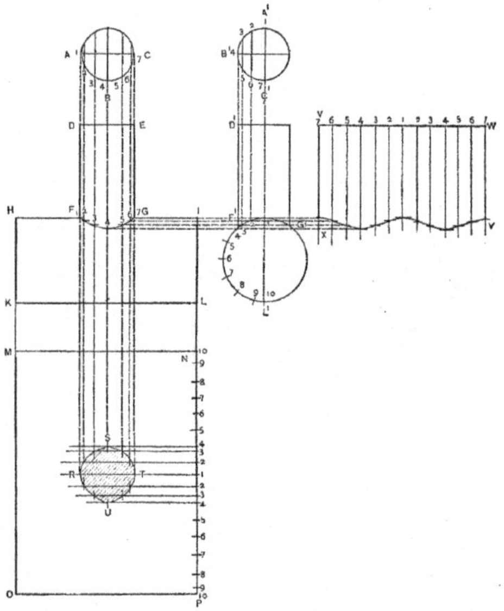


*Fig. 387.—A* T*-Joint between Pipes of Different Diameters.*

by intersecting the lines dropped from A B C by corresponding lines carried across from the same points obtained on the profile F' H', by dropping from A B C, as previously explained and all as shown by the dotted lines.

524. *A Joint between Two Pipes of the Same Diameter at Other than Right Angles.*—Let L F D E K I H M of Fig. 389 represent the elevation of two pipes meeting in the angle M H I, for which patterns are required. Draw the profile or section A' B' C' in line with the branch pipe, and the section A B C in line with the main pipe. Space both the profiles into the same number of equal divisions, commencing at the same point in each. Draw lines from these points, which produce until corresponding lines from the two sections intersect, and through the several points of intersection thus obtained draw the lines F G and G H, which is the miter line between the two pipes. For the pattern of the arm proceed as follows: Lay off the stretchout O N opposite the end of the arm, and draw the usual measuring lines at right angles through it, as shown. Place the T-square at right angles with the arm, or, what is the same, parallel with the stretchout line, and, bringing the blade successively against the points in the miter line F G H, cut the corresponding measuring lines. Through the points thus obtained trace the line P R S T, which will form the pattern required. For the pattern of the main pipe proceed as follows: Opposite one end lay off the stretchout, as shown by V Y, and opposite the other end lay off a corresponding line, as shown by U X. Connect U V and X Y. From so many of the points in the stretchout line V Y as correspond to points in the miter line F G H, draw the usual measuring lines. Place the T-square at right angles to them, and draw lines from the points in the miter line F G H, intersecting the corresponding measuring lines. A line traced through these points of intersection, as F' Z H' W, will describe the shape required. The position of the seam in both the arm and the main pipe is determined by the manner of numbering the spaces in the stretchout. In the illustration the seam in the arm is located in the shortest part, or at a point corresponding to 1 of the profile. Accordingly, in numbering the divisions of the stretchout, that number is placed first. In like manner the seam in the main pipe is located at a point opposite the arm. Therefore, in numbering the spaces in the stretchout we commence at 1, which, as will be seen by the profile, represents the part named. If it were desirable to make the seam come on the opposite side of the main pipe from where we have located it—that is, come directly through the opening made to receive the arm—we would commence numbering the stretchout with 7.

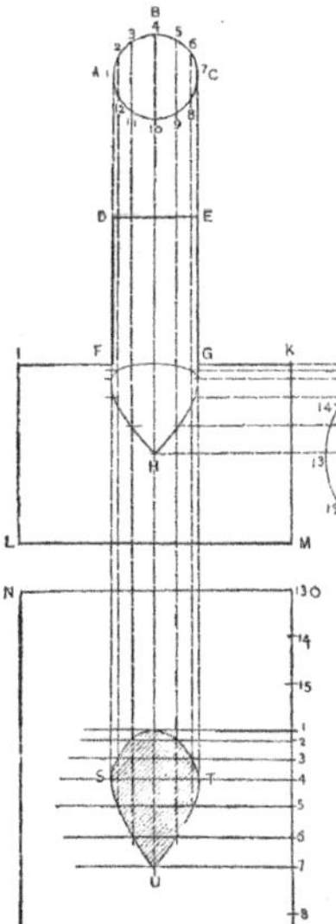


*Fig. 388.—A T-Joint between Pipes of Different Diameters, the Smaller Pipe Setting to One Side of the Larger.*

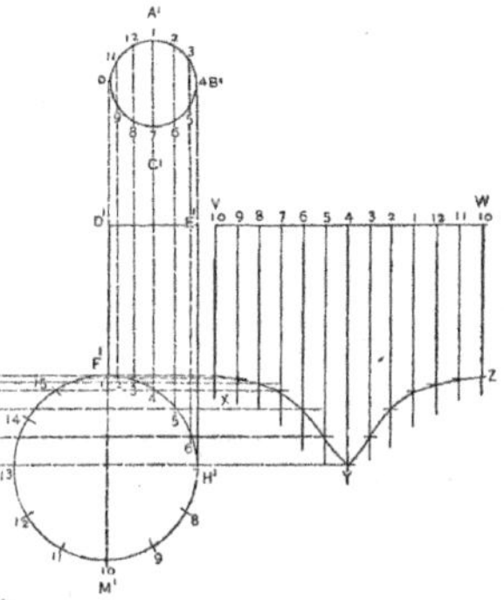

In that case the opening F' W H' Z would appear in two halves, and the shape of the pattern would be as though the present pattern were cut in two on the line 7 and the two pieces were joined together on 1. By this explanation it will be seen that the seams may be located during the operation of describing the pattern wherever desired.

525. *The Joint between Two Pipes of Different Diameters Intersecting at Other than Right Angles.*—Let A B C, Fig. 390, be the size of the smaller pipe, and Y N' Z the size of the larger pipe, and let H L M be the angle at which they are to meet. Draw an elevation of the pipes, as shown by G K I O N M L H, placing the profile of the smaller pipe above and in line with the arm, as shown. Place an end view of the larger pipe in line with that part of the elevation, as shown, and directly above it, their center lines corresponding. Place a second profile of the small pipe, as shown by A' B' C'. Divide both sections of the small pipe into the same number of spaces, commencing at the same point in each. From these points drop lines on to the large pipe, as shown, both in section and elevation. From the points thus obtained upon the profile of the large pipe carry lines across to the left, producing them until they intersect corresponding lines in the elevation. A line traced through these several points of intersection gives the miter line K L, from which the points in the two patterns are to be obtained. For the pattern of the small pipe proceed as follows: Opposite the end lay off a stretchout, at right angles to it, as shown by E F. Through the points in it draw the usual measuring lines, as shown. Bring the T-square to right angles with the pipe, and, placing it successively against the points in the miter line K L, cut the corresponding measuring lines, as shown by the dotted lines. A line traced through the points thus obtained will give the pattern, as indicated. For the pattern of the large pipe proceed as follows: Opposite one end, and at right angles to it, lay off a stretchout, as shown by R S. Draw a corresponding line P T opposite the other end, and connect P R and T S. In order to afford corresponding points for measurement in describing the shape of the opening to be cut in the pattern of the large pipe, in spacing the profile, as shown by Y N' Z, the points 4 3 2 1 2 3 4 are taken, as already established by the lines dropped from the profile of the small pipe. The other points in the profile are taken at convenience, simply for stretchout purposes. In laying off the stretchout R S that number is placed first which represents the point at which it is desired the seam shall come. For the shape of the opening in the pattern, draw measur-

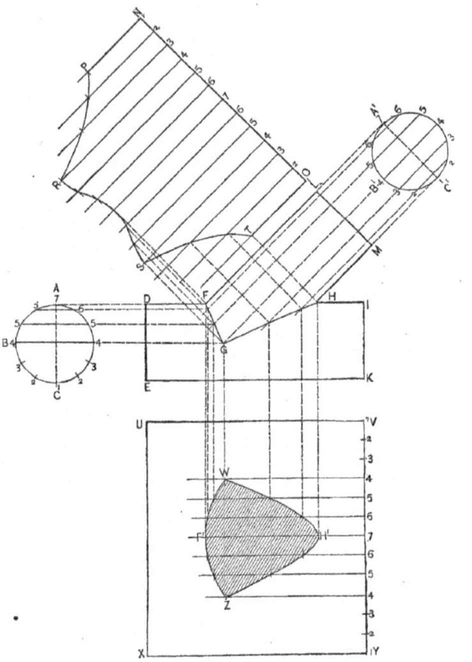


*Fig. 389.—A Joint between Two Pipes of the Same Diameter at Other than Right Angles.*

ing lines from the points 4 3 2 1 2 3 4, as shown, and intersect them by corresponding lines dropped from the miter line. Through the points thus obtained trace the line U V W X, which will represent the shape of the opening required.

526. *A Joint at other than Right Angles between Two Pipes of Different Diameters, the Axis of the Smaller Pipe being Placed to One Side of that of the Larger One.*—In Fig. 391, let C' B' A' be the size of the smaller pipe, and D' E' I' the size of the larger pipe, between which a joint is required at an angle represented by W F K, the smaller pipe to be placed to the side of the larger. Draw an elevation of the pipes, joined as shown by V D G H I K F W. Place a profile or section of the arm in line with it, as shown by C' B' A'. Opposite and in line with the end of the main pipe draw a section of it, as shown by D' E' I'. Directly above this section draw a second profile of the small pipe, as shown by A B C, placing the center of it—relative to the center of the profile of the large pipe—in the same position that the arm is to have in the main pipe. Divide the two profiles of the small pipe into the same number of equal spaces, commencing at the same point in each. From the divisions in C' B' A' drop lines parallel to the lines of the arm indefinitely. From the divisions in A B C drop lines until they cut the profile of the large pipe, as shown by the points in the arc D' E'. From these points carry lines to the left, producing them until they intersect the corresponding lines from C' B' A'. A line traced through these points of intersection, as shown by D E F, will be the miter line between the two pipes. For the pattern of the arm proceed as follows: Lay off a stretchout at right angles to and opposite the end of the arm, as shown by R P, and through the points in it draw the usual measuring lines. Place the T-square at right angles to the arm, and, bringing it successively against the points in the miter line, cut the corresponding measuring lines. A line traced through these points, as shown by U T S, will form the required pattern. For the pattern of the main pipe draw a stretchout line opposite one end of it, as shown by M O, numbering the divisions in it with reference to locating the seam, which can be placed at any point desired. Draw a line corresponding to the stretchout line opposite the other end of the pipe, as shown by L N, and connect L M and N O. Through as many of the points in the stretchout line as correspond to the points forming the miter line D E F in the elevation draw measuring lines, as shown by 1, 2, 3, 4, 5, 6 and 7. Place the T-square at right angles to the main pipe, and, bringing the blade against the points in D E F successively, cut the corresponding measuring lines, all as shown by the dotted lines. A line traced through these points of intersection, as shown by E' D' F', will give the shape of the opening to be cut in the pattern.

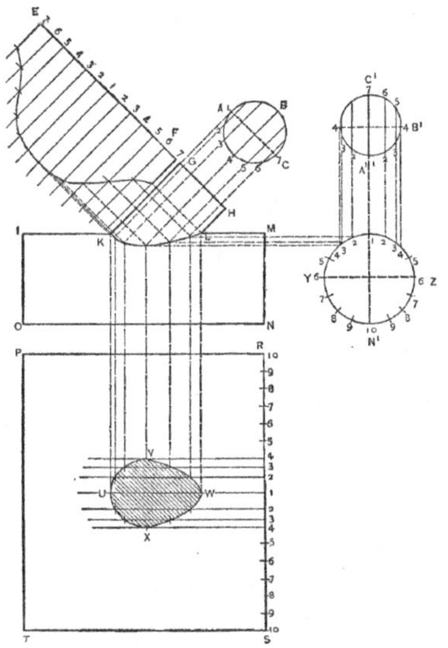

*Fig. 390.—The Joint between Two Pipes of Different Diameters Intersecting at Other than Right Angles.*

527. *The Patterns of a Cylinder (or Pipe) and Cone Meeting at Right Angles to their Axes.*—In Fig. 392, let B G E D F A C be the elevation of the required article. Draw the plan in line with the elevation, making like points correspond in the two views, as shown by M O S T U P N. Draw a section of the pipe in proper position in both elevation and plan, as shown by E M' D and N D' M respectively. Divide these sections of the pipe into any convenient number of equal parts, commencing at the same point in each, as shown by the small figures. From the center of the section of the pipe, as shown in plan, draw a straight line to the center of the plan of the cone, as shown by D' R. From each of the points in the section of the pipe shown

in elevation carry lines parallel to the sides of the pipe, cutting the side of the cone, and for convenience extend them some distance into the figure—for example, until they meet the axis. From the several points of intersection with the side of the cone, as shown by *a b c d e*, drop lines parallel to the axis of the cone, on to the line $D^1$ R of the plan, giving the points $a^1$ $b^1$ $c^1$ $d^1$ $e^1$, and through each of these points, from R as center, describe an arc, as indicated in the engraving. From the points in the profile N $D^1$ M of the pipe in the plan draw lines parallel to the sides of the pipe, producing them until they meet the arcs drawn through corresponding points, as dropped upon $D^1$ R from the elevation, giving the points indicated by $1^1$, $2^1$, $3^1$, $4^1$ and $5^1$. From these points carry lines vertically to the elevation, producing them until they meet the lines drawn from points of corresponding numbers in the profile of the pipe to the axis of the cone, giving the points $1^2$, $2^2$, $3^2$, $4^2$ and $5^2$. A line traced through these points, as shown from G to F, will be the miter line in elevation formed by the junction of the pipe and cone. To describe the patterns proceed as follows: Opposite the end of the pipe, as shown in elevation, and at right angles to it, lay off a stretchout, K H, through the points in which draw the usual measuring lines. Intersect these measuring lines by lines from corresponding points, $1^2$, $2^2$, $3^2$, etc., in the miter line, as produced in the elevation. A line traced through these points of intersection, as shown from L to I, will be the shape of the end pipe to fit against the side of the cone, and the entire pattern of the piece will be as shown by H I L K.

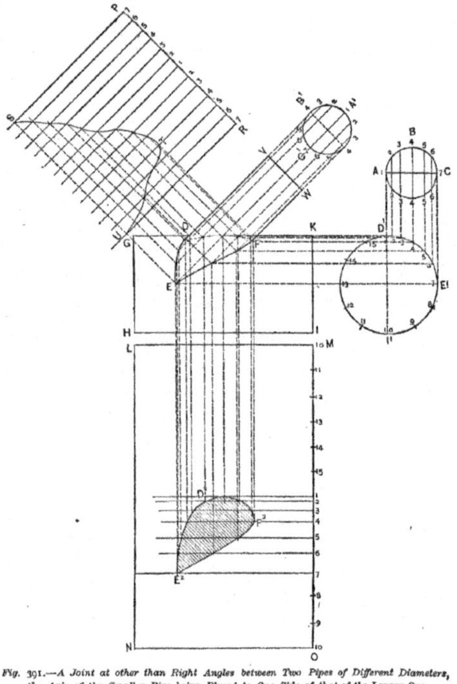


*Fig. 391.—A Joint at other than Right Angles between Two Pipes of Different Diameters, the Axis of the Smaller Pipe being Placed to One Side of that of the Larger One.*

From any convenient point, as $A^1$, Fig. 393, draw $A^1$ $B^1$, in length equal to A B of the elevation. Set off points $e^2$, $d^2$, $c^2$, $b^2$ and $a^2$ in it corresponding to *e*, *d*, *c*, *b* and *a* of A B, Fig. 392. From $A^1$ as center, with radius $A^1$ $B^1$, describe the arc $B^1$ V, upon which lay off the stretchout of the plan of the cone, as indicated by the small figure outside of the pattern. (But one-half of the pattern is shown in the engraving.) From the same center $A^1$ describe arcs corresponding to the points $e^2$, $d^2$, $c^2$, $b^2$ and $a^2$. From the center R of the plan draw lines to the circumference through the points $2^1$, $3^1$, $4^1$, etc., giving the points in the circumference marked $2^3$, $3^3$, $4^3$, etc. Set off corresponding points in the arc $B^1$ V, as shown by $3^4$, $2^4$, $4^4$, $5^4$, etc.

From these points draw lines to the center A', intersecting the arcs drawn from $a^2$, $b^2$, $c^2$, etc. A line traced through these points of intersection, as shown by F' O' G' P', will be the shape of the opening to cut to correspond with the pipe.

528. *The Patterns of a Frustum of a Cone Intersecting a Cylinder, their Axes being at Right Angles.*—Let S P R T in Fig. 394 be the elevation of the cylinder, and *a* G K H *b* the elevation of the frustum. Draw the axis of the cylinder, as shown by A B, which prolong, as shown by C D, on which construct a profile of the cylinder, as shown by C E D F. Produce the sides of the frustum, as shown in the elevation, until they meet in the point L, which is the apex of the cone. Draw the axis L K, which produce in the direction of O, and at any convenient point in the same construct a profile of the cone M O N as it would appear if cut on the line *a b*. In connection with the profile of the cylinder draw

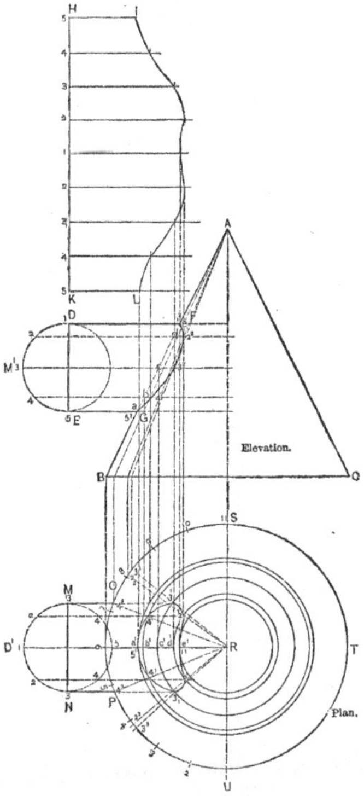


*Fig. 392.—The Patterns of a Cylinder (or Pipe) and Cone Meeting at Right Angles to their Axes.*

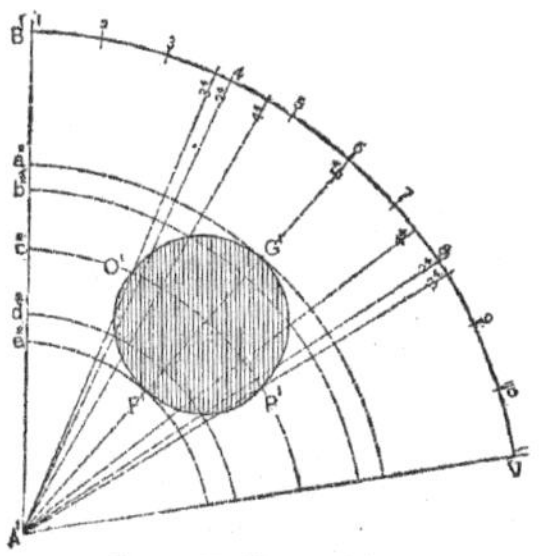

Fig. 393.—Half Pattern of Cone.
*The Patterns of a Cylinder (or Pipe) and Cone Meeting at Right Angles to their Axes.*

a corresponding elevation of the cone, as shown by K' *a*' *b*' K². Produce the sides K' *a*' and K² *b*' until they intersect, thus obtaining the point L', the apex corresponding to L of the elevation. Draw the axis L' E, as shown, which produce in the direction of N', and upon it draw a second profile of the cone taken on the line *a b*, as shown by M' O' N'. Divide the profiles M O N and M' O' N' into the same number of equal parts, commencing at corresponding points in each, as shown. With the T-square set parallel to the axis of the cone, and brought successively against the points in the profile, drop lines to the lines *a b* and *a*' *b*', as shown. Place the T-square against the apex L', and, bringing it successively against the points in *a*' *b*', cut the profile of the cylinder, as shown in K' E K². In like manner place the T-square against the apex L, and draw lines indefinitely through the points in *a b*. Place the T-square parallel to the sides of the cylinder, and, bringing it against the points in the profile K' E K² just described, cut corresponding lines in the elevation, as shown by H K G. A line traced through these points of intersection, as shown by H K G, will form the miter line between the two pieces as it appears in elevation. Continue the lines drawn from K' E K² until they meet the side *a* G of the cone prolonged, as shown from G to Z. From

L as center, and with radius L $a$, describe the arc $b^2$ $a^2$, upon which lay off a stretchout of the profile M O N of the cone. Through each of the points in this stretchout draw lines indefinitely, radiating from L, as shown. Number the points in the stretchout $a^2$ $b^2$ corresponding to the numbers in the profile, commencing with the point occurring where it is desired to have the seam. Set the compasses, with L for center, to L Z as radius, and describe an arc cutting the corresponding lines drawn through the stretchout, as shown by 1, 5 and 1. In like manner reduce the radius to the second point in G Z, and describe an arc cutting 2, 4, 4 and 2. Also bring the pencil to the third point and cut the lines corresponding to it in the same way. Then a line traced through the points thus obtained, as shown by $H^1$ $K^3$ $G^1$, will be the pattern of the cone. At right angles to and opposite one end of the cylinder draw a stretchout, taken from the profile C E D F, as shown by X V. In laying off this stretchout let points in the portion of the profile represented by $K^1$ E $K^2$ correspond to the divisions obtained by dropping lines from the apex of the cone. The other points in the profile may be taken at will, being used only for stretchout purposes. In laying off the stretchout commence at a point corresponding to the place at which the seam is desired to be in the finished work, in this case at 9. Through the points 1, 2, 3, 4 and 5, being those in the portion of the profile over which the cone sets, draw measuring lines, as shown. The usual measuring lines may be dispensed with through the other points. Place the T-square at right angles to the cylinder, and, bringing it successively against the points in the miter line, as shown in the elevation, cut the corresponding measuring lines. Then a line traced through these points of intersection, as shown by $G^2$ $K^4$ $H^2$ $K^5$, will be the opening to be cut in the pattern of the cylinder. Draw U W opposite the other end of the cylinder, as shown in elevation, parallel and equal in length to X V, and connect U X and W V, thus completing the pattern of the cylinder.

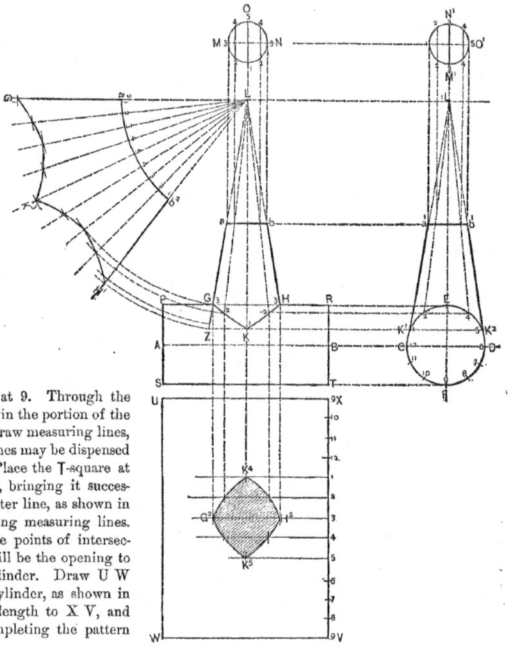


*Fig. 394.—The Patterns of a Frustum of a Cone Intersecting a Cylinder, their Axes being at Right Angles.*

529. *The Frustum of a Cone Intersecting a Cylinder of Greater Diameter than Itself at Other than Right Angles.*—In Fig. 395, E G H F represents an elevation of the cylinder, and M N L K an elevation of the frustum of a cone intersecting it. $F^1$ Z Q represents the profile of the cylinder, or, what is the same, the cylinder in plan. Having drawn the elevation and the plan under it, as shown in the engraving, for the patterns proceed as follows: At any convenient point on the axial line of the cone, as indicated by T O, construct the profile V Y X W, which represents a section through the cone on the line M N. Divide the section V Y X W into any convenient number of equal spaces in the usual manner, as shown by the small figures, 1, 2, 3, 4, etc. From each of the points thus established drop lines parallel with the axis of the cone, cutting the line M N. From the intersections in M N thus obtained drop points parallel with the side G H of the cylinder,

and continue them indefinitely, cutting the line $F^1$ $O^1$, which is drawn through the center of the plan of the cylinder at right angles to the elevation, all as shown in the engraving. Make $Y^1$ $W^1$ equal to Y W of the first section constructed. In like manner measure distances from the center line V X of the first section to the points 2, 3, 4, etc., and set off corresponding spaces in the second section, measuring from $M^1$ $N^1$, upon lines of corresponding numbers dropped from the intersections in M N already described. Then a line traced through these points will represent a view of the upper end of the frustum as it would appear when looked at from a point directly above it. Produce the sides of the frustum K M and L N until they meet in the point O. From O drop a line parallel to the side G H of the cylinder, cutting the line $F^1$ $O^1$ in the point $O^1$, thus establishing the position of the apex of the cone in the plan. From the point $O^1$ thus established draw lines through the several points in the section $M^1$ $Y^1$ $N^1$ $W^1$, which produce until they intersect the plan of the cylinder in points between Z and Q, as shown in the engraving. From O, the apex of the cone in the elevation, draw lines through

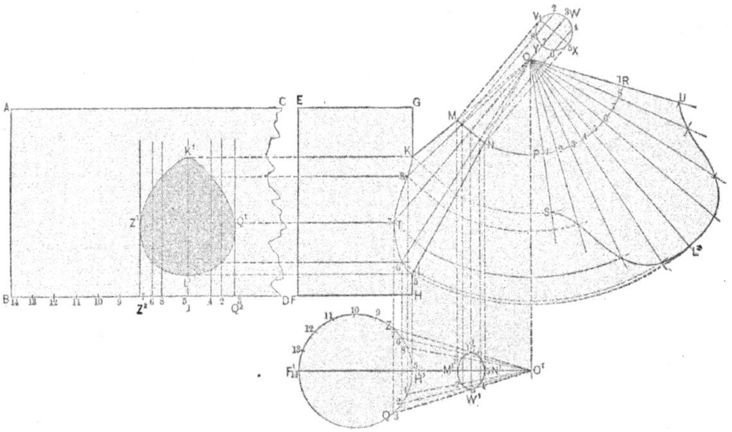


*Fig. 395.—The Frustum of a Cone Intersecting a Cylinder of Greater Diameter than Itself at Other than Right Angles.*

the several points in M N already determined, which produce until they cross G H, the side of the cylinder, and continue them inward indefinitely. Intersect these lines by lines drawn from the points between Z and Q of the plan just determined. Then a line traced through these intersections, as indicated by K T L, will represent the miter between the frustum and cylinder as seen in elevation. With this line determined we are now ready to lay off the patterns, to do which proceed as follows: From O as center, with O N as radius, describe the arc P R, on which set off a stretchout of the section Y V W X in the usual manner. From O, through the several points in P R thus obtained, draw radial lines indefinitely. From the several points in the miter line K T L draw lines at right angles to the axis O T of the cone, producing them until they cut the side N L. From O as center, with radii corresponding to the several points in N L just obtained, describe arcs, which produce until they intersect radial lines of corresponding number drawn through the stretchout P R. Then a line traced through these points of intersection, as indicated by S $L^2$ U, will be the lower line of the pattern sought, and P S $L^2$ U R will be the complete pattern. For the pattern of the cylinder and the opening in it proceed as follows: Draw the line B D at right angles to the cylinder and in line with one end of it, upon which set off a stretchout of the cylinder from the plan $F^1$ Z Q in the usual manner. The points between Z and Q of the plan, as indicated by $Z^2$ $Q^2$ of the pattern, must be made to correspond with the divisions in the plan. From these points lines are to be drawn perpendicular to the stretchout line B D. Then, with the T-square placed at right angles to the cylinder, and brought successively against the points in the miter line K T L, cut lines of corresponding numbers. A line traced through the points of intersection thus formed, as shown by Z' K' Q' L', will be the shape of the required opening in the cylinder.

530. *The Patterns of the Frustum of a Cone Joining a Cylinder of Greater Diameter than Itself at Other than Right Angles, the Axis of the Frustum passing to One Side of the Axis of the Cylinder.*—Let E F H G in Fig. 396 be the elevation of a cylinder, which is to be intersected by a cone or frustum of a cone, D A I C, at the angle F D A in elevation, and which is to be set to one side of the center, all as shown by S O P L M R of the plan. Opposite the end of the frustum, in both elevation and plan, construct a section of it, as shown by T U V W in the elevation and T' U' V' W' in the plan. Divide both of these sections into the same number of equal parts, commencing at corresponding points, and number them as shown by the small figures in the diagram. From the points in T U V W carry lines parallel to the axis of the cone, cutting the line A I, and thence drop them vertically across the plan. From the points in the section U' V' W' draw lines parallel to the axis of the cone, as seen in plan, intersecting the lines dropped from A I described. Through these points of intersection trace a line, as shown by L M. Then L M will show the end of the frustum A I as it appears in plan. Through the points in L M draw lines cutting the plan of the cylinder, as shown from P to R. From the points of intersection between these lines and P R of the plan of the cylinder carry lines vertically, intersecting those in the elevation drawn from the apex X. Then a line traced through these points, as shown by K D C, will be the miter line in elevation. For the pattern of the cylinder lay off the stretchout G' H', in length equal to S O P R of the plan, in which set off points corresponding to the points used in the plan between P and R. The other lines appearing in G' H' are used merely for the purposes of a stretchout.

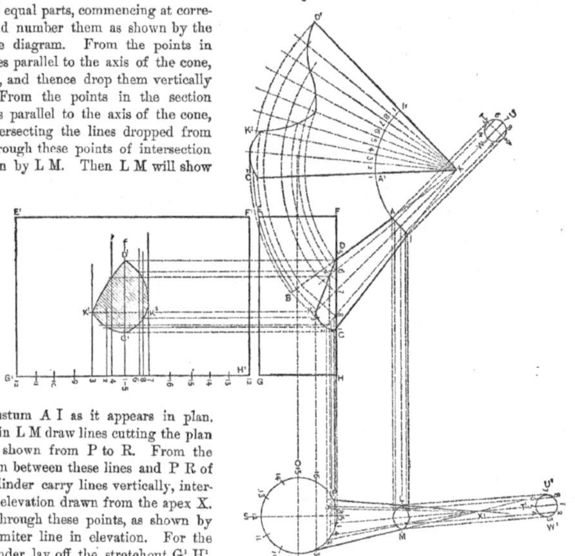

*Fig. 396.—The Patterns of the Frustum of a Cone Joining a Cylinder of Greater Diameter than Itself at Other than Right Angles, the Axis of the Frustum passing to One Side of the Axis of the Cylinder.*

Through these points, specially named above, and which in numbers are from 1 to 7 inclusive, draw lines at right angles to the stretchout line, as shown. With the T-square placed at right angles to the axis of the cylinder, and brought successively against the points in the miter line D K C, cut these lines in the manner indicated by the dotted lines. Then a line traced through these points of intersection, as indicated by D' K' C' K', will be the shape of the opening to be cut in the pattern of the cylinder to correspond with the intersection of the cone. Draw G' E' equal to G E of the elevation, and H' F' equal to H F of the elevation, and connect E' F', thus completing the patterns of the cylinder. For the pattern of the frustum, from any convenient center, as X, with radius X A, describe the arc A' I', upon which lay off a stretchout of the section W T U V, through the points in which, from X, draw radial lines indefinitely. Intersect these lines by arcs drawn from X, with radii corresponding to points in the side A D produced, as shown from D to B. These points are obtained by lines drawn at right angles to the axis of the cone from the several points of intersection between the side D C and the lines drawn from X through the points in A I, all of which is clearly indicated by the dotted lines in the diagram. Through the points of intersection in the pattern thus obtained trace a line, as shown by $C^1$ $K^2$ $D^1$. Connect $D^1$ $I^1$ and $C^1$ $A^1$. Then $D^1$ $C^1$ $A^1$ $I^1$ will be the pattern of the frustum D A I C, mitering with the cylinder at the angle described.

531. *Patterns of a Cylinder Joining a Cone of Greater Diameter than Itself at Other than Right Angles.*—Let B A K in Fig. 397 be the elevation of a right cone, perpendicular to the side of which a cylinder, L S T M, is to be joined. The first operation will be to describe the miter line as it would appear in elevation. Draw the plan U V W of the cylinder, which divide into any convenient number of equal parts, as indicated by the small figures, and from these points drop lines, cutting the side A K of the cone in the points H, F and D, producing them until they cut the axis A X in the points G, E and C. The next step is to construct sections of the cone as it would appear if cut on the lines G H, E F and C D. Draw a second elevation of the cone, as shown by $B^1$ $A^1$ $K^1$, representing the cone turned quarter way around; or, the first may be regarded as a side elevation and this as an end elevation. Draw a plan under the side elevation of the cone, as shown by N R P O, which divide into any convenient number of equal parts, and in like manner draw a corresponding plan under the end elevation, as shown by $R^1$ $P^1$ $O^1$ $N^1$. Divide this second plan into the same number of equal parts, commencing to number them at the same point as in the other plan. From the points 1 to 4 in plan N R P O, carry lines vertically to the base B K, and thence toward the apex A, cutting the lines C D, E F and G H. In like manner, from the same points (1 to 4 inclusive) in the plan $R^1$ $P^1$ $O^1$ $N^1$, carry vertical lines to the base $B^1$ $K^1$, and thence to the apex $A^1$. Place the T-square at right angles to the axes of the two cones, and, bringing it against the points of intersection of the lines from B K with C D, cut corresponding lines in the second elevation, and through the points of intersection thus established trace a line, as shown by $M^1$ $M^7$. Produce the axis $X^1$ $A^1$ to any convenient distance, upon which set off $C^1$ $D^1$, in length equal to C D, in which set off the points corresponding to the points in C D, and through these points draw lines at right angles to $C^1$ $D^1$. Place the T-square parallel to the axis $X^1$ $A^1$, and, bringing it against the several points in $M^1$ $M^7$, cut the lines drawn through $C^1$ $D^1$, as shown, and through the intersections thus established trace a line, as shown by $M^3$ $D^1$ $M^4$. Then $M^3$ $D^1$ $M^4$ is a section of the cone as it would appear if cut on the line C D. In like manner carry lines from E F across to the second elevation and thence parallel to the axis, cutting lines drawn through $E^1$ $F^1$, which with its points is equal to E F, by which to establish the profile $M^5$ $F^1$ $M^6$, which is a section of the cone as it would appear if cut on the line E F. Also use the points in G H in like manner, establishing the profile $M^7$ $H^1$ $M^8$, which represents a section of the cone as it would appear if cut on the line G H. (The lines indicating the operation in connection with the sections corresponding to E F and G H are omitted in the engraving to avoid confusion; the opera-

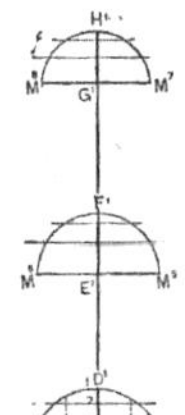

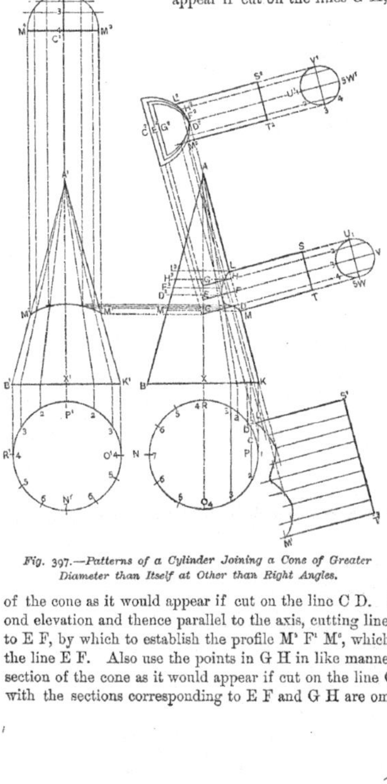

*Fig. 397.—Patterns of a Cylinder Joining a Cone of Greater Diameter than Itself at Other than Right Angles.*

tion is identical with that explained in connection with C D.) Having thus obtained sections of the cone corresponding to the several lines C D, E F, G H, arrange them together in line with the side K A of the cone, placing the points $D^1$ $F^1$ $H^1$ tangent, all as indicated by $C^2$ $D^2$, $E^2$ $F^2$, $G^2$ $H^2$. In connection with these sections draw a plan of the cylinder, as shown by $L^2$ $S^2$ $T^2$ $M^2$, opposite the end of which draw a profile, as indicated by $U^1$ $V^1$ $W^1$, which divide into the same number of equal parts as used in the divisions of the profile U V W, commencing the division at corresponding points in each. From the points in the profile $U^1$ $V^1$ $W^1$ drop lines against the several profiles $C^2$ $D^2$, $E^2$ $F^2$ and $G^2$ $H^2$, arranged together, and thence drop the points back on to the elevation, cutting corresponding lines in it. That is, from the intersection of the line drawn from point 4 in $U^1$ $V^1$ $W^1$ with the profile $C^2$ $D^2$ cut the line C D, which in the elevation corresponds to the point 4 in the profile U V W, and from the intersection of a line drawn from 3 with $E^2$ $F^2$ cut the line E F, and so on, all as indicated by the dotted lines. Then a line traced through these points of intersection, as shown by L M, will be the miter line in elevation, after which the patterns are readily obtained, as follows: For the pattern of the cylinder lay off a stretchout of the profile U V W $S^1$ $T^1$, opposite the end S T, through the points in which draw the usual measuring lines. Place the T-square at right angles to the same, and, bringing it against the points in the miter line L M, cut the corresponding measuring lines. Then a line traced through these points, as shown from $L^1$ to $M^1$, will be the shape of the pattern of the cylinder to fit against the cone. For the pattern of the cone, from any convenient center, as $A^3$ in Fig. 398, with radius A B, describe the arc $B^3$ $K^3$, which in length make equal to the circumference of the plan N R P O. From the apex of the cone, through such points in the miter line L M as do not correspond with lines already drawn, draw lines cutting the base B K, and thence drop them on to the plan N R P O, all as indicated by *a*, *b* and *c*. Set off in the arc $B^3$ $K^3$ points corresponding, as indicated by $a^1$ $b^1$ $c^1$ and $a^2$ $b^2$ $c^2$. From these points draw lines to the center $A^3$, as shown, and also likewise from other points corresponding to points obtained in the miter line in the elevation. From $A^3$ as center, with radii corresponding to the points L, H, F, D and M of the elevation, strike arcs intersecting the lines just drawn, all as shown by $L^4$ $L^5$, $H^4$ $H^5$, $F^4$ $F^5$, etc. Then a line traced through the intersections thus obtained will be the shape of the opening to be cut in the envelope of the cone corresponding to the cylinder.

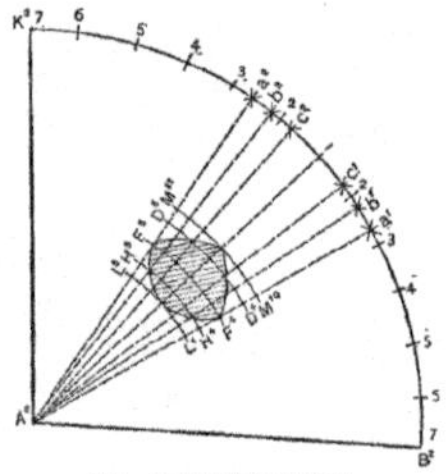

**Fig. 398.—Envelope of Cone.**

***The Patterns of a Cylinder Joining a Cone of Greater Diameter than Itself at Other than Right Angles.***

532. *The Patterns of Two Cones of Unequal Diameter Intersecting at Right Angles to their Axes.*—Let U T V in Fig. 399 be the elevation of a cone, at right angles to the axis of which another cone or frustum of a cone, O F G P, is to miter. Let L K N M be a section of the frustum on the line F G. Let $U^2$ W $V^2$ $W^1$ be a plan of the larger cone at the base. The first step in describing the patterns is to obtain the miter line in the elevation, as shown by the curved line from O to P. With this obtained the development of the pattern is a comparatively simple operation. To obtain the miter line O P we proceed as follows: Divide the profile L K N M into any convenient number of equal parts, as shown by the small figures. Inasmuch as the divisions of this profile are used in the construction of the sections—or, in other words, since sections must be constructed to correspond to certain lines through this profile—it is desirable that each half be divided into the same number of equal parts, as shown in the diagrams. Thus 2 and 2, 3 and 3, 4 and 4 of the opposite sides correspond, and sections, as will be seen in the upper part of the diagram, are made to agree with them. From the points thus obtained in the profile draw lines cutting the end F G of the frustum. Produce the sides O F and P G until they meet in E, which is the apex of the cone. From the points in F G draw lines from E, producing them until they cut the axis of the cone, as shown by A $A^1$ $A^2$. Next construct sections of the cone as it would appear if cut through upon lines corresponding to these points, as A C, $A^1$ B, $A^2$ D. Divide the plan $U^2$ W $V^2$ $W^1$ into any convenient number of parts. From the points thus established carry lines vertically to the base line U V, and thence carry them to the apex T, cutting the lines A C, $A^1$ B, $A^2$ D, all as shown. Through each of the several points of intersection in these lines draw horizontal lines from the axis of the cone to the side, all as shown. At right angles to the lines A C, $A^1$ B, $A^2$ D draw lines to any convenient point, at which to construct the required sections. Upon the lines drawn from the points A, $A^1$, $A^2$, at convenience, locate the points $A^3$, $A^4$, $A^5$. Inasmuch as $A^1$ B is at right angles to the axis of the cone, the section corresponding to it will be a semicircle whose

radius will be equal to A' B. Therefore, from A³ as center, with radius A' B, describe the semicircle S B' R. For the section corresponding to A² D lay off from A⁶ the distances A⁵ S² and A⁵ R², in a line drawn at right

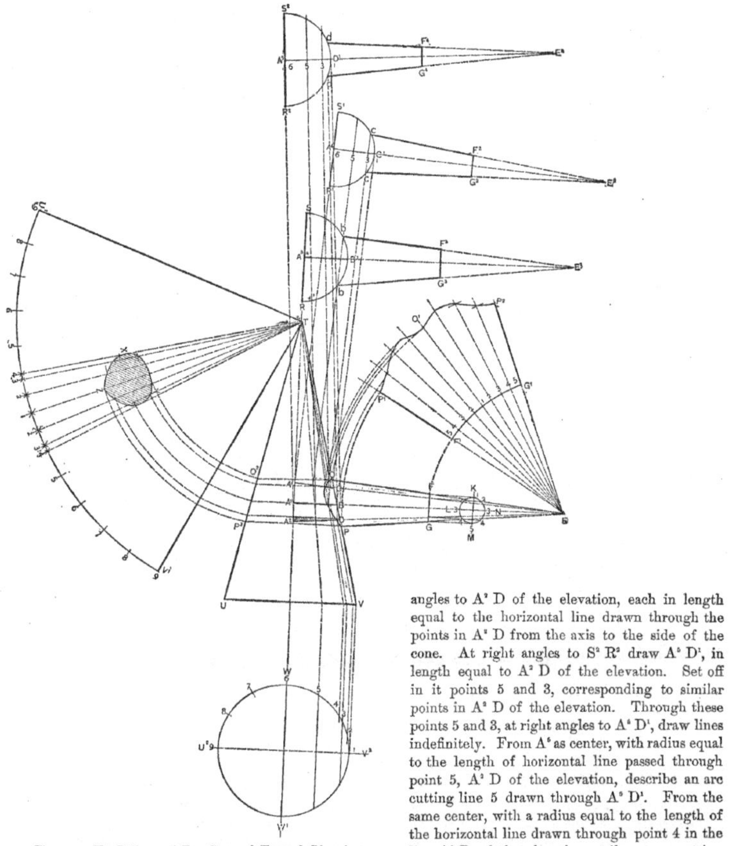

*Fig. 399.—The Patterns of Two Cones of Unequal Diameters Intersecting at Right Angles to their Axes.*

angles to A² D of the elevation, each in length equal to the horizontal line drawn through the points in A² D from the axis to the side of the cone. At right angles to S² R² draw A⁵ D¹, in length equal to A² D of the elevation. Set off in it points 5 and 3, corresponding to similar points in A² D of the elevation. Through these points 5 and 3, at right angles to A⁶ D¹, draw lines indefinitely. From A⁶ as center, with radius equal to the length of horizontal line passed through point 5, A² D of the elevation, describe an arc cutting line 5 drawn through A⁵ D¹. From the same center, with a radius equal to the length of the horizontal line drawn through point 4 in the line A² D of the elevation, strike an arc cutting the line 3. Then a line traced through these points, as shown by S² D¹ R², will be the section of the cone as it would appear if cut on the line A² D of the elevation. In like manner obtain the section S¹ C¹ R¹, corresponding to A C of the elevation. Prolong A⁵ D¹, as shown by $E^3$, making $A^3$ $E^3$ in length equal to $A^1$ E of the elevation. In like manner make $A^4$ $E^2$ and $A^5$ $E^1$ equal to A E and $A^1$ E of the elevation respectively. At right angles to these lines in the sections set off $F^2$ $G^2$, $F^3$ $G^3$, $F^4$ $G^4$, in position corresponding to F G of the elevation. Make the length of $F^2$ $G^2$ equal to the length across the section of the frustum marked 2 2. In like manner make $F^3$ $G^3$ equal to 3 3, and $F^4$ $G^4$ equal to 4 4 of the section. From $E^1$, $E^2$ and $E^3$ respectively, through these points in the several sections, draw lines. From the several points of intersection between the lines drawn from $E^1$, $E^2$, $E^3$ of the sections of the cone, as shown

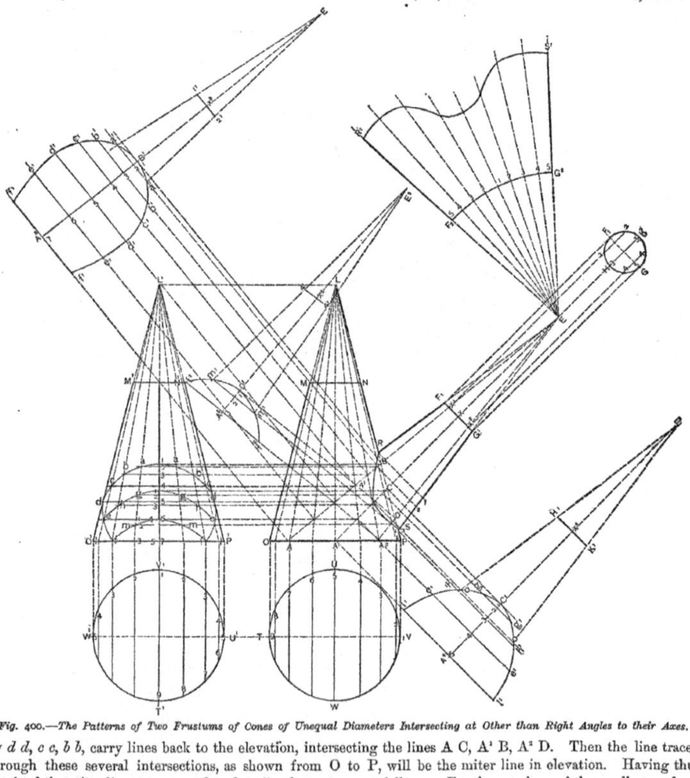

*Fig. 400.—The Patterns of Two Frustums of Cones of Unequal Diameters Intersecting at Other than Right Angles to their Axes.*

by *d d*, *c c*, *b b*, carry lines back to the elevation, intersecting the lines A C, $A^1$ B, $A^2$ D. Then the line traced through these several intersections, as shown from O to P, will be the miter line in elevation. Having thus obtained the miter line, we proceed to describe the patterns as follows : For the envelope of the small cone, from any convenient center, as E, with radius E F, describe the arc $F^1$ $G^1$, upon which set off the stretchout of the section K M N L. Through the points in this arc, from E, draw radial lines indefinitely. From E as center, with radii corresponding to the several points in the miter line O P, cut the corresponding radial lines, as indicated by the dotted lines. Then a line traced through these points of intersection, as shown by P¹ O¹ P², will be the shape of the pattern to fit against the larger cone. For the pattern of the larger cone, from any convenient point, as T, as center, with radius T U, describe the arc V¹ U¹ in length equal to the circumference of the plan U W V² W² of the cone. Upon this arc, V¹ U¹, set off points corresponding to the points from which lines were drawn to the base and thence to the apex. For obtaining measurements in connection with the miter line, through these points, 4 3 2 1 2 3 4, draw lines to the apex T. Intersect these lines by arcs struck from T as center, with radii corresponding to the points in the side of the cone between O² and P³, corresponding to the intersections of the lines drawn from the section L K M N of the frustum. Then a line traced through these intersections, as shown by X Y Z, will be the shape of the opening to be cut in the envelope of the larger cone, over which the smaller cone will fit.

533. *The Patterns of Two Frustums of Cones of Unequal Diameters Intersecting at Other than Right Angles to their Axes.*—In Fig. 400, let M N P O be the side elevation of the larger frustum, and F¹ G¹ S R the side elevation of the smaller, the two joining upon some line to be drawn from R to S. Produce the sides S G¹ and R F¹ until they meet in the point E. At any convenient place in line of the axis of the smaller frustum draw the profile H F K G, corresponding to the end F¹ G¹. Divide this profile into any convenient number of equal parts, as shown by the small figures, 1, 2, 3, etc., and from these divisions, parallel to the axis of the cone, drop points on to F¹ G¹. From the apex E, through these points in F¹ G¹, carry lines, cutting the side N P of the larger frustum, and producing them until they meet the opposite side, or, as in this case, the base O P, all as shown by B A, C A¹ and D A². The next step is to construct sections of the larger frustum as it would appear if cut on each of these lines, from which to obtain points of measurement for determining the miter line from R to S in the elevation. Draw the plan of the base of the larger frustum, as shown by T U V W, and divide one-half of it in the usual manner. From these points carry lines vertically to the base O P of the frustum. Produce the sides O M and P N until they meet in the point L. From the points in the base obtained from the plan carry lines to the apex L, cutting the section line A B, A¹ C and A² D, as shown. Parallel to A B and of the same length, at any convenient point outside of the elevation, draw A³ B¹, and from the points in A B, obtained by intersections with the lines from the base O P to the apex L, draw lines at right angles, cutting it as shown in the points 7, 6, 5, 4, etc. In like manner make A⁴ C¹ equal and parallel to A C, and from the points in A C draw lines at right angles to it, cutting it as shown, giving the points 5, 4, 3, etc. Also make A⁵ D¹ equal to the section line A² D of the elevation, and by drawing lines from the points in it cut A⁵ D¹ in the points 3, 2, 1, etc., as shown. In order to complete these several sections, the width of the frustum through each of the points indicated is to be set off on corresponding lines drawn through A³ B¹, A⁴ C¹ and A⁵ D¹. To obtain the width through these points draw an end elevation of the article, as

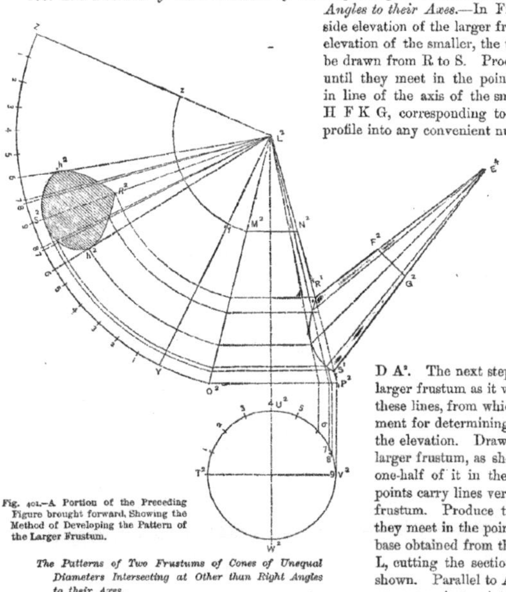


**Fig. 401.—A Portion of the Preceding Figure brought forward, Showing the Method of Developing the Pattern of the Larger Frustum.**

***The Patterns of Two Frustums of Cones of Unequal Diameters Intersecting at Other than Right Angles to their Axes.***

shown by $M^1$ $N^1$ $P^1$ $O^1$. Produce the sides, obtaining the apex $L^1$. Draw a plan and divide it into the same number of spaces as that shown in T U V W, and commence numbering at a corresponding point, all as indicated by $V^1$ $U^1$ $T^1$ $W^1$. From the points in the plan carry lines vertically to the base $O^1$ $P^1$, and thence to the apex $L^1$. Place the blade of the T-square at right angles to the axis of the cone, and, bringing it successively against the points in the section line A B in the side elevation, draw lines cutting the axis of the end elevation, and cutting the lines corresponding in number to the several points in A B, all as shown by *a a*, *b b*, *c c*, etc. Make the length of the lines drawn through $A^1$ $B^1$ equal to the corresponding lines thus obtained, as shown by $a^1$ $a^1$, $b^1$ $b^1$, $c^1$ $c^1$, $d^1$ $d^1$, etc., and through these extremities trace a line, as shown by $f^1$ $B^1$ $f^1$, which will be the section through the cone when cut on the line A B. In like manner obtain $l^1$ $C^1$ $l^1$ and $f^1$ $D^1$ $f^1$. Produce $A^1$ $B^1$, making $B^1$ $E^1$ equal to B E of the elevation, and $B^1$ $X^3$ equal to B $X^2$ of the elevation. In like manner make $C^1$ $E^2$ equal to C E, and $C^1$ $X^4$ equal to C X. Make $D^1$ $E^3$ equal to D E, and $D^1$ $X^6$ equal to D X. Through $X^3$, at right angles to $B^1$ $E^1$, draw a line in length equal to the line 2 2 drawn across the profile F K G H, with which this section corresponds, as shown by $2^1$ $2^1$. In like manner, through $X^4$ draw a line equal to H K, as shown by $H^1$ $K^1$, and through $X^6$ draw $4^1$ $4^1$, in length equal to the line 4 4 drawn through the profile F K G H. From $E^1$, through the extremities of $2^1$ $2^1$, draw lines cutting the section. In like manner draw lines from $E^2$ through the points $H^1$ $K^1$, and from $E^3$, through the points $4^1$ $4^1$. From the points at which these lines meet the profiles of the sections, $a^1$ $a^1$ in the first, *o o* in the second, and $m^1$ $m^1$ in the third, carry lines at right angles to and cutting the corresponding section lines in the elevation. A line traced through the points thus obtained, as shown by R S, is the miter line in elevation formed by the junction of the two frustums. Having thus obtained the miter line in elevation, we proceed to develop the patterns as follows: From the points in R S, at right angles to $A^1$ E, which is the axis of the smaller cone, draw lines cutting the side E S, as shown by the small figures, 1, 2, 3, 4 and 5. These points are to be used in laying off the pattern of the smaller frustum. From any convenient point for center, as E, with radius E $G^1$, describe the arc $F^2$ $G^2$, upon which step off the stretchout of the profile F K H G, numbering the points in the usual manner. Through the points, from the center E, draw radial lines indefinitely. From the same center, E, with radius E 1 (of the points in E S), cut the radial line numbered 1, and in like manner, with radii $E^2$, $E^3$, etc., cut the corresponding numbers of the radial lines. A line, $R^1$ $S^1$, traced through the several points of intersection thus formed will be the larger end of the pattern for the small frustum, thus completing the shape of that piece, all as shown by $R^1$ $S^1$ $G^2$ $F^2$. To avoid confusion of lines, the manner of obtaining the envelope of the large frustum is shown in Fig. 401, which is a duplicate of the side elevation and plan shown in Fig. 400. The miter line $R^1$ $S^1$ and the points in it are obtained by transfer, being the same in all particulars as employed in the operations already described. Similar letters refer to corresponding parts in the several figures. From any convenient point, as $L^2$, with radius $L^2$ $O^2$, describe an arc, as shown by Y Z, and from the same center, with radius $L^2$ $M^2$, describe a second arc, as shown by *y z*. Draw Y *y*, and upon Y Z lay off the stretchout of the plan $U^2$ $V^2$ $W^2$ $T^2$, all as shown. Draw Z *z*. Then Z *z y* Y will be the envelope of the large frustum. Through the points in the miter line $R^1$ $S^1$ draw lines from the apex of the cone to the base, and from the base continue them at right angles to it until they meet the circumference of the plan. Mark corresponding points in the stretchout Y Z, and insert any points which do not correspond with points already fixed therein. From each of the points thus designated draw a line across the envelope already described to the apex, as shown by 6 $L^2$, 7 $L^2$, 8 $L^2$, 9 $L^2$, etc. Also, from the points in the miter line $R^1$ $S^1$ draw lines at right angles to the axis of the frustum, cutting the side $L^2$ $O^2$, as shown. From $L^2$ as center, describe arcs corresponding to each of these points, and cutting the radial lines drawn across the envelope of

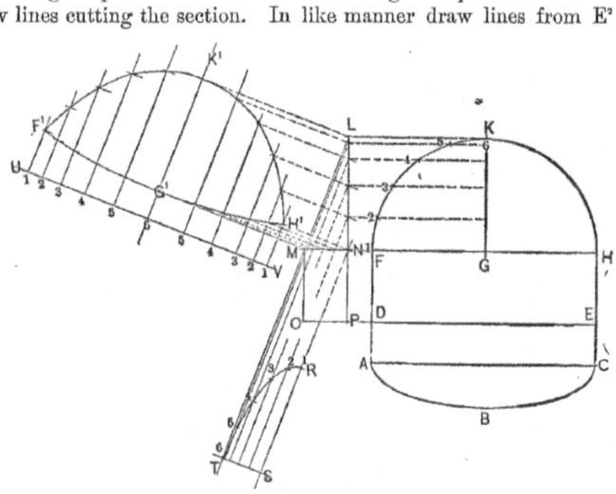


*Fig. 402.—Pattern for a Blower for a Grate.*

the cone. A line traced through the points of intersection between arcs and lines of the same number, as shown by $h^2$ $R^2$ $h^2$ $S^2$, will be the shape of the opening to fit the base of the smaller frustum.

534. *Pattern for a Blower for a Grate.*—The blower shown in Fig. 402 consists of two pieces, the body and the hood. A section through the body, taken horizontally, shows an arc of an ellipse—a shape somewhat more flattened than a segment of a circle. The profile, taken through the blower vertically, shows the body straight, with the hood pitching toward the grate. L M O P is a profile through the blower, taken vertically at its center. A D C is a profile taken horizontally through the line F G H. D F K H E is the elevation. Before it is possible to cut the miters at the top and bottom of the piece F H K, a true stay of these pieces must be obtained, which is shown in connection with the side elevation. To obtain this stay proceed as follows: Divide one-half of the arc F K H into any number of equal spaces. Carry lines from each of these several points to the vertical line L N of the profile, and thence parallel with the line L N indefinitely. Intersect them at right angles by the line T S, located at any convenient point outside of the diagram. With the dividers take the horizontal distance between the points in the arc F K to the line K G, and set them off on the lines of corresponding number, measuring from the line T S. Then a line drawn through the points thus obtained, and as indicated by T R, will be a horizontal section through the correct profile of the inclined portion of the blower. Take the stretchout of the profile T R point by point, and place the spaces on the line U V, which is drawn at right angles to L M. Through the points in U V draw the usual measuring lines at right angles to it. Drop the points from the profile T R on to both the miter lines M N and N L, and thence carry them, at right angles to L M, on to the stretchout lines of corresponding numbers drawn from U V. Then a line traced through the points thus obtained, and as indicated by $F^1$ $K^1$ $H^1$, will be the desired pattern.

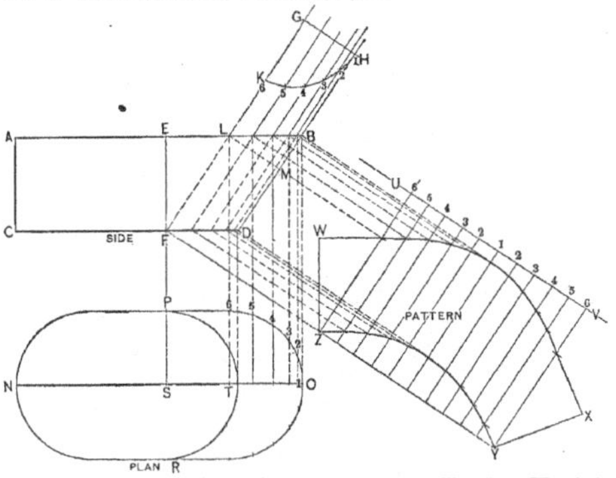


*Fig. 403.—Pattern of the Flaring End of an Oblong Pan. First Case—When both Bottom and Top of the Flaring End are Curved.*

535. *Pattern of the Flaring End of an Oblong Pan. First Case—When both Bottom and Top of the Flaring End are Curved.*—In Fig. 403, A B D C shows in elevation, and N P O R in plan, a vessel of the description indicated. To obtain the patterns, after having correctly drawn the plan and elevation, proceed as follows: Divide half of the boundary line of the bottom into any number of equal spaces, commencing at O, all as shown by the small figures 1, 2, 3, etc., in the plan. From the points thus obtained carry lines vertically until they cut the top line of the elevation, as shown in the points between B and L; also continue the lines downward until they meet the line T O, all as shown. From the points between L and B thus obtained draw lines parallel to B D, producing them upward indefinitely, and continue them downward until they meet the bottom line of the elevation F D, as shown. At right angles to the lines thus drawn, and at any convenient distance from the elevation, draw G H. With the dividers, from the line G H, set off on each of the lines drawn through it the distance from T O, on the lines of corresponding number, to the line representing the plan of the end. In other words, make G K equal to T 6 of the plan. Set off spaces on the other lines corresponding to the distance on like lines in the plan. Through the points thus obtained trace a line, as shown by K H. Then G H K will be the half profile of the end of the vessel at right angles to the line D B. The stretchout of the pattern is to be taken from the profile thus constructed. At right angles to D H, and at any convenient distance from it, lay off U V equal to twice the length of K H, and make the divisions in it correspond with the divisions in K H. From the points in the stretchout thus obtained draw lines at right angles to it indefinitely. With the blade of the T-square set at right angles with D B, and brought

successively against the points in F D, cut lines of corresponding numbers drawn through the stretchout. Then a line traced through these points, as shown by Z Y, will be the pattern of the bottom of the end piece. In like manner, with the T-square in the same position, bring the blade against the points in L B, and cut corresponding lines drawn through the stretchout. What may be called the corner pieces of the pattern are to be added to the portion already obtained as follows: With the dividers take the distance F E of the elevation as radius, and from the point Z of the pattern as center describe an arc. In like manner take the distance E L of the elevation in the dividers, and from the last point in the upper edge of the pattern already obtained, being that of the line 6, describe a second arc, cutting the one first drawn in the point W. Connect W Z, and also draw a line from W to the point in the line 6 already obtained. Add a corresponding corner piece to the other extremity. Then Z W X Y will be the pattern required.

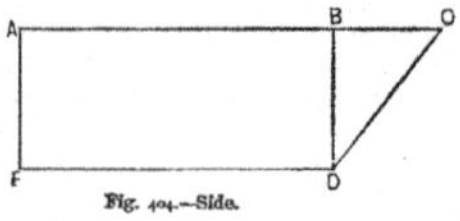


Fig. 404.—Side.

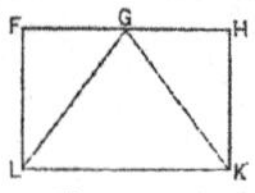


Fig. 405.—End.

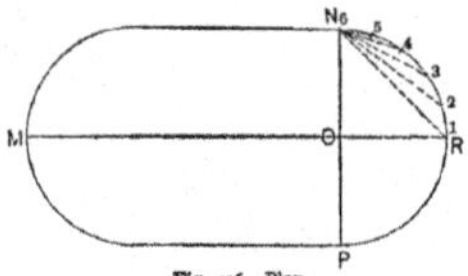


Fig. 406.—Plan.

*Pattern of the Flaring End of an Oblong Pan. Second Case—When Top is Curved and Bottom is Straight.*

536. *Pattern of the Flaring End of an Oblong Pan. Second Case—When Top is Curved and Bottom is Straight.*—In Fig. 404, A C D E represents the side elevation of the article, F H K L in Fig. 405 the end elevation, and M N R P in Fig. 406 the plan or bottom. By inspection of these it will be seen that the shape of the end piece required is such that it may be resolved into three parts or sections. The middle one of these will be flat, or as represented upon the end elevation by G L K. The two side pieces are sections of the envelope of a cone. To obtain the patterns proceed as follows: Divide one-half of the end of the plan into any convenient number of equal spaces, all as shown by small figures 1, 2, 3, 4, etc., in N R. From each of the points thus determined draw lines to the point N, all as shown in the engraving. From the measurements made possible by these lines we next proceed to construct the diagram shown in Fig. 407. Draw A B, in length equal to D B of Fig. 404. At right angles to it draw B C, which produce indefinitely. From B along B C set off spaces equal to the distance from N, Fig. 406, measured to the points in the boundary line of the plan. That is, make B 5 of Fig. 407 equal to N 5 of Fig. 406, and B 4 equal to N 4, and so on. With the measurements to be obtained from this diagram we lay off the patterns as follows: Draw A' D, in length equal to A C of the diagram. Set off points in A' D to represent the length of the lines in the diagram drawn from A, or, in other words, make A' 2 equal to A 2 of the diagram, and so on. From A' as center, with radius A' D, describe the arc D E indefinitely. In like manner, from the same center, with radius A' 2, describe a corresponding arc, and proceed in this way with each of the other points lying in the line A' D. From A', and at any convenient angle, draw A' E, letting E fall in the arc D E, already mentioned. From E, stepping from one arc to another, lay off the stretchout of N R of Fig. 406, all as shown by E F of the pattern. Connect A F. Then A' F E will be one section of the required pattern. From E as center, with radius E A', describe the arc A' G indefinitely. Make the chord A' G equal to L K of the end elevation, Fig. 405. Connect G E. Then A' E G will be a second section of the pattern. To this add E G H, equal to E A' F. Then F E H G A' will be the pattern sought.

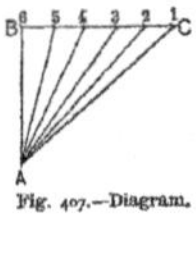


Fig. 407.—Diagram.

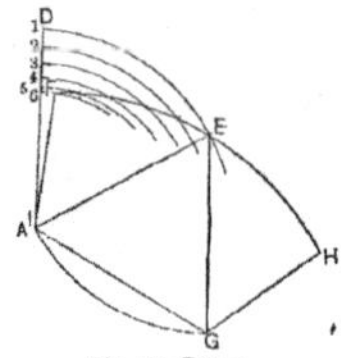


Fig. 408.—Pattern.

*Pattern of the Flaring End of an Oblong Pan. Second Case—When Top is Curved and Bottom is Straight.*

537. *Patterns for a Soapmaker's Float.*—Fig. 409 represents a soapmaker's float as commonly constructed in some places. The part A O B, or the bottom, is to be regarded as raised work, and shaped by means of the raising hammer without regard to any rules. The sides are to be considered as parts of two cones having elliptical bases, the short diameters of which are alike, but the long diameters of which vary. Thus in the plan, Fig. 410, L D' M represents the half of the base of an elliptical cone, the short diameter of which is equal to

L M, and the half of the long diameter of which is equal to K' D'. By thus resolving the envelope of the vessel into sections of cones, the development of patterns becomes, comparatively speaking, a simple operation. First develop that part of the pattern which corresponds to A E F D of the section, Fig. 410. To do this proceed as follows: Drop the point A on to the center line of the plan, as shown by A'. As the curve D' L is the quarter of a perfect ellipse, it becomes necessary to locate the point P so that the curve A' P shall be a section of the same elliptical cone as D' L. This may be determined in this manner: Connect D' L by a straight line, as shown. Then from A' draw a line parallel to D' L, cutting the short diameter, which point of intersection will be the point P required. With this point determined, draw the curve A' P, being a portion of the regular ellipse, by any convenient method.

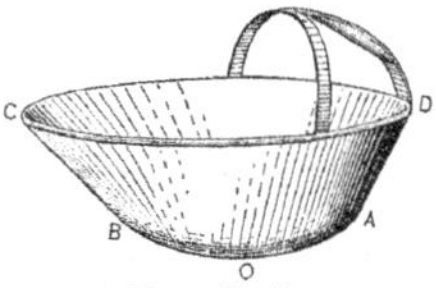

Fig. 409.—Elevation.
*Soapmaker's Float.*

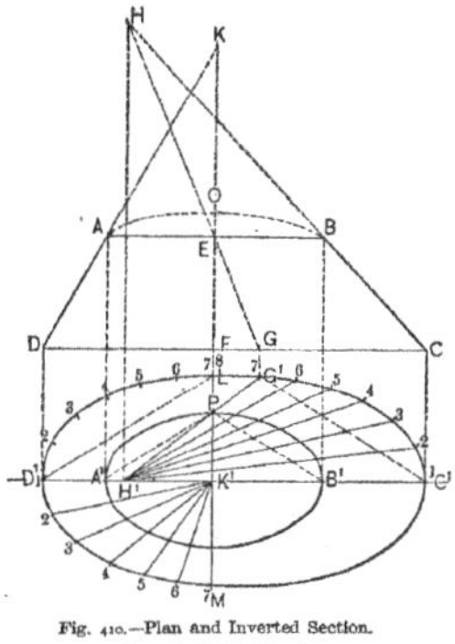

Fig. 410.—Plan and Inverted Section.
*Soapmaker's Float.*

Produce the line F E of the elevation in the direction of K indefinitely. In like manner produce D A of the elevation until it reaches F E produced in the point K. Then D K F may be regarded as the section of a half cone, of which that part of the vessel indicated by A E F D is a portion, and K F the perpendicular hight. Next, divide both halves on the plan L D' M into the same number of equal parts, as shown by the small figures 1, 2, 3, etc., running both ways from D'. Construct the diagram shown in Fig. 412 by drawing the line D K' of indefinite length, and the line K' K at right angles to it, making K' K in length equal to F K, Fig. 410. Establish the point K² by making the distance K' K² equal to E F of Fig. 410. Draw K² A parallel to K' D. From each of the points 2, 3, 4, etc. of the plan, draw lines to the center K', and set off distances equal to these lines upon the line K' D of Fig. 412, measuring from K' toward D. From each of the points thus obtained draw lines to the point K, cutting A K². With the foot of the compasses in the point K, and the other brought successively against the points 1, 2, 3, etc., in the line D K' and the line A K², describe arcs, producing them indefinitely. Take in the dividers a space equal to the divisions 1, 2, 3, 4, etc., of Fig. 410, and, commencing at the point *a* in the first arc (Fig. 412), step to the second arc, and thence to the third arc, and thus continue stepping from one arc to another until the entire stretchout of the half plan has been laid off, as shown in Fig. 412. The same operation is to be repeated upon the arcs drawn from the points in the line A K². It may be shortened, however, by drawing radial lines to the point K from the several points determined in the first set of arcs. Then a line traced through the several points of intersection thus obtained, as shown by *b c* and *a d*, will be the boundary lines of the pattern. The pattern for the other end of the article is to be, in the main, developed in the same manner as we have described. There is, however, a slightly different case arising, to which we shall give attention, without repeating that portion of the description which would coincide with what has been stated. The points P and B' being established, connect them by a true ellipse, half of the long diameter of which is K' B', and half of the short diameter of which is K' P. It now becomes necessary to obtain a curve between the points L and C', which shall be a section of the same elliptical cone as B' P. To do this proceed as follows: Connect the points P and B' by means of a straight line. From the point C' draw a line parallel to P B', and

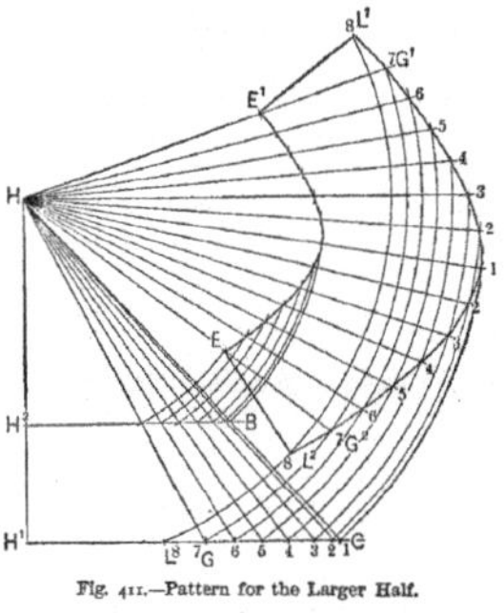

Fig. 411.—Pattern for the Larger Half.
*Soapmaker's Float.*

produce it until it cuts the line L $G^1$, which is a straight line drawn at right angles to L M. Then $G^1$ becomes a point in the lower base of the cone corresponding to the point P in the upper base. Draw the line $G^1$ P, and continue it until it intersects the long diameter in $H^1$. Drop the point $G^1$ vertical from the plan on to the base line D C of the elevation, as indicated by the point G. Draw a line through the points G and E, which produce indefinitely in the direction of H. In like manner produce the side C B of the cone until it intersects G E produced in the point H. Then it will be found that the point H of the elevation and the point $H^1$ of the plan coincide, as indicated by the line H $H^1$. The operation of developing the pattern from this stage forward is the same as in the previous case, save only in the matter of the triangular piece indicated by G E F of the elevation. After completing the other portions of the pattern, this triangular piece is added as follows: The distance $H^1$ L in Fig. 411 is to be set off on the line $H^1$ C in the same manner as the other points—*i. e.*, $H^1$ L of Fig. 411 is equal to $H^1$ L of Fig. 410. Then L is to be treated in the same manner as the other point, an arc being struck from it, as indicated in the engraving, by which to determine the corresponding point $L^2$ in the outline of the pattern. $L^2$ $G^2$ becomes equal to L $G^1$ of the plan, Fig. 410. From $L^2$, Fig. 411, draw a line to E. Then E $L^2$ $G^2$ will be the pattern of the triangular piece indicated in Fig. 410 by E F G. It is to be added upon the opposite end of the pattern in like manner, as indicated by $E^1$ $G^1$ $L^1$.

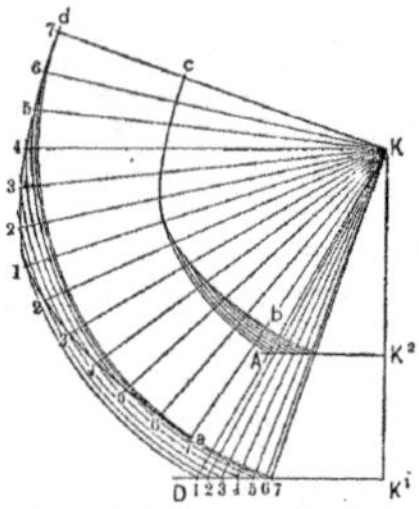

**Fig. 412.—Pattern for the Smaller Half.**
*Soapmaker's Float.*

538. *A Square Return Miter, or a Miter at Right Angles, as in a Cornice at the Corner of a Building.*—In Fig. 413, let A B D C represent a cornice at the corner of the building for which a miter at right angles is desired. As has been elsewhere explained, the process of cutting a miter, when applied to a right angle, admits of certain abbreviations not employed in the use of other angles. The demonstration here introduced is calculated to show the method of obtaining the pattern for a square miter with the least possible labor. Divide the profile A B into any convenient number of parts, as shown by the small figures. At right angles to the lines of the molding, and in convenient proximity to it, lay off the stretchout E F, through the points in which, parallel to the lines of the cornice, draw measuring lines in the usual manner, producing them far enough to intercept lines dropped vertically from points in A B. Place the T-square at right angles to the cornice, or, what is the same, parallel to the stretchout line, and, bringing it successively against points in the profile A B, cut measuring lines of corresponding numbers. Then a line traced through these points, as shown by G H, will be the pattern sought.

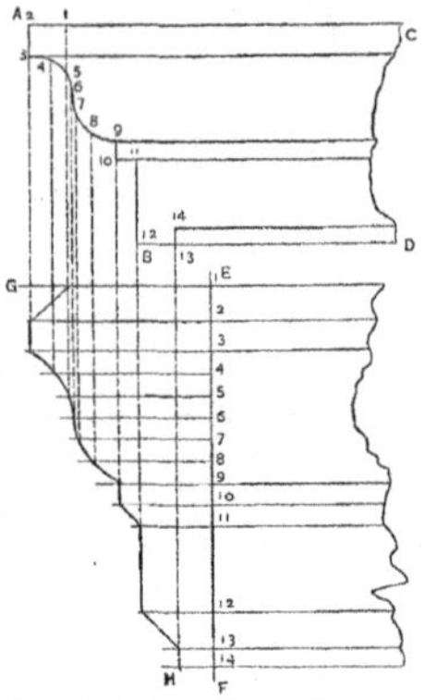

*Fig. 413.—A Square Return Miter, or a Miter at Right Angles, as in a Cornice at the Corner of a Building.*

539. *A Return Miter at Other than a Right Angle, as in a Cornice at the Corner of a Building.*—In Fig. 414, let A B C D be a section of the cornice, of which a pattern is to be cut forming a miter in the angle, shown in plan by G H K. As remarked in the previous demonstration, the cutting of a miter at other than a right angle demands certain work not necessary in the case of a right angle; therefore the demonstration which follows is applicable in all cases save that of a right angle. Construct a plan of the required miter, as indicated by E F L K H G, and draw the miter line F H. By inspection it will be evident that F H is in reality the only line in the plan constructed which is used, therefore the work may be abbreviated to the extent of laying off simply the line F H, its inclination being determined by any means most convenient. The full plan, however is here introduced in order to show the requirements of that line. It must be drawn as though the complete plan were represented. Divide the profile A B in the usual manner into any convenient number of parts, and from the points thus obtained drop lines vertically on to the miter line in the plan F H, as shown. At right angles to one arm of the cornice, as shown in plan—in this case at right angles to E F—lay off a

stretchout of the profile, as shown by N M, through the points in which draw the usual measuring lines, as indicated. Place the T-square parallel to this line, or, what is the same, at right angles to E T, and, bringing it successively against the points F H, cut measuring lines of corresponding numbers. Then a line traced through the points thus obtained, as shown by O P, will be the pattern sought. It is evident that the stretchout M N could with equal propriety be laid off at right angles to F L, the general rule in miter cutting being that the stretchout must be laid off at right angles to the molding the pattern of which is being produced. By the operation shown above, M O P N represents a pattern of a portion of the molding shown in plan by E F H G. If for any reason the stretchout had been laid off at right angles to F L, the pattern produced would have represented a portion of the molding shown in plan by F L K H. But since these two pieces are alike, all necessary results are accomplished in performing the operation once, and therefore it is performed at such a place as is most convenient, which, of course, is where the T-square can be used from adjacent sides of the board.

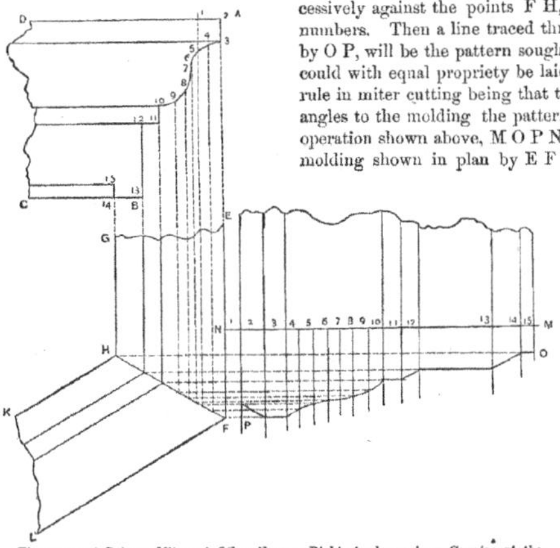

*Fig. 414.—A Return Miter at Other than a Right Angle, as in a Cornice at the Corner of a Building.*

539. *A Butt Miter against a Plain Surface shown in Elevation.*—Let C D in Fig. 415 be the profile of a cornice, and A B the angle or inclination of the surface in elevation against which the cornice miters. Let A K L B be the length of the cornice for which the pattern is desired. Space the profile in the usual manner, and from the points draw lines cutting the miter line A B. At right angles to the cornice lay off, on any convenient line, as E F, a stretchout of the profile C D, through the points in which draw the usual measuring lines, all as indicated by the small figures. Placing the T-square at right angles to the lines of the cornice, or, what is the same, parallel to the stretchout line, bring it successively against the points in the miter line A B and cut corresponding measuring lines, as indicated by the dotted lines. A line traced through these points, as indicated by H G, will be the pattern required.

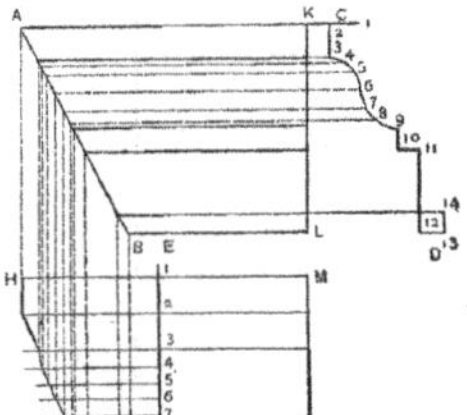

*Fig. 415.—A Butt Miter against a Plain Surface shown in Elevation.*

540. *A Butt Miter against a Regular Curved Surface.*—In Fig. 416, let A B be the profile of any cornice, a butt miter in which is to be cut to fit it against a surface, the profile of which is a regular curve, as shown by C D. Space the profile in the usual manner, and through the points draw lines cutting C D. At right angles to the line of cornice lay off the stretchout L M, as shown, through the points in which draw measuring lines in the usual manner. Place the T-square parallel to the stretchout line, or, what is the same, at right angles to the lines of the cornice, and, bringing it against the several points in C D, cut the corresponding measuring lines, as shown. In the event of a wide space, as shown by $a^1$ $b^1$ in the elevation, two methods are at the choice of the pattern cutter. One is to divide this space in the profile in the usual manner, as though it was a molding from which to obtain a number of points approximating to the curve. The other method is as given in the engraving. Transfer to the pattern $G^1$ a point corresponding to G of

the elevation, the center by which the curve C D was struck, as indicated by the line and arrow point. Then from G' as center, with the same radius as used to strike the curve in the elevation, strike the arc *a b*, extending from the measuring line 11 to measuring line 12. A line traced through the several points of intersection, together with the arc struck from the center G', as above explained, all as shown by E F, will be the shape of the required pattern.

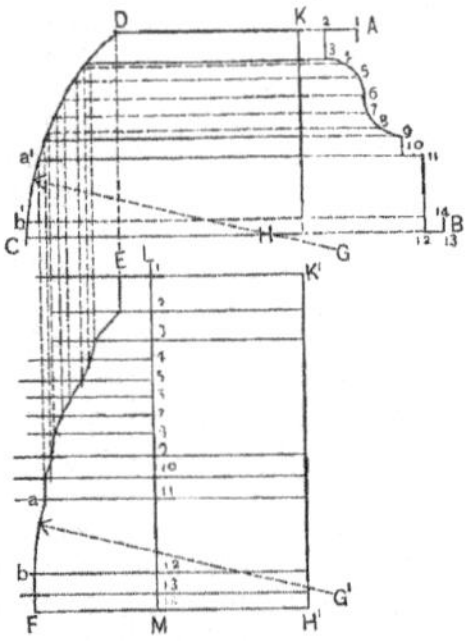


*Fig.* 416.—*A Butt Miter against a Regular Curved Surface.*

541. *A Butt Miter against a Plain Surface shown in Plan.*—Let C D in Fig. 417 be the profile of the cornice which is required to miter against a vertical surface standing at any angle with the lines of the cornice, the angle being shown in plan by A B. Draw the profile C D, corresponding to the lines of the cornice, all as indicated. Space in the usual manner, and through the points draw lines cutting the miter line A B. At any convenient point at right angles to the lines of the cornice, lay off the stretchout E F of the profile C D, through the points in which draw measuring lines in the usual manner. Placing the T-square at right angles to the cornice, or, what is the same, parallel to the stretchout line E F, bring it successively against the points in A B and cut the corresponding measuring lines. A line traced through the points of intersection thus obtained, shown by K G, will be the pattern required.

542. *A Butt Miter of a Molding Inclined in Elevation against a Plain Surface Oblique in Plan.*—Let A B in Fig. 418 be the profile of a given cornice, and let E D C F represent the rake or incline of the cornice as seen in elevation. Let G H represent the angle of the intersecting surface in plan. The first step in developing the pattern will be to obtain miter lines in the elevation, as shown by E F. For this purpose draw the profile A B in connection with the raking cornice, which space in the usual manner, as indicated by the small figures. Draw a duplicate of this profile, as shown by A' B', placing it in a horizontal position, with points corresponding to those shown in the raking cornice. Space the profile A' B' into the same number of parts as A B, and through the points thus obtained carry lines parallel to the lines of the cornice, as seen in plan, cutting the miter line G H, as shown. In like manner draw lines through the points in A B, carrying them parallel to the lines of the raking cornice in the direction of E F indefinitely, as shown. Place the T-square at right angles to the lines of the cornice, as shown in plan, and, bringing it against the points of intersection in the line G H, carry lines vertically, cutting corresponding lines in the inclined cornice drawn from the profile A B. Through the points of intersection thus obtained trace a line, as shown from E to F. Then this profile E F will be the miter line in elevation, formed by a cornice of the profile A B meeting a surface in the angle shown by G H in the plan. At right angles to the raking cornice lay off a stretchout upon any line, as K L, and through the points draw the usual measuring lines, all as shown. Place the T-square at right angles to the lines of the raking cornice, and, bringing it against the several points in the profile E F, cut corresponding measuring lines drawn

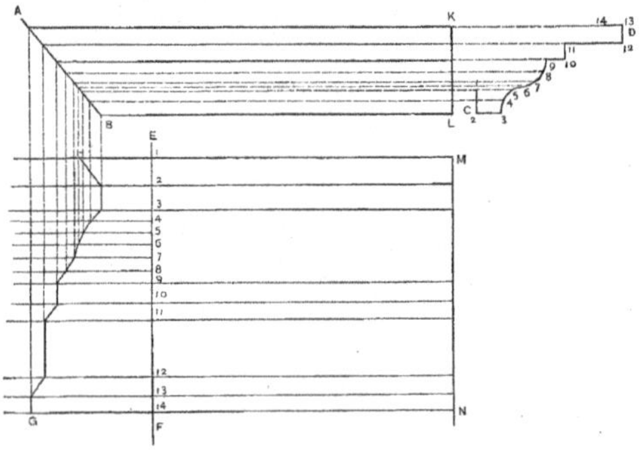


*Fig.* 417.—*A Butt Miter against a Plain Surface shown in Plan.*

from the stretchout K L. A line traced through these points of intersection, as shown from M to N, will be the pattern required.

543. *A Butt Miter against an Irregular or Molded Surface.*—Let B A in Fig. 419 be the profile of a cornice, against which a molding of the profile, shown by G H, is to miter, the latter meeting it at an angle, as indicated by C D. Draw the profile B A ; also construct an elevation of the cornice meeting it, as shown by C D F E, in line with which draw the profile G H. Divide G H in the usual manner into any number of convenient parts, and through the points draw lines parallel to the lines of the inclined molding, cutting the profile B A, all as indicated by the dotted lines. At right angles to the lines of the inclined molding lay off a stretchout, M N, in the usual manner, through the points in which draw measuring lines. Place the T-square at right angles to the lines of the inclined molding, or, what is the same, parallel to the stretchout line, and, bringing it against the points of intersection formed by the lines drawn from the profile G H across the profile B A, cut the corresponding measuring lines. In the event of any angles or points occurring in the profile B A which are not met by lines drawn from the points in G H, additional lines from these points must be drawn, cutting the profile G H, in order to establish corresponding points in the stretchout. Thus the points 3 and 13 in the profile G H are inserted after spacing the profile, as above described, because the points with which they correspond in the profile B E are angles which must be clearly indicated in the pattern to be cut. Having thus cut the measuring lines corresponding to the points in the profile B A, draw a line through the points of intersection, as shown by O P. Then O P will be the shape of the pattern of the incline cornice to miter against the profile A B.

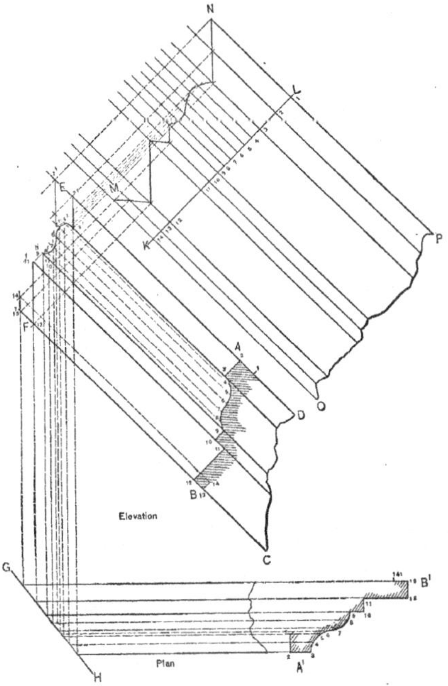


*Fig. 418.—A Butt Miter of a Molding Inclined in Elevation against a Plain Surface Oblique in Plan.*

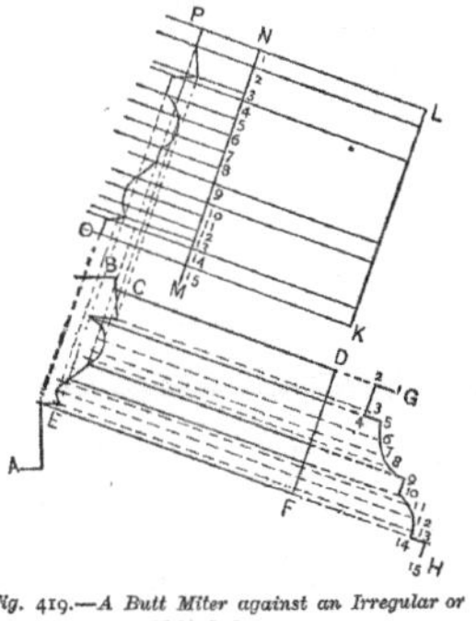


*Fig. 419.—A Butt Miter against an Irregular or Molded Surface.*

544. *Miter between Two Moldings of Different Profiles.*—To construct a square miter between moldings of dissimilar profiles requires two distinct operations. The miter upon each piece is to be cut as it would appear when intersected by the other molding. Let the profiles A B and A' B' in Figs. 420 and 421 be of the

same hight, but differing in members, between which a square miter is to be formed. Proceed as follows: Draw E F, a duplicate of A' B', in line with A B. Divide A B into any convenient number of parts in the usual manner, from which carry lines horizontally against E F, and thereby construct an elevation of the molding as it would appear if intersected by F E, all as shown by F C D E. For the pattern of this piece, at right angles to its lines lay off a stretchout, G H, of the profile A B, through the points in which draw the usual measuring lines. Bring the T-square against the points of intersection in the line E F, and cut the corresponding measuring lines. Then a line traced through these points, as shown by E' F', will give the shape of the cut to fit the molding against the profile E F. For the other piece proceed in the same manner, reversing the order of the profiles. Draw M N, a duplicate of A B, in line with A' B'. Divide A' B' in the usual manner. Through the points draw lines cutting M N, thereby constructing an elevation, K M N L, of the piece the pattern of which is sought. At right angles to this piece lay off the stretchout O P of the profile A' B', through the points in which draw measuring lines, as shown. With the T-square at right angles to the lines K M N L, and brought against the points in M N, cut corresponding measuring lines drawn through O P. A line traced through these points, as shown by M' N', will be the shape of the piece required to fit against the profile M N. In the event of the points obtained by spacing the profiles A B and A' B' not meeting all the points in the profiles F E and M N necessary to be marked in the pattern, then lines must be drawn backward from such points in profiles M N and E F, cutting the profile A' B' or A B, as the case may be. Corresponding points are then to be inserted in the stretchouts, through which measuring lines are to be drawn, which in turn are to be intersected by lines dropped from the points. An illustration of this occurs in point No. 6½ in Fig. 421. It will be seen that this point is absolutely essential to the shape of the pattern.

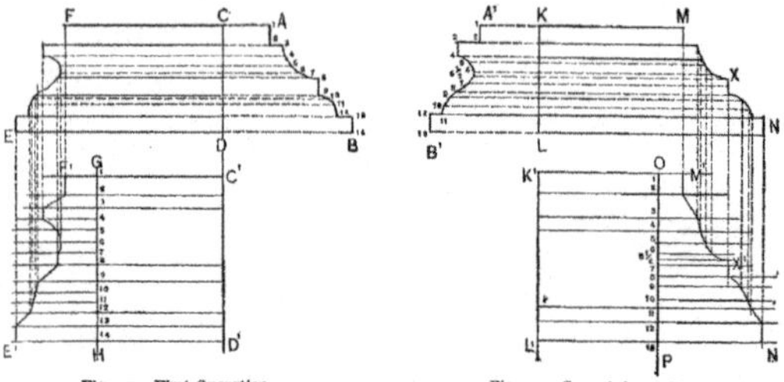


**Fig. 420.—First Operation.** **Fig. 421.—Second Operation.**

***Miter between Two Moldings of Different Profiles.***

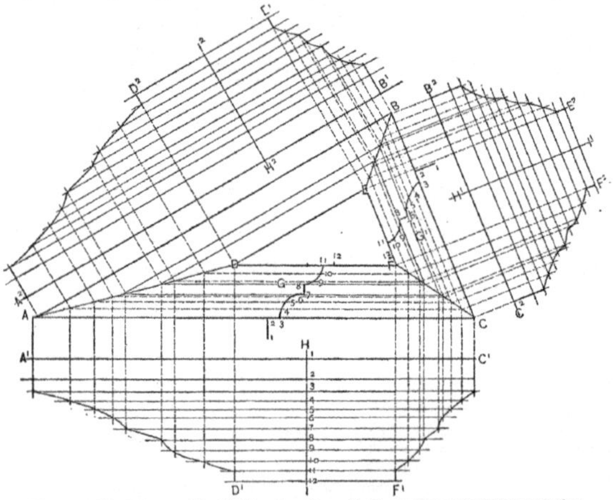


***Fig. 422.—The Patterns of the Moldings bounding a Panel, the Shape of which is a Scalene Triangle.***

Therefore, after spacing the profile a line is drawn from X back to A' B', forming the point No. 6½. In turn this point is transferred to the stretchout O P, also marked 6½, from which a measuring line is drawn in the same manner as through the other points in the stretchout, upon which a point from X is dropped, as shown by X'. In actual practice such expedients as this must be resorted to in almost every case, because usually there is less correspondence between the members of dissimilar profiles, between which a miter is required, than in the illustration here given. By this means profiles, however unlike, can be joined.

545. *The Patterns of the Moldings bounding a Panel, the Shape of which is a Scalene Triangle.*—In Fig. 422, let D E F be the elevation of a triangular panel or other article, surrounding which is a molding of a profile, shown at G and G'. Construct an elevation of the panel, as shown by A B C, and draw the miter lines A D, B E, C F. For the patterns of the several sides proceed as follows: Draw a profile, G, placing it, relative to the side D F, in the position corresponding to the molding to be constructed. Divide it into any convenient number of parts in the usual manner, and through these points draw lines, as shown, cutting the miter lines F C and A D. In like manner place the profile G' in a corresponding position. Divide it into the same number of parts, and draw lines intersecting those drawn from the first profile in the line F to C, also cutting the line E B. By this operation we have points in the three miter lines A D, E B, F C, from which to lay off the pattern in the usual manner. At right angles to each of the three sides, at convenient points, draw stretchout lines, as shown by H I, H' I' and H' I', through the points in which draw the usual measuring lines. With the T-square parallel to each of the several stretchout lines, or, what is the same, at right angles to the respective sides, bringing the blade successively against the points in the several miter lines, cut the corresponding measuring lines, all as indicated by the dotted lines. Then lines traced through the points of intersection thus obtained will describe the patterns required. A' C' F' D' will be the pattern for the side, A D F C of the elevation, and likewise C' B' E' F' is the pattern for the side, described by similar letters.

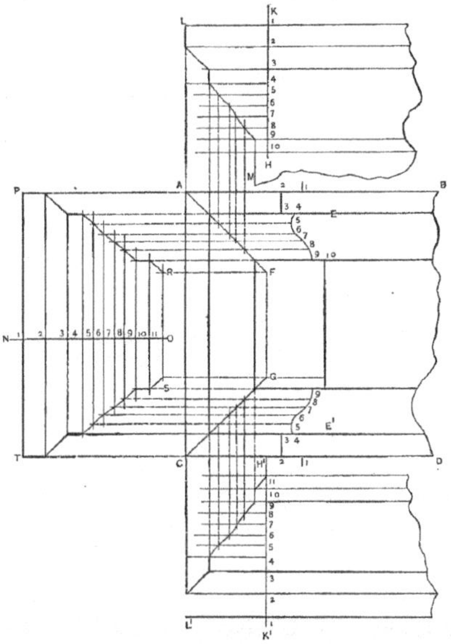


*Fig. 423.—A Face Miter, or Miter at Right Angles, as in the Molding Around a Panel.*

546. *A Face Miter, or Miter at Right Angles, as in the Molding Around a Panel.*—In Fig. 423, let A B C D represent any panel, around which a molding is to be carried of the profile E and E'. The miters required in this case are of the nature commonly known as "face" miters, which in the process of pattern cutting require substantially the same steps as indicated in the preceding problem for a miter at any angle other than a right angle around a panel. That is to say, by reason of the position in which the profile is shown, it is necessary to drop points against a miter line, and thence carry them to the measuring lines, in order to develop the pattern. For the patterns, therefore, we proceed as follows: Draw profiles in opposite sides of the panel, as shown by E and E', or, what is the same, draw a section of the panel as is shown by the lines across its width. Divide the two profiles in the usual manner into the same number of parts. Through

the angles of the panel draw miter lines, as shown by A F and C G. From the points in the profile already determined, draw lines parallel to the lines of the molding, cutting these miter lines, as shown. For the pattern of the side corresponding to A B, lay off a stretchout at right angles to it, as shown by H K, through which draw measuring lines in the usual manner. Place the T-square at right angles to A B, or, what is the same, parallel to the stretchout line H K, and, bringing it successively against the several points in the miter line A F, cut measuring lines of corresponding number. Then a line traced through these points, as shown by L M, will be the pattern sought. In like manner, by dropping points from the profile E' on to the miter line C G, the pattern for the opposite side may be obtained, all as shown by L' H'. So far as the molding bounding the panel is concerned, these two patterns correspond in all particulars, the only difference being that an allowance is made for a seam in connection with the lower piece, whereas a flat surface to form the panel itself is shown attached to the upper piece. The two patterns are presented, in order to show the convenience of working from both sides in producing the two pieces, instead of copying one from the other. The pattern of the end piece is derived from the two miter lines A F and C G, from the points already established in them. It is quite as easy to describe the pattern by this means as to copy it from the pattern first obtained. For the pattern of the end piece, at right angles to the end of the panel A C, lay off a stretchout of the profile, as shown by N O, through the points in which draw measuring lines in the usual manner, producing them sufficiently far in each direction to intercept lines dropped from the points in the two miter lines. Place the T-square at right angles to A C, and, bringing it successively against points in A F and C G, cut measuring lines of corresponding numbers. Then lines traced through the intersections thus formed, as shown by P R and S T, will be the shape of the pattern of the end piece.

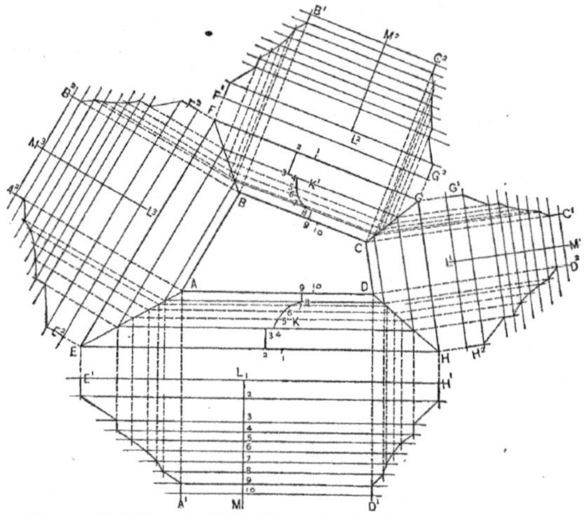


*Fig. 424.—The Patterns of a Molding Mitering Around an Irregular Four-sided Figure.*

547. *The Patterns of a Molding Mitering Around an Irregular Four-sided Figure.*—In Fig. 424, let A B C D be the elevation of an irregular four-sided figure, to which a molding is to be fitted of the profile shown by K and K'. Place the profile in two of the sides, as shown, and construct an elevation of the molding as it would appear when finished, as shown by E F G H. Draw the several miter lines B F, C G, D H and A E. Divide the two profiles into the same number of parts in the usual manner, through the points in which draw lines parallel to the lines of the molding in which they occur, cutting the miter lines, as shown. At right angles to each of the several sides lay off a stretchout from the profile, as shown by L M, L' M', L² M, L³ M'. Through the several points in these several stretchouts draw measuring lines in the usual manner, producing them until they are equal in length to the respective sides, the pattern of which is to be cut. Placing the T-square at right angles to the lines of the several sides, or, what is the same, parallel to the stretchout lines, bring it against the points in the miter lines, cutting the corresponding measuring lines, all as indicated by the

dotted lines. Then the lines traced through these points of intersection will give the several patterns required. Thus $E^1$ $H^1$ $D^1$ $A^1$ will be the pattern of the side E H D A of the elevation, and $H^2$ $D^2$ $C^1$ $G^2$ will be the pattern of the side H D C G, and so on for the others.

548. *The Patterns of Simple Gable Miters.*—In Fig. 425, let A B K R be the elevation of the miters of a cornice at the foot and peak of a gable. The conditions of the elevation are established by the requirements of the work. Let H be the profile of the molding, as shown at the extremity of the horizontal part. Draw the miter line B C, separating the horizontal part from the raking part and the miter line K L at the top. Divide the profile H in the usual manner into any convenient number of equal parts. Place the T-square parallel to the lines in the horizontal molding, and, bringing it successively against the points in the profile, cut the miter line B C, as shown. At right angles to the lines of the horizontal cornice draw the stretchout E F, through the points in which draw the usual measuring lines, as shown. Reverse the T-square, letting the blade lie parallel to the stretchout line E F, and, bringing it against the several points of the profile H, cut the corresponding measuring lines. Then a line traced through these points of intersection, as shown from G to V, will be the pattern of the end of the horizontal cornice mitering with the return. In like manner, with the T-square in the same position, bring it against the points in the miter line B C, and cut the corresponding measuring lines drawn through the stretchout E F. Then a line traced through the points of intersection thus obtained, as shown by T U, will be the pattern of the end of the horizontal cornice mitering against the raking cornice. At right angles to the lines of the raking cornice draw a duplicate profile, as shown by $H^1$, which divide into any convenient number of equal parts, all as indicated by the small figures. Through these points draw lines cutting the miter line B C, and also the miter line K L at the top. At right angles to the lines of the raking cornice draw the stretchout line $E^1$ $F^1$ equal to the profile $H^1$, through the points in which draw the usual measuring lines, as shown. Place the T-square parallel to this stretchout line, and, bringing it successively against the points in B C and K L, cut the corresponding measuring lines, all as indicated by the dotted lines. Through the points thus obtained trace lines, as indicated by M N and O P. Then M N will be the pattern for the bottom of the raking cornice mitering against the horizontal, and O P will be the pattern for the top of the raking cornice. The pattern shown at G V will also be the pattern for the return mitering against A D of the elevation, it being necessary only to establish its length, which may be done from a plan drawn in connection with the elevation or from actual measurements of the work.

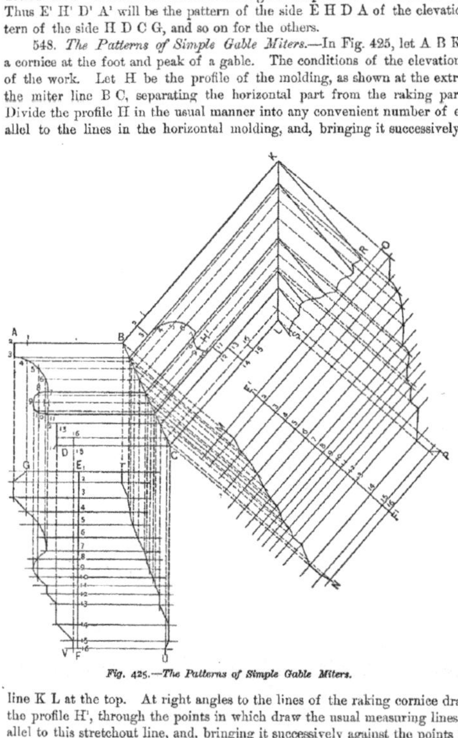

*Fig. 425.—The Patterns of Simple Gable Miters.*

549. *To Ascertain the Profile of a Horizontal Molding Adapted to Miter with a Given Inclined Molding at Right Angles in Plan, and the Several Miter Patterns Involved.*—In the elevation B O E D, and plan

G H K, of Fig. 426, is presented one of the sets of conditions which necessitate a change of profile, in either the horizontal or raking molding, in order to accomplish a miter joint at the point indicated by I H in the plan. In other words, the conditions are such that with a given profile, as shown by A' in the raking molding, the horizontal molding forming the return will require to be modified, as shown by the profile A², in order to form a miter upon the line I H in the plan; or, if A² is established, A' will have to be constructed to correspond with A². The reason for this is quite obvious. The distance across the raking molding at right angles to its lines is greater than the corresponding distance across the return molding at right angles to its lines; therefore the projection in the cornice, as shown by the profile A², must be distributed through a smaller space than is shown in the profile A'. In this problem we assume that the pitch of the raking cornice B C is established and that the profile A is given, and from these parts it is required to develop the modified profile. We have the choice of placing the normal profile in the horizontal return and making the raking profile correspond with it, or of placing the normal profile in the raking molding and making the profile of the horizontal molding agree with it. Although the principle upon which these operations is performed is identical in both, the demonstration will be made clearer if each is fully illustrated independent of the other. In this problem and the following one, therefore, we show the several steps necessary to take in modifying the profile, and in cutting the several patterns required to form the structure indicated by the elevation and plan. First we will assume that the normal profile occurs in the raking cornice, and that the horizontal profile is to be modified to suit it. We then proceed as follows: Draw a representation of the normal profile in the raking cornice, as shown by A', placing it to correspond to the lines of the cornice, as shown. Draw another profile corresponding to it in all parts, directly above or below the foot of the raking cornice, in line with the face of the new profile to be constructed, placing this profile A so that it shall correspond with the lines of the horizontal cornice. Divide the profiles A and A' into the same number of parts, and through the points thus obtained draw lines, those from A' being parallel to the lines of the raking cornice, and those from A intersecting them vertically. Through these points of intersection trace a line, which gives the modified profile, as shown by A². Then A² is the profile of the horizontal return, indicated by G H I F in the plan. It is also the elevation of the miter line I H of the plan for the several patterns involved. We therefore proceed as follows: At any convenient point at right angles to the lines of the raking cornice lay off the stretchout M N of the profile A', through the points in which draw measuring lines in the usual manner. Place the T-square at right angles to the lines of the raking cornice, and,

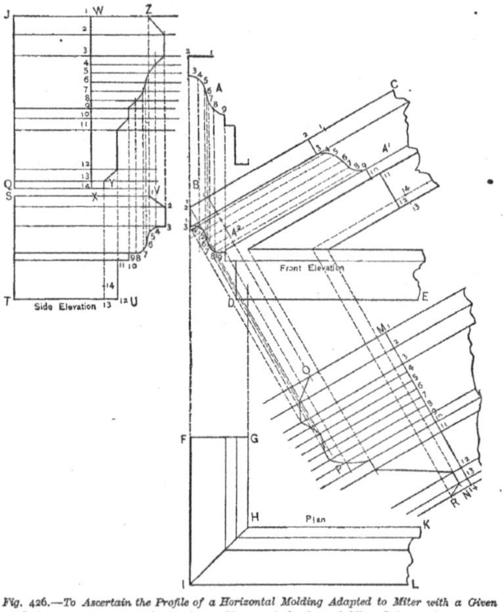


*Fig. 426.—To Ascertain the Profile of a Horizontal Molding Adapted to Miter with a Given Inclined Molding at Right Angles in Plan, and the Several Miter Patterns Involved.*

bringing it successively against the points in the profile A³, cut the corresponding measuring lines just described. Through the points of intersection trace a line, as shown by O P R. Then O P R will be the shape of the lower end of the raking cornice mitering against the return. For the pattern of the return proceed as follows. Construct a side elevation of the return, as shown by S V U T, making the profile V U to correspond to the profile A³ of the elevation, all as shown by B D. Let the length of the return correspond to the return as shown in the plan F G I. In the profile V U set off points corresponding to the points in the profile A³, as shown from B to D. At right angles to the elevation of the return lay off a stretchout of V U, or, what is the same, of the profile A³, as shown by W X, through the points in which draw measuring lines in the usual manner. Placing the T-square parallel to this stretchout line, and bringing it successively against the points in V U, cut the corresponding measuring lines. Then a line traced through these points of intersection, as usual, from Y to Z, will be the pattern of the horizontal return.

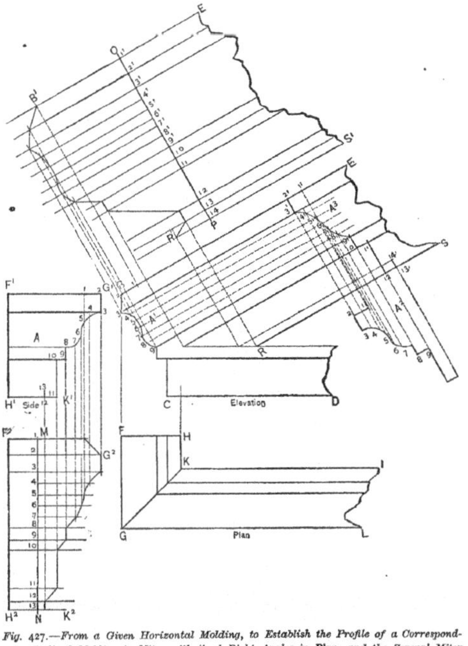


*Fig. 427.—From a Given Horizontal Molding, to Establish the Profile of a Corresponding Inclined Molding to Miter with it at Right Angles in Plan, and the Several Miter Patterns Involved.*

550. *From a Given Horizontal Molding, to Establish the Profile of a Corresponding Inclined Molding to Miter with it at Right Angles in Plan, and the Several Miter Patterns Involved.*—The conditions shown in this problem are similar to those in the one just demonstrated. In this, however, the normal profile is given to the horizontal return, and the profile or the raking cornice is modified to correspond with it. To obtain the new profile we proceed as follows: Divide the normal profile A¹, Fig. 426, into any convenient number of parts in the usual manner, and from these points carry lines parallel to the lines of the raking cornice indefinitely. At any convenient point outside of the raking cornice, and at right angles to its lines, construct a duplicate of the normal profile, as shown by A², which divide into like number of spaces. With the T-square at right angles to the lines of the raking cornice, and brought successively against the several points in this profile, cut corresponding lines drawn through the cornice from the profile A¹. Then a line traced through these points of intersection, as shown by A³, will be the profile of the raking cornice. For the pattern of the foot of the raking cornice mitering against the return, take the stretchout of the profile A³ and lay it off on any line at right angles to the raking cornice, as shown by P O. Through the points in this stretchout line draw the usual measuring lines, as shown. With the T-square at right angles to the lines of the raking cornice, or, what is the same, parallel to the stretchout line, and, bringing it successively against the points in the profile A¹, which is also an elevation of the miter, cut the measuring lines drawn through the stretchout P O. Then a

line traced through the points of intersection, as shown by $B^1 R^1$, will be the miter pattern of the foot of the raking cornice. For the pattern of the return proceed as follows: Construct an elevation of the return, as shown by $F^1 G^1 K^1 H^1$, in dimensions making it correspond to F G K H of the plan. Space the profile A of the elevation of the return in the same manner as $A^1$. At right angles to the lines in the return cornice draw any straight line, as M N, on which lay off a stretchout of the profile A, through the points in which draw measuring lines in the usual manner. With the T-square at right angles to the lines of the return cornice, and bringing it successively against the points in the profile A, cut the corresponding measuring lines. In like manner draw a line corresponding to $F^1 H^1$ of the side elevation. Through the points of intersection obtained from the profile trace a line, as shown by $G^2 K^2$. Then $F^2 G^2 K^2 H^2$ will be the pattern of the horizontal return to miter with the raking cornice, as described.

551. *In a Broken Pediment to Ascertain the Profile of the Horizontal Return at the Top, Together with its Miters.*—Still another change of profile in connection with gable and pediment cornices occurs in constructions commonly known as "broken pediments." Whether the normal profile be placed in the horizontal return at the foot of the gable or in the raking cornice, a third profile is to be constructed by which to cut the patterns and establish the shape of the return occurring at the top. In Fig. 428, C B D represents a section of a broken pediment, of which the normal profile is $A^1$. The profile for the return cornice at the foot of the gable, as shown by B C, is to be obtained by the rule just explained in Fig. 427. The profile for the return at the top of the raking cornice, as shown by $A^2$, is to be obtained in the following manner. Divide the profile $A^1$ of the raking cornice into any convenient number of parts in the usual manner, and through these points draw lines parallel to the lines of the cornice indefinitely. At any convenient point outside of the cornice, and in a vertical line with the point at which the new profile is to be constructed, draw a duplicate of the profile of the raking cornice, as shown by A, which space into the same number of parts as $A^1$, already described. From the points in A draw lines vertically, intersecting lines drawn from $A^1$. Then a line traced through these several points of intersection, as shown by $A^2$, will constitute the profile of the horizontal return at the top and also the miter line as shown in elevation. As before remarked, it matters not whether the normal profile occurs in a horizontal cornice at the base or in a raking cornice, a change still remains to be made at the top. In which-

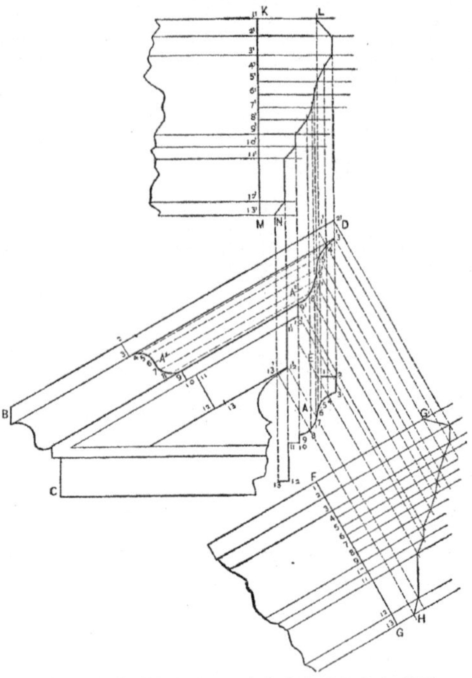


*Fig. 428.—In a Broken Pediment to Ascertain the Profile of the Horizontal Return at the Top, Together with its Miters.*

ever way the profile occurs the steps to be taken are the same as above described. If the normal profile were in the horizontal cornice at the foot of the gable and the modified profile in the position of $A^1$, it would be immaterial whether the normal profile or a duplicate of the modified profile were in the place of A by which to obtain the intersecting lines. It is obvious that it can make no difference which is employed, for what we have to deal with is the projection only, which is the same in both cases. In this connection it may be remarked that the normal profile may be located in the horizontal return at the top of the other profiles, established by working from or reversing the several steps here described. We are led to allude to the possible modifications of the plan here suggested by reason of the demands made upon pattern cutters in cornice work, owing to the whimsicalities of modern designers. For the patterns of the several parts shown in the elevation proceed as follows: At right angles to the lines of the raking cornice lay off a stretchout of the profile of the raking cornice A, as shown by F G, through the points in which draw measuring lines in the usual manner. Place the T-square at right angles to the lines of the raking cornice, and, bringing the blade successively against the points in the profile $A^2$, which is also the miter line in the elevation, cut the corresponding measuring lines, and through these points of intersection trace a line, as shown by G H. Then G H will be the pattern of the top of the raking cornice to miter against the horizontal return. For the horizontal return the usual method would be to construct an elevation of it in a manner similar to that described for the return at the foot of the gable in the preceding demonstrations; the equivalent of this, however, can be done in a way to save a considerable portion of the labor. Draw the line K M perpendicular to the lines of the horizontal return, as it would be if shown in elevation. Upon K M lay off a stretchout of the profile $A^1$, all as shown by the small figures, and through the points draw the usual measuring lines. With the T-square parallel to the stretchout line K M, bring the blade successively against the points in the profile $A^2$, which is also the miter line in elevation, cutting the corresponding measuring lines. Through these points of

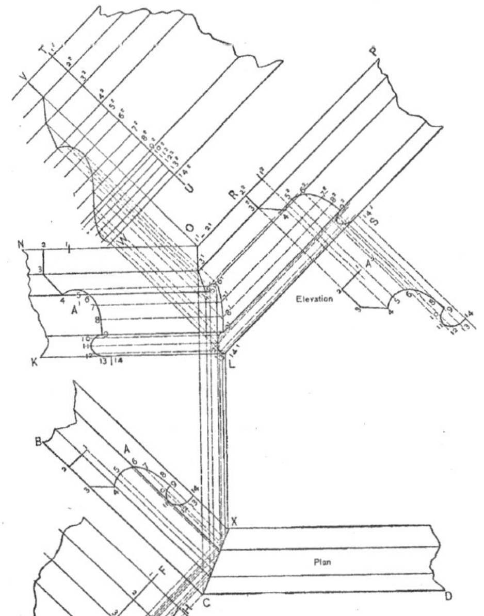


*Fig. 429.—From a Given Horizontal Profile, to Establish the Profile for a Corresponding Inclined Molding to Miter with it at an Octagon Angle in Plan, and the Several Miter Patterns Involved.*

intersection trace a line, as shown by N L, which will be the pattern of the end of the horizontal return to miter against the gable cornice, as shown.

552. *From a Given Horizontal Profile, to Establish the Profile for a Corresponding Inclined Molding to Miter with it at an Octagon Angle in Plan, and the Several Miter Patterns Involved.*—Another example wherein is required a change of profile in order to produce a miter between the parts is shown in Fig. 429. In this case the angle shown in plan between the abutting members is that of an octagon, as indicated by B C D. To produce the modified profile and to describe the patterns we proceed as follows: In the side B C draw the profile A, as indicated, and in the corresponding side, as shown in elevation by N O L K, draw a duplicate profile, as shown by A¹. Divide both of these profiles into the same number of parts, and from the points carry lines parallel to the lines of molding in the respective views. Then produce the lines drawn from A until they meet the miter line C X. From the points thus obtained in C X carry lines vertically until they meet those drawn through A¹, intersecting in points as shown from O L. Through these points of intersection draw the line O L, which will be the miter line in elevation corresponding to the miter line C X of the plan. From the points in O L carry lines up the raking molding in the direction of P M indefinitely. At any convenient point outside of the raking cornice draw a duplicate of the normal profile, as shown by A², placing its vertical line at right angles to the lines of the raking cornice. Divide the profile A² into the same number of spaces as employed in A and A¹, and from these points carry lines at right angles to the lines of the raking cornice, intersecting those drawn from the points in O L. Trace a line through these intersections, as shown from R to S. Then R S will be the required profile of a raking cornice to miter against a level cornice of the profile A at an angle indicated by C X in the plan, or an octagon angle. For the pattern of the level cornice, at right angles to the arm B C in the plan lay off a stretchout of the profile A, as shown by E F, through the points in which draw the usual measuring line. With the T-square at right angles to B C, bringing the blade successively against the several points in X C, cut corresponding measuring lines drawn through E F. Then a line traced through these points, as shown from H to G, will be the required pattern of the horizontal cornice. In like manner, for the pattern of the raking cornice, at right angles

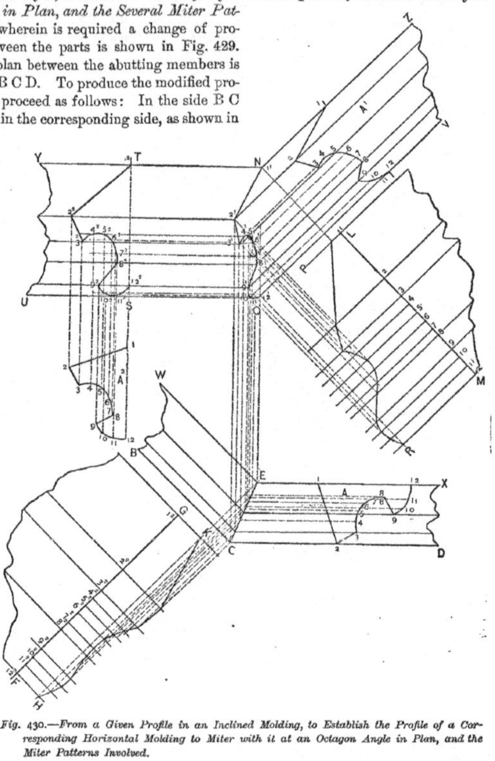


*Fig. 430.—From a Given Profile in an Inclined Molding, to Establish the Profile of a Corresponding Horizontal Molding to Miter with it at an Octagon Angle in Plan, and the Miter Patterns Involved.*

to its lines lay off a stretchout of the profile R S, as shown by U T, through the points in which draw measuring lines in the usual manner. With the T-square at right angles to the lines of the raking cornice, and brought successively against the points in the miter line O L, as shown in elevation, cut the corresponding measuring lines. Then a line traced through the points thus obtained, as shown by W V, will be the required pattern for the raking cornice.

553. *From a Given Profile in an Inclined Molding, to Establish the Profile of a Corresponding Horizontal Molding to Miter with it at an Octagon Angle in Plan, and the Miter Patterns Involved.*—In Fig. 430, let B C D be the angle in plan at which the two sections are to join, and U O V the angle in elevation. To form a miter between moldings meeting under these conditions a change of profile is required. To obtain the modified profile and the miter line in elevation proceed as follows : Draw the normal stay A with its vertical side parallel to the lines in the plan of the arm corresponding to the front of the elevation, all as shown by E X D C. Draw a duplicate of the normal profile in correct position in the elevation, as shown by $A^1$. Divide both of these stays into the same number of parts, and through the points draw lines parallel, in the one case with the lines in the plan and in the other with the lines of the raking cornice, all as indicated by the dotted lines. From the points in the miter line of the plan C E, obtained by the lines drawn through the stay A, carry lines vertically intersecting the lines drawn from $A^1$. Then a line traced through the intersections thus obtained, as shown from N to O, will be the miter line in elevation. From the points in N O carry lines horizontally along the arm of the horizontal molding N O U Y, as shown. At any convenient point outside of this arm, either above or below it, draw a duplicate of the normal stay, as shown by $A^2$, which divide into the same number of parts as before, and from the points carry lines vertically intersecting the lines drawn from N O, just described. Then a line traced through these points of intersection, as shown by T S, will give the modified profile. For the patterns of the parts proceed as follows: For the pattern of the arm Y N O U, at right angles to the same as shown in plan by W E C B, lay off on any straight line, as G F, a stretchout of the profile T S, all as shown by the small figures $1^2$, $2^2$, $3^2$, etc. Through these points draw measuring lines in the usual manner. With the T-square parallel to the stretchout line, and brought against the points of the miter line E C in plan, cut corresponding measuring lines, as indicated by the dotted lines, and through these points of intersection trace a line, as shown by K H. Then K H will be the shape of the end of Y N O U to miter against the raking molding. For the pattern of the raking molding, at right angles to the arm N Z V O in the elevation lay out a stretchout, L M, from the profile $A^1$. Through the points in this stretchout draw measuring lines in the usual manner. Place the T-square parallel to the stretchout line, or, what is the same, at right angles to the arm N Z V O, and, bringing it against the several points in the miter line in elevation N O, cut corresponding measuring lines, as indicated by the dotted lines. Then a line traced through these points of intersection, as shown by P R, will be the shape of the cut on the arm N Z V O to miter against the horizontal molding.

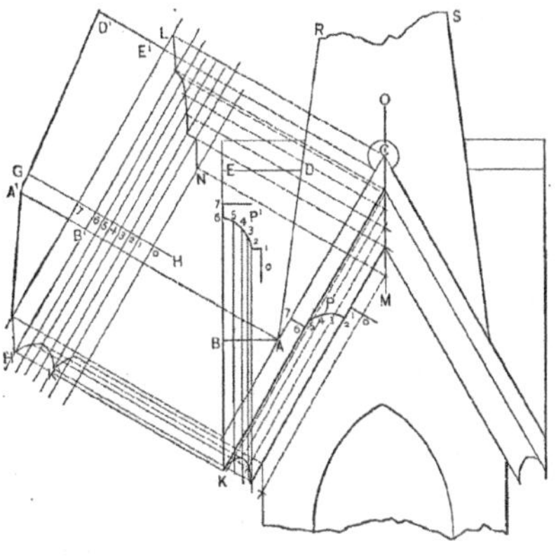


*Fig. 431.—Patterns for the Moldings and Roof Pieces in the Gable of a Square Pinnacle.*

554. *Patterns for the Moldings and Roof Pieces in the Gables of a Square Pinnacle.*—Fig. 431 shows

the elevation of one of four similar gables occurring in a square pinnacle. The profile of the molding is shown by P. The first step is to obtain the miter line shown at K, from which to measure for the pattern. Draw the profile P in the molding, as shown, placing it so that its members will correspond with the lines of the molding. Draw a second profile, $P^1$, in the side view of the gable, placing it, as shown in the engraving, so that its members will coincide with the line of the side view, and also with the first profile already drawn. Space both of these profiles into the same number of parts in the usual manner, and through the points thus obtained draw lines parallel to the lines of the elevation, as shown. Trace a line through these intersections. Then K is the line in elevation upon which the moldings will miter. Draw the miter line O M for the top of the gable, as shown. Upon any line, as G H, drawn at right angles to the line of the gable in elevation, lay off a stretchout of the profile, as shown by the small figures. Through these points draw measuring lines, as shown. Place the T-square parallel to the stretchout line, or, what is the same, at right angles to the line of the gable, and, bringing it successively against the several points in O M and the miter line K, cut the corresponding measuring lines, as shown. Make $E^1$ $D^1$ equal to E D of the side view of the gable, and set it off at right angles to $E^1$ $B^1$. In like manner, at right angles to the same line, set off $A^1$ $B^1$ equal to A B of the side view. Draw the line indicated by $A^1$ $D^1$, as shown, and trace lines through the intersection of points dropped from the elevation on to the measuring lines, thus completing the patterns.

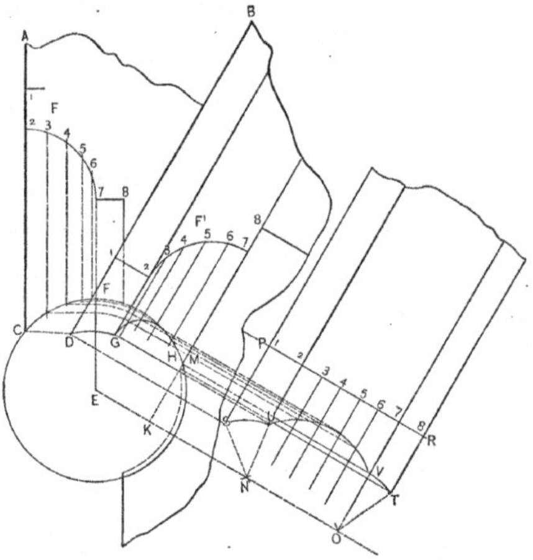

*Fig. 432.—The Pattern for the Miter Between the Moldings of Adjacent Gables Upon a Square Shaft, Formed by Means of a Ball.*

555. *The Pattern for the Miter Between the Moldings of Adjacent Gables Upon a Square Shaft, Formed by Means of a Ball.*—In Fig. 432, let A C be one of the gables in profile and B D the other in elevation, the moldings forming a joint against a ball, the center of which is shown at E. For the patterns we proceed as follows : Place the profile in each gable as shown by F and $F^1$, locating them in such a manner with regard to their respective positions that corresponding points in each shall fall upon the same lines. Divide each of these profiles into the same number of equal parts, as indicated by the small figures. From the points thus obtained in F drop lines vertically, meeting the profile of the ball, as shown from C to F. From the center E of the ball erect a vertical line, as shown by E F. From the points in C F already obtained carry lines horizontally, cutting E F, as shown, and thence continue them, by arcs struck from E as center, until they meet corresponding points dropped from the profile $F^1$ by lines parallel to the gable in elevation. Through the intersections thus obtained trace a line, as indicated by G H. Then G H will be the miter line in elevation. At right angles to the gable lay off a stretchout of the profile at any convenient place, as shown by P R, through the points in which draw the usual measuring lines. Place the T-square parallel to the stretchout line, or, what is the same, at right angles to the lines of the gable, and, bringing it successively against the points in the miter line G H, cut the corresponding measuring lines. Since the surface against which the two moldings miter is that of a sphere, the pattern representing the space between the points 1 and 2 of the profile, and also between 7 and 8 of the profile, will necessarily be an arc of a circle. Therefore in the pattern the line running from S to U, and also the line from V to T, must be struck from centers which are to be found. By inspection of the elevation it will be seen that the space S U is equal to that of D G struck from the center E. Set the dividers,

therefore, to E D or E G of the elevation, and from S and U respectively as centers, strike arcs, which will be found to intersect at N. Then N is the center by which to describe the arc S U. By further inspection it will be seen that the lines corresponding to 7 and 8 of the stretchout meet the profile of the ball at M. Continue this line indefinitely in the direction of K. From E, at right angles to it, draw the line E K. Then K M is a radius of the arc to be described from V to T. Set the dividers to K M, and from V and T respectively as centers, strike arcs which will intersect in the point O. From O, with the same radius, describe the arc V T. Trace a line through the points from U to V. Then S U V T is the pattern for the gable molding to fit against the ball, as shown.

556. *The Miter Between the Moldings of Adjacent Gables upon a Square Shaft, the Gables being of Different Pitches.*—The problem presented in Figs. 433 and 434 is one occasionally arising in what is known as pinnacle work among cornice makers. The figures represent the side and end elevations of a pinnacle which is rectangular, but not square. Each of its faces is finished with a gable, which miter against each other at the eaves, and which are of the same hight in the line of their ridges, as indicated by L M and L' M'. Whatever profile is given to the molding in one face of such a structure, the profile of the gable in the adjacent face will require some modification in order to form a miter. Let A be the profile of the molding in the side. For the miter line in elevation and the pattern we proceed as follows : Draw a duplicate of A, placing it in a vertical position directly below or above the point at which the two moldings are to meet, as shown by A'. Divide both of these profiles into the same number of parts, as indicated by the small figures, and through these points draw lines intersecting in the points from H to K. Then a line traced through these intersections, as shown by H K, will be the miter line in elevation. At right angles to the lines of the molding, as shown in elevation, lay off a stretchout of the profile A, as shown by B C, through the points in which draw the usual measuring lines. Place the T-square at right angles to the lines of the molding, or, what is the same, parallel to the stretchout line, and, bringing it against the several points in the miter line H K, cut corresponding measuring lines. Then a line traced through these points, as shown by D E, will be the shape of the cut at the foot of the side gable to miter against the adjacent gable. For the pattern of the piece to miter against the one just obtained, and belonging to the adjacent end, transfer the profile H K, reversing it, as shown by K' H' in Fig. 434, or, in lieu

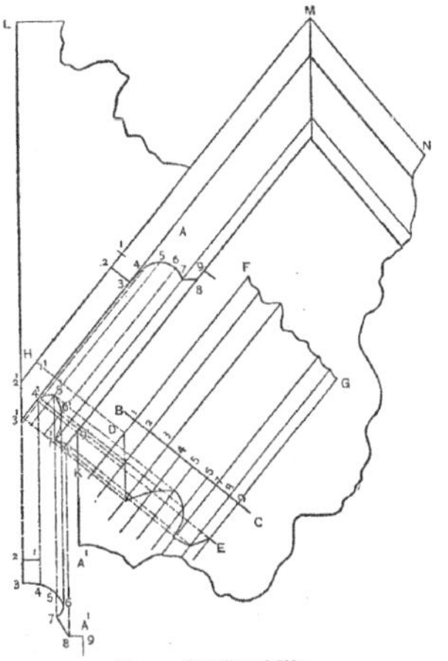


Fig. 433.—Elevation of Side.
*The Miter Between the Moldings of Adjacent Gables Upon a Square Shaft, the Gables being of Different Pitches.*

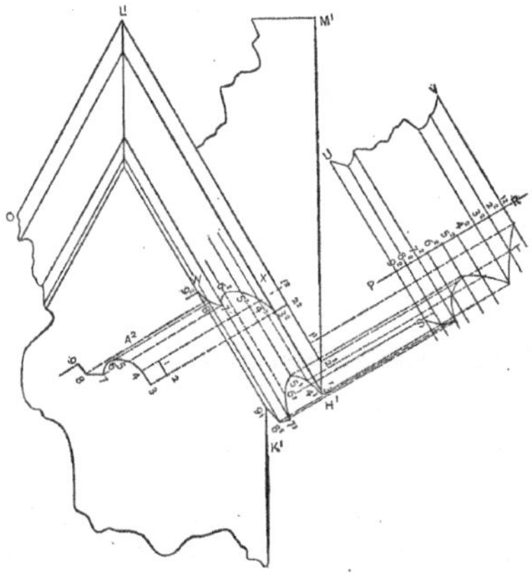


Fig. 434.—Elevation of End.
*The Miter Between the Moldings of Adjacent Gables Upon a Square Shaft, the Gables being of Different Pitches.*

of this, repeat the operation by which H K was obtained. At right angles to the line of the raking cornice in the end elevation, draw a duplicate of the normal profile, as shown by A², which divide into the same number of equal parts as in the other case, and through the points carry lines across the cornice, as shown, intersecting

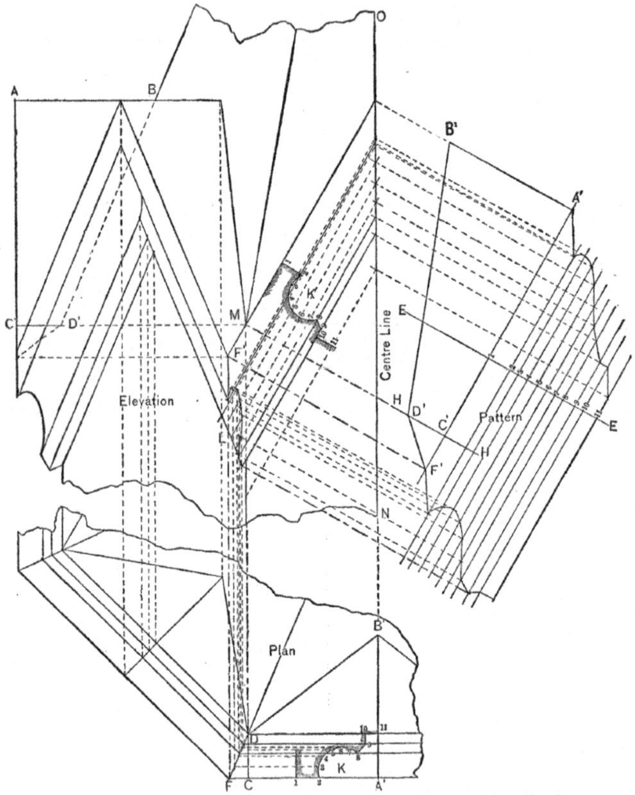


*Fig. 435.—Pattern for the Moldings and Roof Pieces in the Gables of an Octagon Pinnacle.*

these lines by lines drawn parallel to the lines of the cornice through the points in K' H'. Then a line traced through these points of intersection will form the modified profile, as shown by W X. For the pattern we proceed as follows: At right angles to the lines of the raking cornice lay off a stretchout of the profile W X, as shown by P R, through the points in which draw measuring lines in the usual manner. With the T-square at right angles to the lines of the raking cornice, or, what is the same, parallel to the stretchout line P R, bringing it successively against the points in K' H', cut corresponding measuring lines. Then a line traced through these points of intersection, as shown from S to T, will be the pattern for the foot of the gable cornice on the end elevation.

557. *Pattern for the Moldings and Roof Pieces in the Gables of an Octagon Pinnacle.*—Fig. 435 shows a partial elevation and a portion of the plan of an octagon pinnacle having equal gables on all sides. The first step in developing the patterns is to obtain a miter line at the foot of the gable, as shown by L. To do this proceed as follows: Draw the profile K, as shown, placing it so that it shall correspond in all its parts with the lines of the molding in elevation. Number spaces in it in the usual manner, as shown, and through the points draw lines parallel to the lines of the gable toward L, as shown. In the plan place a duplicate stay or profile, K, so drawn that its parts shall in all respects correspond to the position of those of the profile in the elevation. Divide it into the same number of spaces, and through the points in it draw lines parallel to the lines of the plan, cutting the line D F, representing the plan of the miter. From the points in D F thus obtained carry lines vertically, intersecting corresponding lines drawn through the profile in the elevation. A line traced through the several points of intersection, as shown by L, will be the line of miter in elevation between the moldings of the adjacent gables. The center line O N forms the miter line for the top of the gable. For the pattern proceed as follows: Upon any line, as E E, drawn at right angles to the lines of the gable, lay off a stretchout of the profile, as shown by the small figures. Through the points of the stretchout draw the usual measuring lines. Place the T-square at right angles to the lines of the gable, and, bringing the blade successively against the points in the two miter lines above described, cut the corresponding measuring lines, as shown. Lines traced through the points of intersection thus obtained will give the pattern of the molding. The roof piece may be added by setting off A' B' at right angles to A' C', equal in length to A B of the side view. In like manner set off D' C' equal to C D of the side view. Then draw F' D' B', thus completing the pattern.

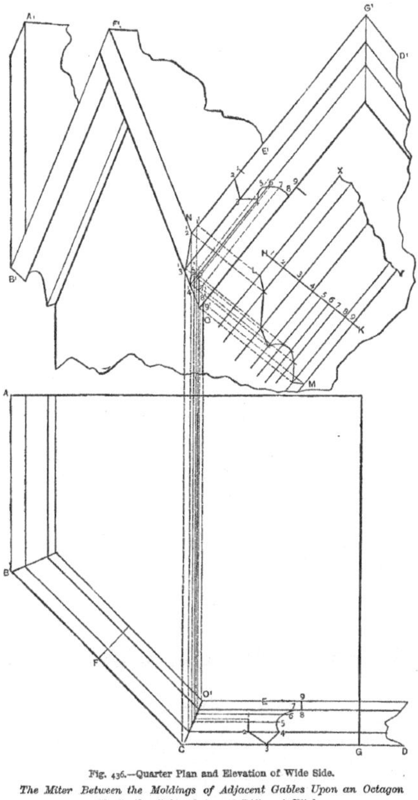


Fig. 436.—Quarter Plan and Elevation of Wide Side.
*The Miter Between the Moldings of Adjacent Gables Upon an Octagon Shaft, the Gables being of Different Pitches.*

558. *The Miter Between the Moldings of Adjacent Gables Upon an Octagon Shaft, the Gables being of Different Pitches.*—The problem illustrated in Figs. 436 and 437 resembles that presented in Section 556, save

that the angle is not a right angle. The elevations represent an octagon pinnacle of unequal sides, and the problem is to cut the miter at the eaves occurring between adjacent gables of the same hight, but of different widths. Let A' B' F' O G' D' of Fig. 436 be a correct elevation, and A B C G be a quarter plan of the structure. In that portion of the plan corresponding to the part of the elevation shown to the front, draw the normal profile E, placing its vertical side parallel to the lines of the plan. Divide it into any convenient number of spaces, and through these points draw lines parallel to the lines of the plan, cutting the miter line C O', as shown. In like manner place a duplicate of the normal profile, as shown by E' in the elevation. Divide it into the same number of equal parts, and through the points draw lines parallel to the lines of the raking cornice, which produce in the direction of N O indefinitely. Bring the T-square against the points in C O', and with it erect vertical lines, cutting the lines drawn through E', as shown from N to O. Then a line, N O, traced through these points of intersection will be the miter line in elevation. For the pattern of the foot of the gable shown in elevation proceed as follows: At right angles to the lines of the gable cornice lay off a stretchout of the profile E', as shown by H K, through the points in which draw the usual measuring lines. Placing the T-square at right angles to the lines of the cornice, or, what is the same, parallel to the stretchout line, and bringing it against the several points in N O, cut corresponding measuring lines. Then a line traced through the points of intersection thus obtained, as shown from L to M, will be the pattern of the foot of the gable shown in elevation. For the modified profile and the pattern of the gable piece forming the narrow side, proceed as follows: Transfer the miter line N O, reversing it, to the foot of the gable of the narrow side, as shown by R P in Fig. 437, and through the points carry lines parallel to the lines of the gable cornice indefinitely, as shown. Draw a duplicate of the normal profile at any convenient point outside of the gable cornice, as shown by E', placing its vertical side at right angles to the lines of the cornice. Divide E² into the same number of parts as used in the other profiles, and through the points draw lines at right angles to the lines of the cornice, intersecting the lines drawn from P R. Through these points trace a line, as indicated by E², which will be the modified profile. Take the stretchout of E² and lay it off on any straight line drawn at right angles to the lines of the cornice, as S T, and through the points in it draw the usual measuring lines. Place the T-square at right angles to the lines of the gable cornice, and, bringing it against the points in P R, cut the measuring lines, as indicated by the dotted lines. Then a line traced through these points of intersection, as shown by U T, will be the pattern of the foot of the side gable.

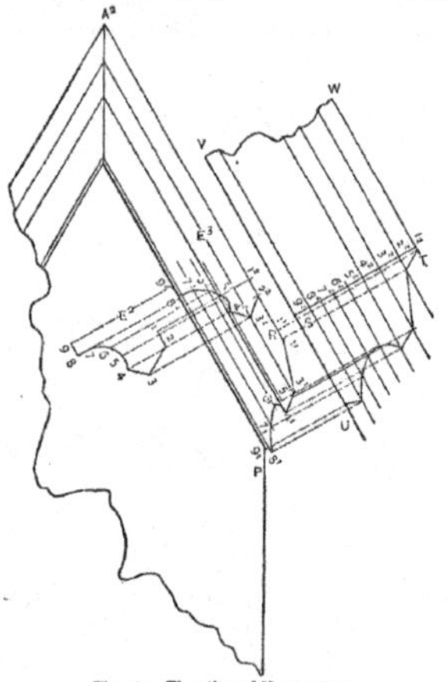


Fig. 437.—Elevation of Narrow Side.
*The Miter Between the Moldings of Adjacent Gables Upon an Octagon Shaft, the Gables being of Different Pitches.*

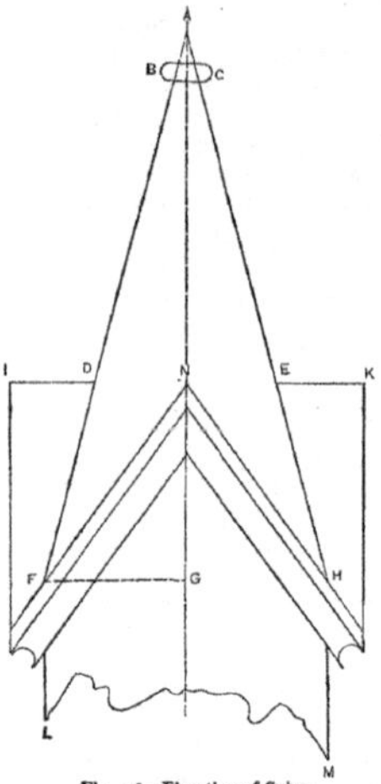


Fig. 438.—Elevation of Spire.
*The Pattern of a Square Spire Mitering Upon Four Gables.*

559. *The Pattern of a Square Spire Mitering Upon Four Gables.*—In Fig. 438, let B F H C be the elevation of a square spire which is required to miter over four equal gables in a pinnacle, the plan of which is also square. Produce D B and E C until they meet in A, which will be the apex of the pyramid of which the

spire is a section. Draw the axis A G, and at right angles to it, opposite the lowest point of contact between the spire and the gable, as F, draw F G. Then F G will represent the half width of one of the sides of the pyramid at the base, and A F will represent the length of a side through the center. From any convenient point, as $A^1$ in Fig. 439, draw $A^1$ $F^1$, in length equal to A F. From $F^1$ set off, perpendicular to $A^1$ $F^1$, on each side a space equal to F G of the elevation, as shown by $F^1$ $G^1$ and $F^1$ $G^2$. From $G^1$ and $G^2$ draw lines to $A^1$, as shown. From $A^1$ as center, and with $A^1$ $G^2$ as radius, describe an arc, as shown by $G^2$ O, in length equal to three spaces of the extent of $G^1$ $G^2$, as shown by $G^2$ *g*, *g* $g^1$ and $g^1$ O. Draw $g^1$ $A^1$, *g* $A^1$ and O $A^1$. Make $A^1$ $B^1$ equal to A B of the elevation, and through $B^1$ draw a perpendicular to $A^1$ $F^1$, as shown. Draw lines corresponding to it through the other sections of the pattern. Make $A^1$ $D^1$ equal to A D, and draw $D^1$ $G^1$ and $D^1$ $G^2$. Set the compasses to $G^2$ $D^1$, and from $G^2$ and *g* as centers, describe arcs intersecting at *d*. Draw *d g* and *d* $G^2$, as shown. Repeat the same operation in the other sections of the pattern, thus completing the required shape.

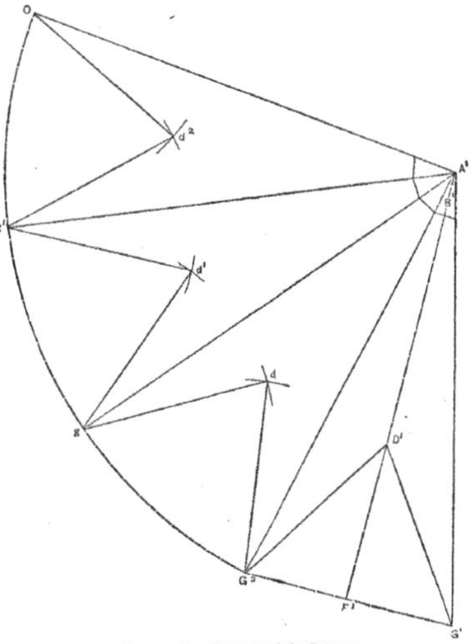


Fig. 439.—Development of the Pattern.
*The Pattern of a Square Spire Mitering Upon Four Gables.*

560. *The Pattern of a Conical Spire Mitering Upon Four Gables.*—Let D B O in Fig. 440 be the elevation of a pinnacle having four equal gables, down upon which a conical spire is required to be mitered, as shown. Produce the sides of the spire until they meet in the apex D. Also continue the side E F downward to any convenient point below the junction between the spire and the gables, as shown by H, which point is to be considered the base of the spire. Let $B^2$ K L M be the plan of the pinnacle. The diagonal lines $B^2$ L and M K represent the angles between the gables, while R S and T U represent the ridges of the gables over which the spire is to be fitted. Through the point H in the elevation draw a line to the center of the cone, and at right angles to the axis, as shown by H C. This will represent the half diameter or radius of the spire at the base. With radius C H, and from center $A^2$ of the plan, describe a circle, as shown, which will represent the spire in plan. At any convenient distance from the elevation, and to one side, construct a diagonal section corresponding to the line $B^2$ $A^2$ in the plan, as shown. Draw $D^1$ $A^1$ corresponding to the axis of the spire prolonged. At right angles to it, opposite the point B, set off $A^1$ $B^1$ equal to $A^2$ $B^2$ of the plan. Opposite F of the elevation establish $F^2$ in the line $D^1$ $A^1$, and draw $F^2$ $B^1$. Draw $C^1$ $H^2$ equal to and opposite C H in elevation. Draw $H^2$ $D^1$, cutting $B^1$ $F^2$ in $G^2$. From $G^2$, at right angles to the axis, carry a line across the elevation, thus establishing the point G in the side of the spire. To describe the pattern, draw any straight line, as $D^2$ $H^1$, Fig. 441, in length equal to the side D H of the spire, and in it set off points corresponding to the points marked on the side of the spire. Thus make $D^2$ $E^1$ equal to D E, $D^2$ $F^1$ equal to D F, and $D^2$ $G^1$ equal to D G of the elevation. From $D^2$ as center, with $D^2$ $E^1$ as radius, describe the arc $E^1$ $I^1$ indefinitely. From the same center, with $D^2$ $H^1$ as radius, describe the arc $H^1$ V indefinitely. Divide one-quarter of the plan of cone into any number of equal spaces, and set off corresponding spaces from $H^1$ on the arc, as shown. Through the last of these points, as shown by 9, draw a line to the center $D^2$, which in the pattern will be the same as R $A^2$ of the plan. In like manner, through the middle point, draw a line, as shown by 5 $D^2$, which will correspond to $B^2$ A of the plan. From $D^2$, with $D^2$ $F^1$ as radius, describe the arc $F^1$ X, and from the same center, with $D^2$ $G^1$ as radius, describe the arc $G^1$ W. Then the arc drawn from $F^1$ will represent points which meet the tops of the gables, and that drawn from $G^1$ will represent the points in the angles between. Then the lines drawn from $g^2$

to F' and $g$ to $f$ will be the required cut. On the arc H' V step off additional spaces, corresponding to the stretchout of a quarter of the plan of the cone, as shown from H' to 9, making four in all, and also mark the middle point (5) in each. Draw the line V D², and also the corresponding intermediate lines. Complete the pattern by drawing the diagonal lines corresponding to $g$ $f$ and $g$ F' already described, all as shown in the engraving.

561. *The Pattern of an Octagon Spire Mitering Upon Four Gables.*—In Fig. 442, let B E Z C be the elevation of an octagon spire, mitering down upon four gables occurring upon a square shaft. Continue the side lines until they intersect in the apex A. Draw the center line A H, from which set off the perpendicular H G, which shows the half width of one of the sides at the point G. Continue the side B E in the direction of F. Draw P F at right angles to the axis A H of the spire, thus establishing the point F, which shows the length of the sides fitting into the angle between the gables of the shaft. Draw A' F' in Fig. 443 equal to A F of the elevation, and set off points on it corresponding to points in A F. Thus make A' B' equal to A B, A' D' equal to A D, and A' E' equal to A E of the elevation, etc. Through E' draw a perpendicular equal in length to the width of a side at the point E, or double to G H, as shown in the elevation, placing one-half on each side from E', all as shown by L K. From L and K draw lines to A', as shown. From A' as center, with A' L as radius, describe an arc, as shown by L U, indefinite in length. Set the dividers to the space L K, and step off spaces from L, as L Y, Y X and X U, until as many sides are set off as are required in one piece—in this case four. Draw the lines A' Y, A' X and A' U. By inspection of the elevation it will be seen that one-half the sides will be notched at the bottom to fit over the gables, while the others will be pointed to reach down into the angle between the gables. From the point D', which, as will be seen by D in the elevation, corresponds to the top of the

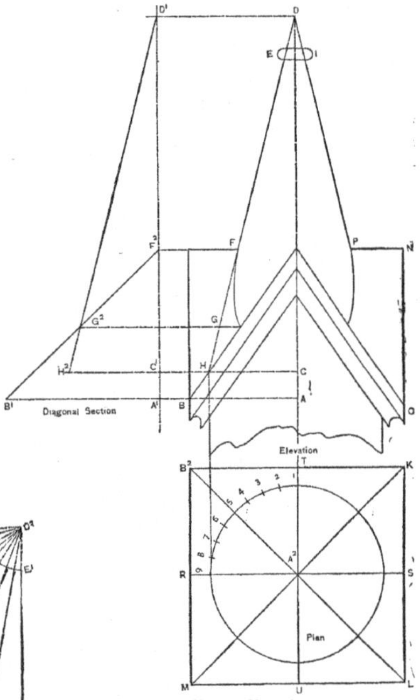


Fig. 440.—Elevation, Plan and Section.
*The Pattern of a Conical Spire Mitering Upon Four Gables.*

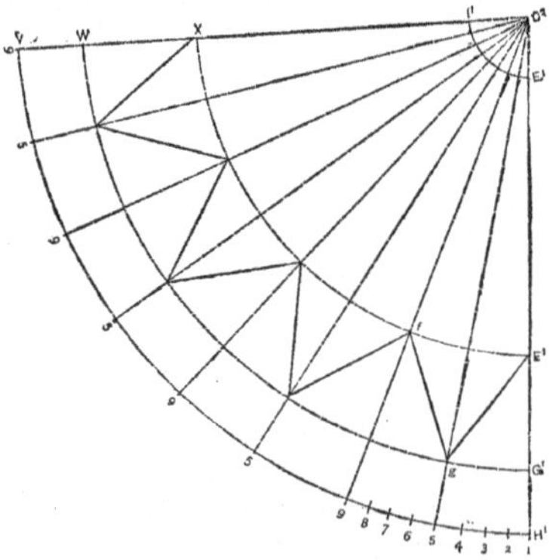

Fig. 441.—Development of the Pattern.
*The Pattern of a Conical Spire Mitering Upon Four Gables.*

gable, draw lines to the points L and K, which gives the pattern for the notch in the first section. Set the dividers to L D' as radius, and from X and Y as centers, describe arcs intersecting at W. Draw W X and W Y. For the pattern of the point place one leg of the dividers at L, and, bringing the other to F', describe an arc, as shown by F' M. With the same radius, from Y as center, describe a second arc intersecting the first at M. Draw M Y and M L. Using the same radius, and U and X as centers, establish the point V. Draw V X and V U, thus completing the pattern.

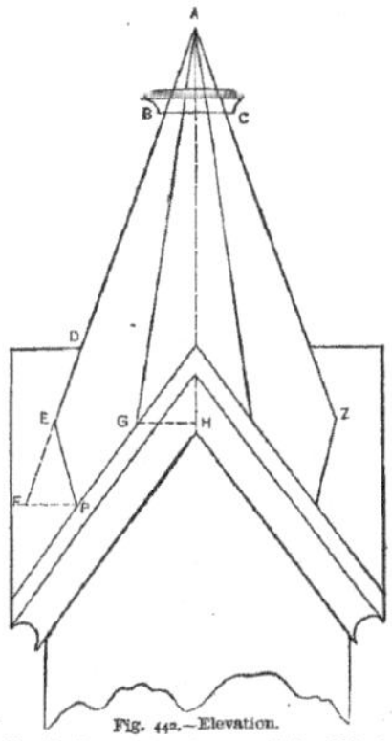


Fig. 442.—Elevation.
*The Pattern of an Octagon Spire Mitering Upon Four Gables.*

562. *The Pattern of an Octagon Spire Mitering Upon Eight Gables.*—Let A C I in Fig. 444 be the elevation of the spire, and M N O P the plan. From the point G, which represents the lowest point of the angle between the gables, to H, which represents the highest point of the gables corresponding to T in the plan, draw the line G H, cutting C Q in the point D. Draw any line, as A' W in Fig. 445, upon which to construct the pattern. Make A' Z equal to A C of the elevation, and A' W equal to A D of the elevation. Through W draw the perpendicular 1 V, as shown. From W set off W V equal to E F of the elevation, and likewise set off W 1 of the same length. Draw A' V and A' 1. Set the dividers to A' 1 as radius, and from A' as center, describe the arc 1 U indefinitely. Set the dividers to 1 V, and step off as many spaces on the arc as it is desired to have sides in the pattern—in this case four—as shown. Draw the lines A' U, A' 3 and A' 2, which represent the lines of bend in the pattern. Draw Z V and Z 1 in the first section of the pattern. Set the dividers to Z V, and from 1 and 2 as centers, describe intersecting arcs, as shown by Z'. In like manner describe similar intersecting arcs at the points Z² and Z³. Draw lines from these points to the points 1, 2, 3 and U, as shown, thus completing the pattern.

563. *The Pattern of a Conical Spire Mitering Upon Eight Gables.* Let E H C N I in Fig. 446 be the elevation of a pinnacle having eight equal gables, upon which the conical spire E F P I is to be fitted. Produce the sides F E and P I until they meet in the point D, which is the apex of the spire. Produce the side E F, continuing it downward until it meets the line of the vertical side of the shaft in G, which point is to be considered as representing the base of the spire. Let A H' S K M N' T U be the plan of the pinnacle, which for convenience draw in line with the shaft. From G in the elevation obtain the point G² in the plan, as shown by the dotted line. With radius B G², from B as center, describe the circle, as shown, which will represent a plan of the spire on a line through its base corresponding to the point G in the elevation. Divide an eighth of this circle into any number of equal parts, as shown by 1, 2, 3, 4 and 5, which spaces are to be used in measuring off the arc describing the pattern further on. To one side of the elevation, construct a diagonal section on the line A B of the plan, as shown. From the point H in the elevation, which corresponds to A in the plan, draw a horizontal line, as shown by B' A', in length equal to B A of the plan. From B' erect a perpendicular, upon which locate the point D', corresponding in hight to D. Mark the point F', corresponding to F of the elevation, and draw A' F'. Set off C' G' equal to and opposite C G in the elevation. Draw D' G', intersecting A' F' in the point R'. Bring the T-square against the point R', the blade being at right angles to the axis of the spire, and draw R' R⁴, cutting the line E G at R. The point R' is of use in drawing the elevation, showing the extreme point of intersection between the spire and gable, but is not essential in cutting the pattern. Draw any line, as D² G², Fig. 447, upon which to lay off the several points in the side of the cone. Make D² E² equal to D E of the elevation, D² F² equal

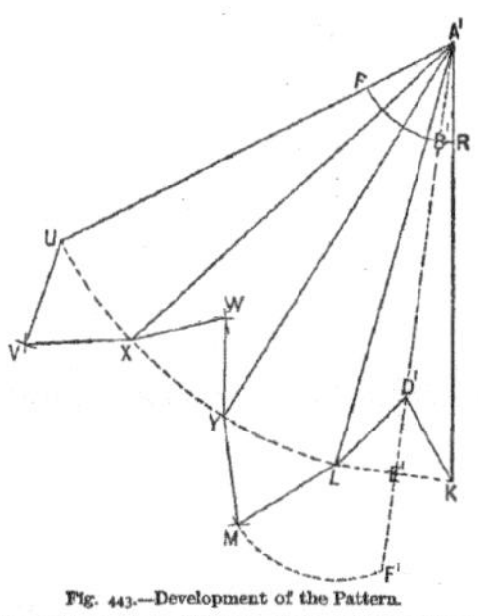


Fig. 443.—Development of the Pattern.
*The Pattern of an Octagon Spire Mitering Upon Four Gables.*

to D F, $D^2$ $R^2$ equal to D R, and $D^2$ $G^2$ equal to D G. From $D^2$ as center, describe the arc $E^2$ $I^2$ indefinitely. Also from the same center, and with $D^2$ $F^2$, $D^2$ $R^2$ and $D^2$ $G^2$ respectively as radii, describe the arcs $F^2$ X, $R^2$ W and $G^2$ V indefinitely. On the arc $G^2$ V step off spaces corresponding to one-eighth of the plan, as shown by 1, 2, 3, 4 and 5, as already described. Draw the line $D^2$ 5, as shown, and also draw a line from $D^2$ to the middle point in the stretchout of the eighth of the circle, as shown by $D^2$ 3. By inspection of the elevation it will be seen that the arc $F^2$ X represents the line at which the gables cut into the cone, and that the arc $R^2$ W represents the line of points in the base of the cone to fit down between the gables. Therefore from $F^2$ draw $F^2$ *g*, and from a point in $D^2$ 5 corresponding to $F^2$, as *f*, draw *f g*. Then $D^2$ *f g* $F^2$ will be one eighth of the required pattern. Set the dividers to 1 5 on the arc $G^2$ V, and step off seven additional spaces, as shown by 5, 5, 5, etc., V. Draw V $D^2$. Also draw the dotted lines $D^2$ 5. From center point in each space, as indicated by 3, draw lines to $D^2$, as shown. Then draw the lines *f g* and *g f*, thus completing the pattern.

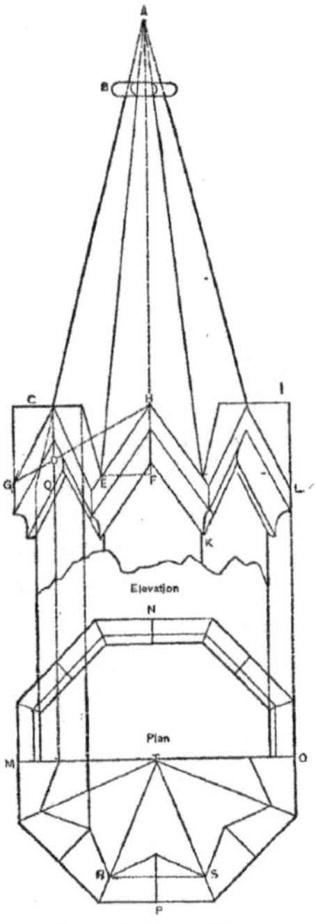


Fig. 444.—Elevation and Plan.
*The Pattern of an Octagon Spire Mitering Upon Eight Gables.*

**564.** *The Gore Piece in a Molding Forming the Transition from an Octagon to a Square.*—In Fig. 448, let A B C D be a half plan of the figure at the outside, and E F G H K L be a half of the inner plan. M O R U P N is an elevation of the article, O R being the normal profile. F B G in the plan represents the transition piece by which the figure is changed from a square to an octagon, the pattern for which is desired. Draw the elevation and plan in such relative position that the profile O R shall fall directly over the miter line B F. Divide O R into any convenient number of spaces, and through these points drop lines cutting B F, all as indicated by the dotted lines. From the points in B F carry lines across the transition piece, parallel to F G, as shown, cutting the miter line G B, and from G B prolong them indefinitely. At right angles to the lines drawn through the transition piece just described, construct a duplicate of the normal profile, as shown by X Y, which divide into the same number of parts as the original profile O R, and from these points, at right angles to the lines drawn across the transition piece, produce lines cutting lines of corresponding numbers. A line traced through the points of intersection thus formed, as shown by $O^1$ $R^1$, will be the profile of the transition piece. At right angles to F G and opposite the point B, draw V $B^1$, upon which lay off a stretchout of the profile $O^1$ $R^1$. Through the points in $B^1$ V draw the usual measuring lines. Place the T-square parallel to this stretchout line, and, bringing it against the several points in the miter lines F B and G B, cut corresponding measuring lines, as shown. Then the lines traced through these points of intersection, as shown by $F^1$ $B^1$ and $G^1$ $B^1$, will be the required pattern.

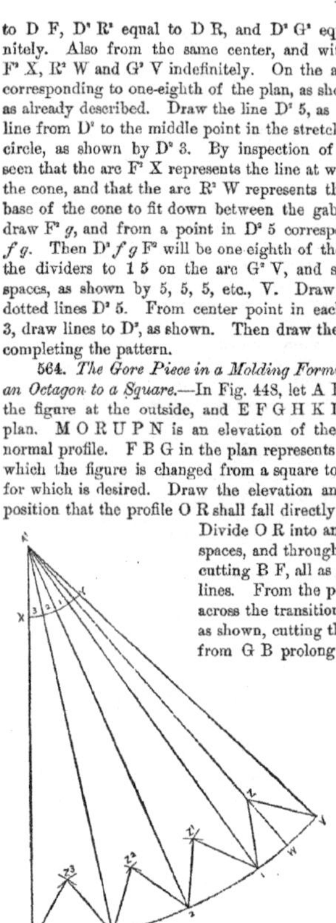


Fig. 445.—Development of the Pattern.
*The Pattern of an Octagon Spire Mitering Upon Eight Gables.*

565. *A Gore Piece Forming the Transition from an Octagon to a Square.*—In Fig. 449, let F F F F represent the square plan of a base, and A A A A a portion of the plan of an octagon shaft which is to be fitted to it. Let D P C be the elevation of the gore piece which is required to form the transition between the two shapes. The rule which follows may be used for whatsoever profile the gore piece is required to be—in the present case an ogee. Since the outline of the gore piece, as shown by C D in elevation, is not a profile of the molding which forms the transition, it will be necessary to construct the profile of it before commencing the pattern. To do this proceed as follows: Divide the profile of the gore piece C D, as it appears in the elevation, in the usual manner, into any convenient number of spaces, as shown by 1, 2, 3, 4, etc. From the points thus obtained drop lines down upon one side of the plan A F, which is to be placed so that corresponding parts of plan and elevation shall agree. Continue the lines from the side A F across the corner, as shown, all parallel to A A of the octagon. Bisect the angle A F A by the line E F, and upon it number the lines drawn across the corner to correspond with the numbers of the points in the elevation, from which they were derived. Draw G H parallel to P D, and at convenient distance from it. Cut G H by lines drawn at right angles to it from the points in the profile C D, as shown by the connecting dotted lines. G H then may

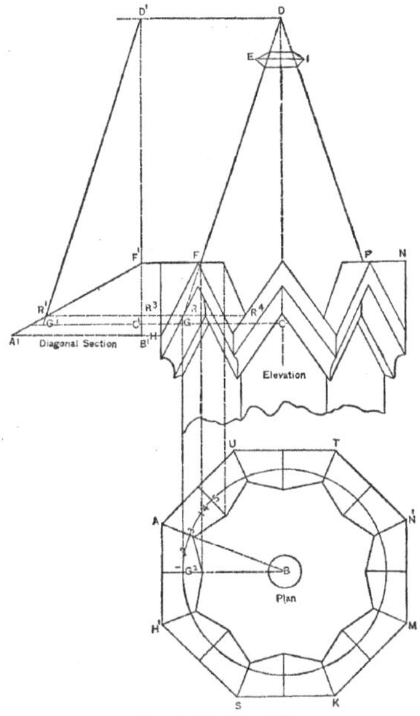


Fig. 446.—Elevation and Plan.
*The Pattern of a Conical Spire Mitering Upon Eight Gables.*

be considered to represent the point F in the plan, or 11 of the numbers on the line E F. From G H, on each of the several lines drawn through it, lay off a distance equal to the space from 11 on E F to the corresponding number on the same line. Thus lay off 11 1 from G H equal to 11 1 on E K, and 11 2 equal to 11 2 of E K, and so on for each of the lines through G H. Then a line traced through these points, as shown by I H, will be the profile of the gore piece, or the shape of its section when cut by the line E F. Prolong E F, as shown by K L, and lay off on the latter a stretchout of the profile I H, the spaces of which must be taken from point to point as they occur, so as to have points in the stretchout corresponding to the points on the miter lines A F, A F, previously derived from C D. Through the points thus obtained draw the usual measuring lines, as shown. Place the T-square at right

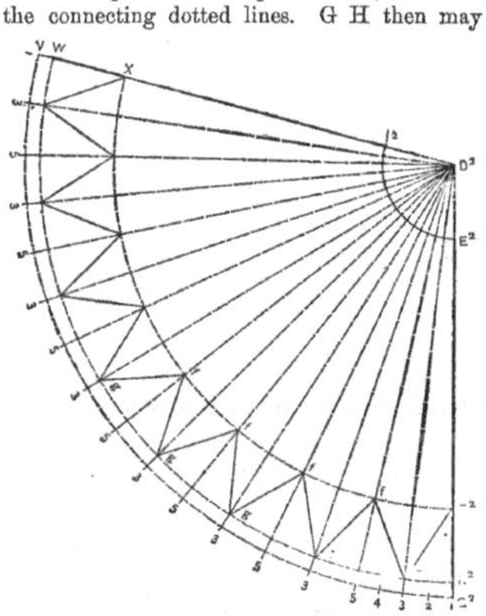

Fig. 447.—Development of the Pattern.
*The Pattern of a Conical Spire Mitering Upon Eight Gables.*

angles to the measuring lines, or, what is the same, parallel to E K, and, bringing it against the points in A F and F A, cut the corresponding lines drawn through the stretchout. Lines traced through these points, as shown, will constitute the pattern.

566. *The Blank for a Curved Molding.*—Figs. 450 and 451 are introduced at this place in order to show the principle upon which blanks for curved moldings are struck. In Fig. 450, A C E D represents the elevation, for example, of a wash basin, in which the sides are made bulging, as shown by the curved line from E to C. If the sides were straight, as shown by the straight line from E to C, the pattern would be easily described; it would be simply the envelope of a section of a right cone. The patterns for curved moldings are cut upon exactly the same principle. The blanks are described in the same manner as though the article was to be formed up of a straight flare and not molded at all, save that additional width is

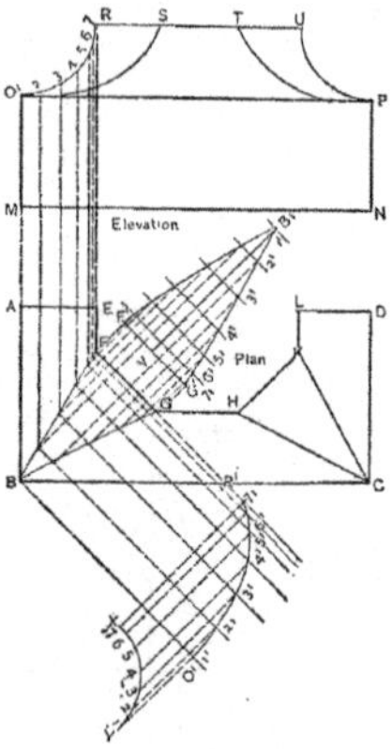


*Fig. 448.—The Gore Piece in a Molding Forming the Transition from an Octagon to a Square.*

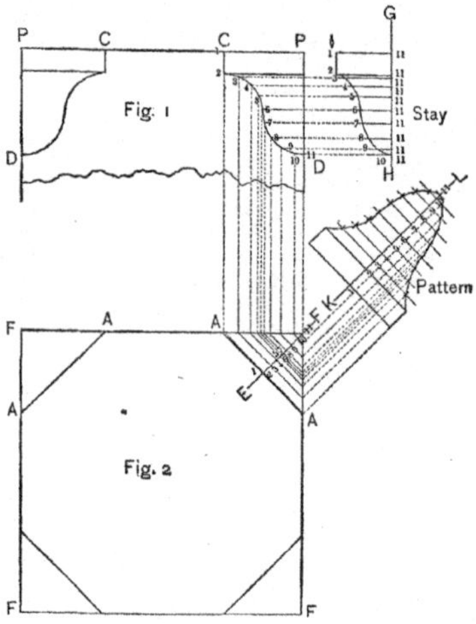


*Fig. 449.—A Gore Piece Forming the Transition from an Octagon to a Square.*

given to it to compensate for the metal which is taken up by the form. Therefore, to describe the pattern of the blank from which to make a curved molding corresponding to the elevation A C E D, proceed in the same manner as though the side E C were to be straight. Through the center of the article draw the line B F indefinitely, and draw a line through the points C and E of one of the sides, which produce until it meets B F in the point F. Then F E will be the radius of the inside of the pattern. The radius of the outside is to be obtained by increasing F C an amount equal to the excess of the curved line E C over the straight line E C, as shown by the distance C S. Then F S is the radius of the outside of the pattern. In Fig. 451 the same operation is shown, applied to an ogee profile. The center line is

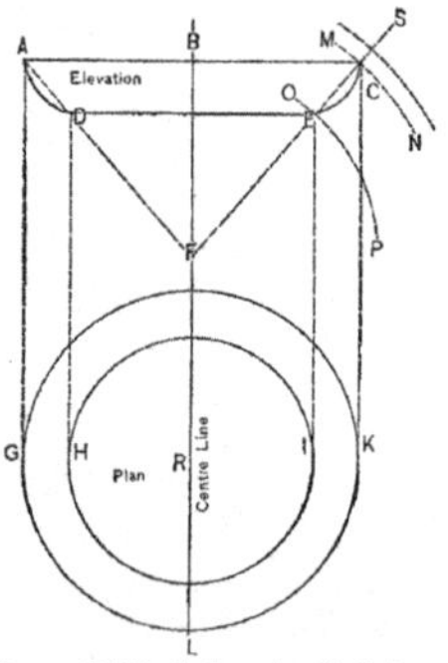


Fig. 450.—Obtaining the Sweep for a Simple Cove.
*The Blank for a Curved Molding.*

drawn. A line through the points of the profile is produced until it cuts the center line in the point F. Then F is the center by which the arcs containing the pattern are to be struck. The distance between the two points E and C is extended to correspond to the stretchout of the profile. In cutting blanks for any metal molding, there must necessarily be some discretion exercised by the mechanic. Some sheets of iron will form more readily than others; in some there is more stretch than in others, while the thickness has much to do with the operation of imparting the form to the blank. In certain kinds of metal it is found necessary to make arbitrary allowance, more or less in amount, to overcome difficulties peculiar to the material in hand. In the demonstration above given we have simply indicated the principles; allowances are to be made as circumstances require.

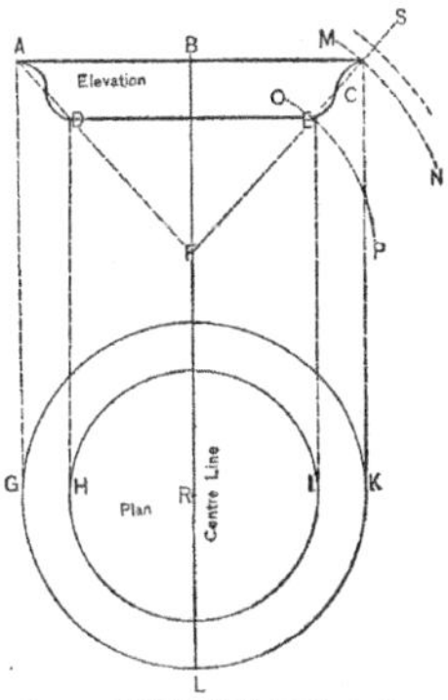


Fig. 451.—Obtaining the Sweep for an Ogee.
*The Blank for a Curved Molding.*

567. *A Plain Window Cap and its Several Patterns.*—One of the simplest articles to be made belonging to cornice work, and also one of the commonest for which patterns are demanded, is a plain window cap, yet in its various features it combines several of the most important principles in pattern cutting. In Fig. 452 we show the elevation and section of a very plain semicircular window cap, with corbels and keystone, of a style in quite general use. By inspection of the engravings it will be seen that a considerable portion of the patterns may be derived directly from these two views without any further drawing. The frame strip B and the roof A, for width are measured upon the section, and for length upon the lines in the elevation corresponding to them. The two flat face strips C and D are taken directly from the elevation. Set the dividers to the radii employed in striking them in the drawing, and lay off corresponding shapes upon the sheet of iron. The face of the keystone must be transferred and shown in the flat, as indicated in the diagram at the right. This is a simple operation and may be performed as follows: Upon any straight line, as P R, which shall be the center of the true face, lay off the length of the face taken from the section. Through the points thus obtained lay off the width of the face at top and bottom, as taken from the elevation. Then draw the connecting lines. The patterns for the diamond ornament on the keystone are developed by means of the side elevation or section and two cross sections, taken through points corresponding with the junction of the miter lines. This operation is clearly shown by the diagram entitled "Ornament on Keystone," and needs no further explanation. The sides of the corbel are pricked direct from the sectional view. One side extends back on to the frame, as indicated by the dotted lines in the section, while the other extends back only to the face of the wall. The face of the corbel is set off upon the stretchout by measurement from the profile, as shown by the diagram entitled "Face of Corbel." The method of developing the pattern for the curve molding is also clearly shown in the diagram bearing that name. Through the center M, by which the elevation lines of the molding were struck, draw a horizontal line, M F, at convenience. Upon any point in it outside of the elevation, as H, erect a perpendicular, H I, which in length make equal to the

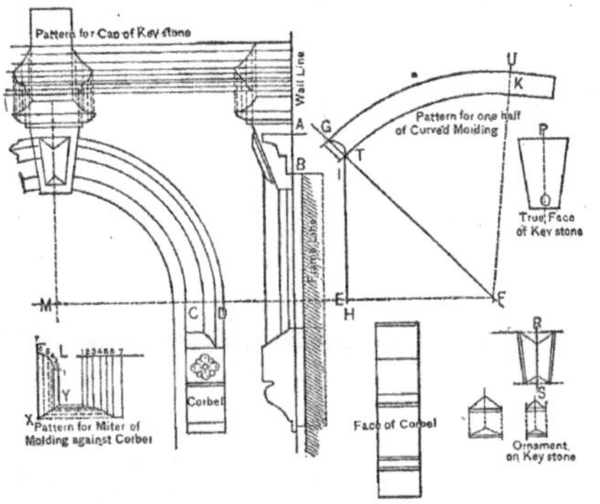


*Fig. 452.—A Plain Window Cap and its Several Patterns.*

radius by which the inner line of the molding was drawn. Upon I construct a profile of the curved molding, as shown by G T, placing it as it would appear if the cap were cut through the center line which H I may be supposed to represent. Through the extreme points in the profile of the curve draw a line, G T, as shown, which prolong until it cuts the line M F in the point F. Then F T will be the radius of the inner edge of the blank for the curved molding. From T, on the line T G, lay off a stretchout of the profile, thus obtaining the point G. Then F G will be the radius of the outer line of the pattern. The straight portion K is added by drawing a line across the pattern through the center by which it is struck, as shown by the dotted line U F, the lines of the straight part being produced to the required length at right angles to it. In the construction of window caps it is usual to trim the lower end of the cap C D after the several parts are joined, in order to fit it against the sloping top of the corbel. For this purpose a miter pattern is shown at L corresponding to the bottom of the window cap as it is to be cut.

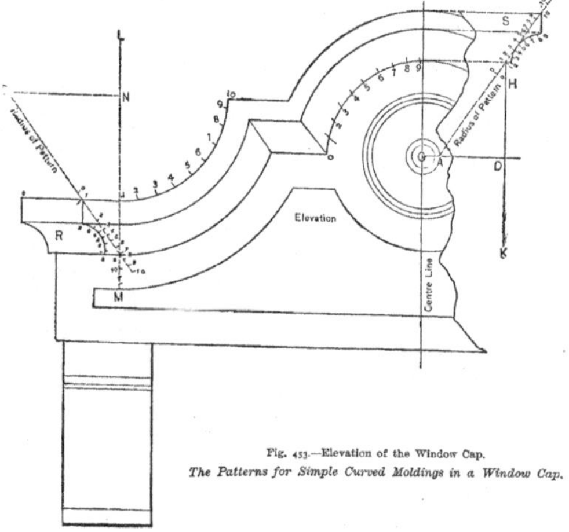


Fig. 453.—Elevation of the Window Cap.
*The Patterns for Simple Curved Moldings in a Window Cap.*

This is used simply as a gauge by which to scribe the line to which the cap is to be trimmed, in order to fit the corbel. Many workmen cut by eye without this pattern and with good results. Lay off a duplicate of the profile of the cap, as shown at L, which divide into any convenient number of equal parts. Draw a miter line, as shown by *x y*, corresponding to the pitch of the top of the corbel. Drop lines from the point of the profile against the miter line, and then with the T-square placed at right angles to the lines drawn from the profile, cut corresponding measuring lines in a stretchout previously laid off, as shown. A line traced through these points in the measuring lines will be a pattern of the shape to which the end C D must be cut to fit the top of the corbel. The molding around the top of the keystone consists of three pieces, joined by simple square miters. Divide the profile of the cap molding into any convenient number of parts. At right angles to the lines of the molding lay off a stretchout, and through the points in the stretchout draw measuring lines in the usual manner. With the T-square placed at right angles to these measuring lines, and brought successively against the points in the profile, cut the measuring lines. A line traced through these points will give the shape of the required patterns. By extending the measuring lines through the stretchout across the space over the elevation and section, and by dividing each of the several profiles of the cap molding shown in the elevation and section and dropping points from all, the two patterns required are produced complete, all as indicated in the engraving. Flanges and joints are to be added at the discretion of the mechanic.

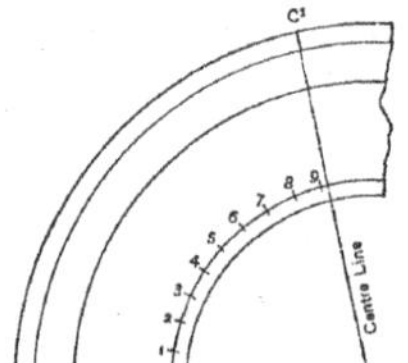


Fig. 454.—Laying out the Blank for Center Piece.
*The Patterns for Simple Curved Moldings in a Window Cap.*

568. *The Patterns for Simple Curved Moldings in a Window Cap.*—In Fig. 453 we show the elevation

of a window cap, in the construction of which two curved moldings are required of the same profile, but of opposite sweeps. The only features peculiar to this cap are the patterns for these curved moldings, which, therefore, are the only parts we shall describe. The patterns for the returns, the corbels and the face portion of the cap are obtained in the same manner as corresponding portions of other caps elsewhere described. The profiles S and R have fillets, and are to be constructed with riveting edges, the whole of which it is possible to raise in one piece. The method of developing the pattern for the blank is the same for both curves. The two pieces will raise to the form by the same dies or rolls, it being necessary only to reverse them in the machine. For the patterns proceed as follows: From G, the center by which the curve in the middle part of the cap is struck, draw A D at right angles to the center line of the cap, as shown. At convenient distance from the center line, and parallel to it through D, draw H K. Draw the profile S, placing it on H K in such position that its several members shall be as far removed from the point D as the corresponding members of the molding in elevation are removed from the center G. The dotted lines running from the profile to the elevation show the correspondence of the parts. By this arrangement it will be seen that the profile S, in connection with the line H K, represents a section of the structure about to be constructed, of which A D is the center line. The principle to be employed in striking the pattern is simply that which would be used in obtaining the envelope of the frustum of a cone. The general average of the profile is to be taken in establishing the section of the cone, or, in other words, a line is passed through its extreme points. Draw a line through the profile in this manner and prolong it until it intersects A D in the point A, all as shown by C A. Then A is the apex of the cone, of which the profile S may be considered a section. Divide the profile S, as in ordinary practice for stretchouts, into any number of spaces, all as shown by the small figures. Transfer the stretchout of the profile S on to the line A C, commencing at the point 1, as shown, letting the extra width extend in the direction of C. From any convenient center, as A' in Fig. 454, with radius A' C', describe the pattern, making the

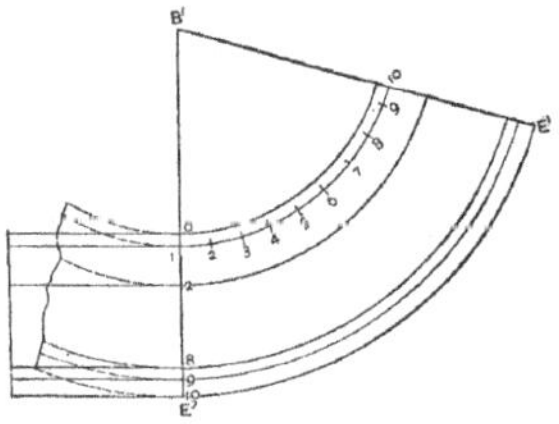


Fig. 455.—Laying out the Blank for Side Piece.
*The Patterns for Simple Curved Moldings in a Window Cap.*

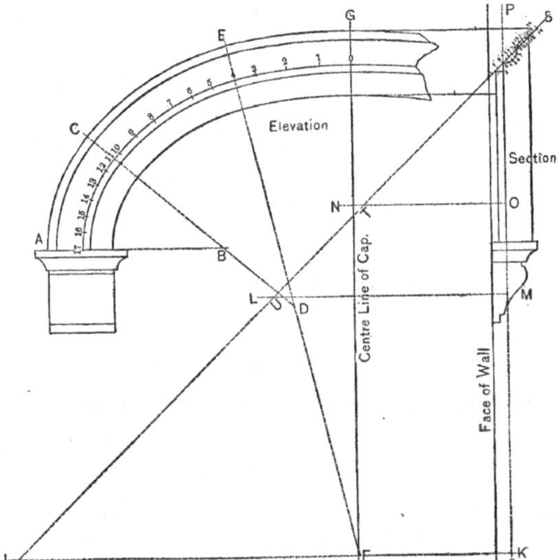


Fig. 456.—Elevation and Section of Cap.
*The Patterns for Elliptical Curved Moldings in a Window Cap.*

length of the arc equal to the length of the corresponding arc in the elevation, all as shown by the spaces and numbers. For the pattern of the curved molding forming the end portion of the cap proceed in the same general manner. Draw a profile, R, as shown, placing it against the line L M drawn through the center F, by which the curve in the elevation is struck. Through F draw the perpendicular F B indefinitely. Through the average of the profile R, as before explained, draw the line E B, cutting F B in the point B, as shown. Lay off the stretchout of the profile upon this line, commencing at the point 1, in the same manner as explained in the previous operation. From any convenient point, as B' in Fig. 455, with radius B E, describe the pattern, as shown from E' to E², which in length make equal to the arc representing the same curve in the elevation, all as shown by the measurements indicated by the small figures. The straight portion forming the end of this molding, as shown in the elevation, is added by drawing, at right angles to the line E² B', a continuation of the lines of the molding of the required length, as shown in the pattern. Upon this end of the pattern a square miter is to be cut by the ordinary rule for such purposes, to join to the return at the end of the cap.

569. *The Patterns for Elliptical Curved Moldings in a Window Cap.*—In Fig. 456 we show the elevation and vertical section of a window cap elliptical in shape, the face of which is molded. The patterns for the corbels, roof and frame strips have no peculiar features about them, and therefore will not be described in this connection. For the pattern of the elliptical curved molding we proceed as follows: In drawing the elevation certain centers were employed, or if the elevation was struck after the manner of drawing an ellipse—with string and pencil or by trammel—then the centers of approximate arcs must be obtained, the number of which, in either case, depends upon the accuracy with which the elliptical curve has been drawn. Having in this way determined the centers B, D and F, by which the respective sections of the elevation are or may be struck, use them in obtaining patterns as follows: Through the center F, from which the arc forming the middle part of the cap is drawn, and at right angles to the center line of the cap G H, draw the line I K indefinitely. Through the average of the profile, as indicated, producing the line until it meets I K, draw S R. Divide the profile in the usual manner and lay off the stretchout, as indicated by the small figures. Then R S is the radius of the pattern of the middle section of the cap. From K, through the section, erect K P, as a common basis of measurement by which to obtain the radii of the other portions. With the dividers, measuring down from the profile, lay off on P K distances equal to the length of the radius A B, as shown by the point O, and C D, as shown by the point M. Through these points O and M, at right angles to P K, draw lines cutting S R in the points T and U. Then U S is the radius for the pattern of the second section of the curve, and T S the radius of the pattern for the third section of the curve. In order to obtain the correct length of the pattern, not only as regards the whole piece, but also as regards the length of each arc constituting the curve, step off the length of the curved molding with the dividers, as shown in the elevation, numbering the spaces as indicated. As a matter both of convenience and accuracy, the spaces used in measuring the arcs are greater in the one of larger radius and are diminished in those of shorter radii, as will be noticed by examination of the diagram. To lay off the pattern after the radii are obtained as above described, proceed as follows: Draw any straight line, as G' H' in Fig. 457, from any point in which, as F',

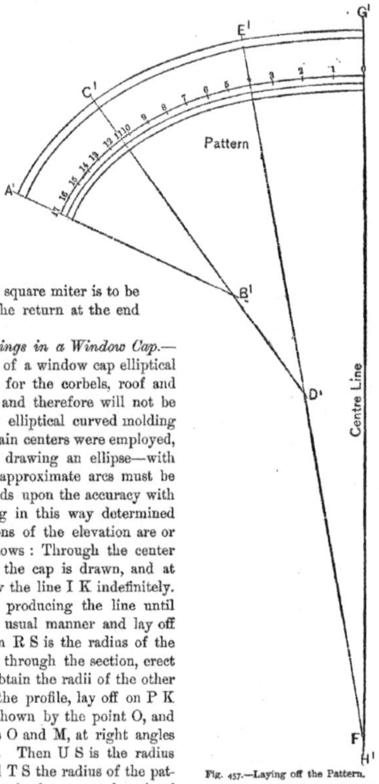

Fig. 457.—Laying off the Pattern. *The Patterns for Elliptical Curved Moldings in a Window Cap.*

with radius equal to R S, as shown by $F^1$ $E^1$, describe an arc, as shown by $E^1$ $G^1$; and likewise, from the same center describe other arcs corresponding to other points in the stretchout of the profile. Make the length of the arc $E^1$ $G^1$ equal to the length of the corresponding arc in the elevation. From $E^1$ to the center $F^1$, by which this arc was struck, draw $E^1$ $F^1$. Set the dividers to the distance U S as radius, with which, measuring from $E^1$ along the line $E^1$ $F^1$, establish $D^1$ as center, from which describe arcs corresponding to the points in the profile, as shown from $E^1$ to $C^1$, which in turn make equal to the length of the corresponding arc in the elevation, all as shown by the small figures. From $C^1$ draw the line $C^1$ $D^1$ to the center by which this arc was struck. Set the dividers to the distance F S in the elevation, and measure from $C^1$ along the line $C^1$ $D^1$. Establish the point $B^1$ for center, from which strike arcs corresponding to those already described in the other section of the pattern. Make the length equal to the length of the corresponding sections in the elevation, and draw the line $A^1$ $B^1$. Then $A^1$ $C^1$ $E^1$ $G^1$ is the half pattern corresponding to A C E G of the elevation.

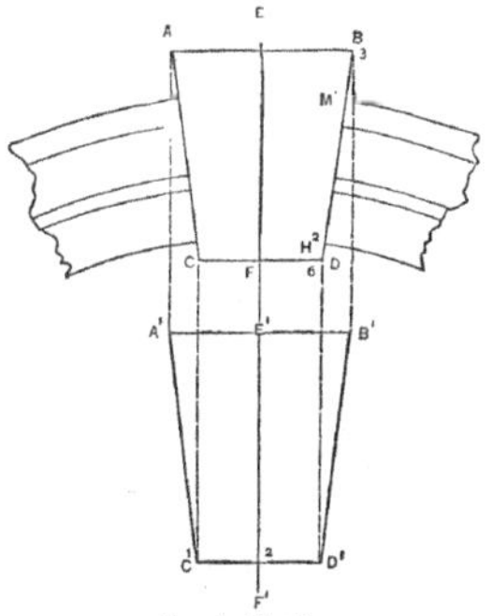

Fig. 458.—Elevation.
*Patterns of the Face and Side of a Plain Tapering Keystone.*

570. *Patterns of the Face and Side of a Plain Tapering Keystone.*—Let A B D C in Fig. 458 be the elevation of the face of a keystone, and G $E^2$ $F^2$ K of Fig. 459 a section of the same on its center line. For the true face and side, or, in other words, for the pattern of the face and side, proceed as follows: Through the center of the face draw E F, which prolong indefinitely. Through any convenient point in E F prolonged, as $E^1$, and at right angles to it, draw $A^1$ $B^1$, equal to A B of the elevation. Set off $E^1$ $F^1$ equal to $E^2$ $F^2$ of the sectional view, and through $F^1$, at right angles to $E^1$ $F^1$, draw $C^1$ $D^1$, in length equal to C D, as indicated by the dotted lines. Connect $A^1$ $C^1$ and $B^1$ $D^1$. Then $A^1$ $B^1$ $C^1$ $D^1$ will be the pattern for face of keystone. For the side we proceed as follows: Produce H K of Fig. 459 indefinitely, as shown by L $K^1$, and at any convenient point erect the perpendicular $L^1$ $E^3$, letting the point $E^3$ fall directly under $E^2$ of the side view, as shown by the dotted lines. Make L $K^1$ equal to B D of the elevation, and from $K^1$ erect the perpendicular $K^1$ $F^3$ equal to K $F^2$ of the sectional view, as shown by the dotted lines. Connect $E^3$ and $F^3$. Then $E^3$ $G^1$ $K^1$ $F^3$ will be the outline of the side of the keystone. Make L M equal to B $M^1$ of the elevation, and make L $H^1$ equal to B $H^2$ of the elevation. Then with M $H^1$ as a basis of measurement draw $N^1$ as a duplicate of the profile N in the side view, thus completing the pattern of the side.

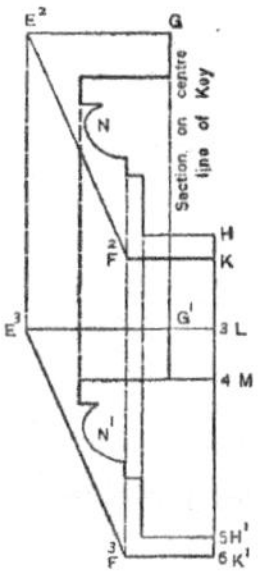

Fig. 459.—Section.
*Patterns of the Face and Side of a Plain Tapering Keystone.*

571. *Patterns for a Keystone with Sink in Face.*—In Fig. 460, E A B F represents the face of a keystone, as for a window cap, fitting over a molding, as shown in profile by M N O. In the face there is a sink, shown by G H D C, extending through the length of the keystone. L K R S represents the profile of the face of the keystone, and K T represents the profile of the sink in the face. By the conditions as thus described it will be seen that the face of the keystone tapers, that its profile is irregular, that the profile of the sink in the face does not correspond to the profile of the face, and that the sink also tapers, being wider at the top than at the bottom. For the several patterns involved proceed as follows: Divide the profile of the face K R into any convenient number of spaces, and from the points thus obtained carry lines across the face of the keystone, as shown. Since K R represents the profile of the face, a stretchout taken from it is to be used by which to locate the measuring lines upon which to drop points from the face piece. At right angles to the keystone lay off a stretchout of K R, as shown by $K^2$ $R^1$, through which draw the usual measuring lines. Placing the T-square parallel to the stretchout line, and, bringing it successively against the points in the lines C D and B A bounding the face strip, cut the corresponding measuring lines. Then a line traced through these points, as shown by $C^3$ $A^2$ $B^2$ $D^3$ will be the pattern for this part. For the pattern of the sink piece, as shown in elevation by G D C H, the profile K T is to be used. The usual method would be to divide K T into equal

spaces, carrying lines across the face; but since this would result in confusion, we have used the same points as established in K R, which are quite as convenient for use as the others mentioned, save that in laying off the stretchout each individual space must be measured by the dividers. At right angles to the line H D of the keystone lay off a stretchout of K T, as shown by $K^1$ $T^1$, through the points in which draw the usual measuring lines. Place the T-square at right angles to the lines across the face of the keystone, and, bringing it successively against the points in the lines G H and C D, forming the sides of the sink, cut the corresponding measuring lines drawn through $K^1$ $T^1$. Then lines traced through these points, as indicated by $G^1$ $H^1$ and $C^1$ $D^1$, will form the pattern of the required sink piece. For the pattern of the strip forming the sides of the sink in the face of the keystone, at any convenient place in line with the side view of the bracket, lay off a space equal to the side strips, as shown in the face by C D. Transfer to that line the several points in C D, as determined by the lines crossing it drawn from the profile, all as indicated by $C^2$ $D^2$. Through the points in $C^2$ $D^2$ draw measuring lines in the usual manner. Place the T-square at right angles to these measuring lines, and, bringing it successively against the several points in the profiles K R and K T, cut the corresponding measuring lines, as shown. Then a line traced through these points, as indicated by $K^2$ $R^2$ and $K^3$ $T^2$, will be the pattern of the strip required. In this connection it is proper to remark that while using the same points in the profile K T as we use in K R, although a matter of some inconvenience in describing the pattern of the sink strip, mention of which was made above, it would be still more inconvenient in describing the pattern last explained if the points of the two profiles were not derived from the same source. In other words, if the points in the profile K T were established arbitrarily and were entirely independent of those in profile K R, it would necessitate two sets of measuring lines drawn through the stretchout $C^2$ $D^2$, resulting in great confusion. For the side of the keystone we proceed in the same manner as described in connection with the sink strip just explained. Lay off $A^1$ $B^1$, in length equal to the side A B of the keystone, putting into $A^1$ $B^1$ all the points occurring in A B, through which draw measuring lines in the usual manner. Place the T-square at right angles to these measuring lines, and, bringing it successively against the points in the profile K R, also against points in the molding N and O, and likewise against L M and P S of the back, cut corresponding measuring lines, as shown. Then a line traced through these points of intersection, as shown by $N^1$ $M^1$ $L^1$ $K^1$ $R^2$ $S^1$ $P^1$ $O^1$, will be the outline of the required pattern, with the exception of that part lying between $N^1$ and $O^1$, which make a duplicate of N O. By examination of the points in $A^1$ $B^1$ and the lines drawn through the same, making comparison with the points in A B, it will be seen that in order to locate all the points in the profile of the molding M N O P, two additional points, as shown by $N^1$ and $O^1$, have been marked in $A^1$ $B^1$, corresponding to the points of intersection between the extreme lines of the molding itself

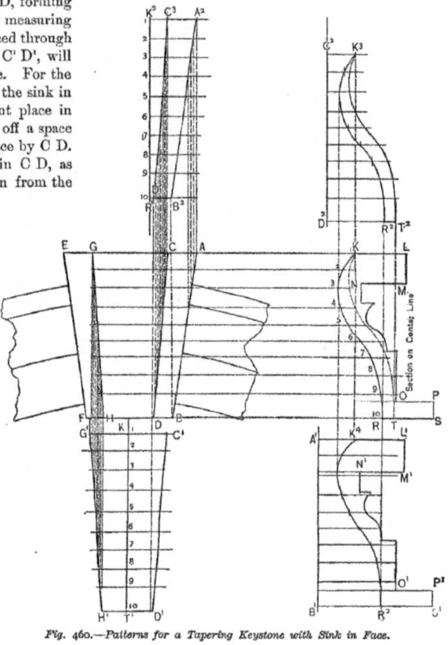


*Fig. 460.—Patterns for a Tapering Keystone with Sink in Face.*

and the side A B, as shown in the elevation by the curved lines of the molding. In practice it is frequently necessary, in operations of this character, to introduce extra points.

572. *The Patterns for a Raking Bracket.*—In Fig. 461 of the accompanying engravings, L P Q represents the normal profile of a bracket, corresponding to which a raking bracket is to be constructed. K O $P^1$ L represents the face view of the raking bracket as it is required to be. In the side view the dotted line U D

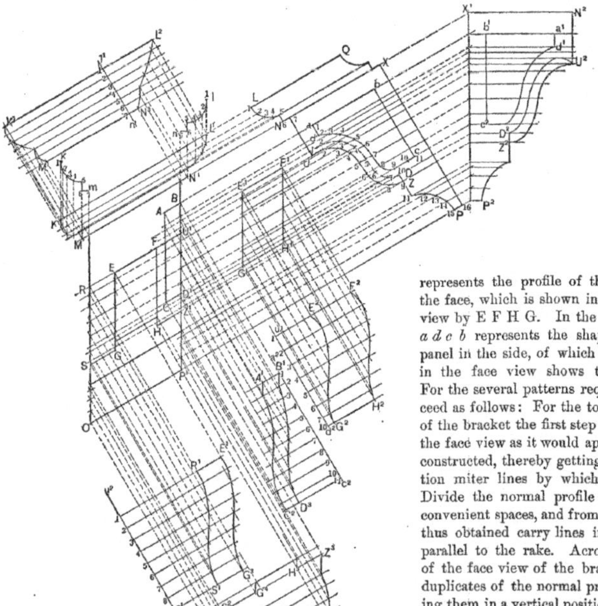

**Fig. 461.—Elevation and Shapes of the Principal Parts.**
*The Patterns for a Raking Bracket.*

represents the profile of the sink in the face, which is shown in the front view by E F H G. In the side view *a d c b* represents the shape of the panel in the side, of which A B D C in the face view shows the depth. For the several patterns required proceed as follows: For the top molding of the bracket the first step is to draw the face view as it would appear when constructed, thereby getting in elevation miter lines by which to work. Divide the normal profile L N into convenient spaces, and from the points thus obtained carry lines indefinitely parallel to the rake. Across the top of the face view of the bracket draw duplicates of the normal profile, placing them in a vertical position directly above where the new sides are required to be, as shown by *n l* and *k m*. Divide these two profiles into the same number of parts employed in dividing the normal profile, and from these points drop lines vertically, intersecting those drawn from L N. Then a line traced through these points of intersection, as shown by $L^1$ $N^4$ and K M, will be respectively the profile of the molding on the upper side and on the lower side of the bracket. At right angles to the line of the rake lay off a stretchout of the profile *n l*, or, in other words, the normal profile, as shown by $L^1$ $N^1$, and through the points in it draw the usual measuring lines. With the blade of the T-square at right angles to the lines of the rake, and brought successively against the several points in the profile $N^1$ $L^1$ and R M, cut the measuring lines drawn through the stretchout. Then a line traced through the points of intersection thus obtained, as shown by $L^2$ $N^3$ and $K^1$ $M^2$, will be the shape of the ends of the molding forming the front of the bracket head. By observation it is evident that in forming this molding the normal profile is to be used as a stay, which is to be placed at right angles to the lines of the molding. For the return moldings, forming the sides of the bracket heads, a duplicate of the profile $L^1 N^1$ is transferred to any convenient place, as shown by $L^3 N^4$ in Fig. 462. By this a representation of the side of the head is drawn, making $N^4 X^2$ equal to N X of the side view of the bracket. Space the profile of the ends of this side view into any convenient number of parts, as shown by the small figures in $L^3 N^4$ and $Q^1 X^2$. At right angles to the lines of the molding lay off a stretchout of these profiles, as shown by *q x*, and through the points in it draw the usual measuring lines. With the T-square at right angles to the lines in the molding, and brought successively against the points in the profiles $L^3 N^4$ and $Q^1 X^2$, cut the corresponding measuring lines. Then lines traced through these points of intersection, as shown by $L^4 N^5$ and $Q^2 X^3$, will form the pattern. The pattern for the return molding of the head occurring on the lower side of the bracket is obtained in the same manner. A duplicate of the profile K M of the face view of the bracket is drawn at any convenient place, as shown by $K^2 M^2$ in Fig. 463. The proper length is given to the molding by measuring upon the side view of the bracket, and a duplicate profile is drawn at the opposite end. Space the profile $K^2 M^2$ into any convenient number of parts, as indicated by the small figures, and in like manner into the same number of parts divide the profile $K^3 M^3$. At right angles to the lines of the molding lay off a stretchout of these profiles, as shown by $K^1 M^1$, through which draw the usual measuring lines. With the blade of the T-square at right angles to the lines of the molding, and brought successively against the several points in the profiles $K^2 M^2$ and $K^3 M^3$, cut the corresponding measuring lines. Then a line traced through these points of intersection, as shown by $K^5 M^5$ and $K^4 M^4$, will constitute the pattern of the return molding, or the lower side of the bracket. For the patterns of the several pieces forming the face of the bracket, the profile, as shown in the side, is divided into any convenient number of spaces, and through the points thus obtained lines are drawn parallel to the lines of the rake, crossing the face of the bracket; stretchouts are taken from the several profiles in the side view and laid out at right angles to the lines of the rake, through which the usual measuring lines are drawn. Points in the several pieces composing the face are then dropped upon these measuring lines, giving points of intersection through which lines are traced constituting the several patterns. For the strip R E G S, forming the face at the side of the sink, the profile U Z is subdivided, as indicated by the small figures, and lines from these points are carried across R E G S, as shown. At right angles to the lines of the rake a stretchout of the profile U Z is laid off, as shown by $u^2 z^2$, through the points in which the usual measuring lines are drawn. With the T-square placed at right angles to the lines of the rake, and brought successively against the points in the sides R S and E G, the corresponding measuring lines are cut. Then lines traced through these points of intersection, as shown by $R^1 S^1$ and $E^1 G^1$, form the pattern for that piece. For the piece forming the face of the bracket below the sink, as shown in the elevation by S O $P^1 Z^1$, proceed in like manner. The profile Z P in the side view is divided into any convenient number of parts, and through the points lines are drawn, crossing the face as shown. A stretchout, as indicated by $d^2 p$, is laid off at right angles to the lines of the rake, through which the usual measuring lines are drawn. The T-square is then placed at right angles to the lines of the rake, and brought against the several points in the sides S O and $Z^1 P^1$, by which the corresponding measuring lines are cut. In like manner it is brought against the points G and H, by which the shape of the part extending up to meet the sink is determined. Then lines traced through these several points of intersection, as shown by $H^2 Z^2 P^2 O^1 S^2 G^4$, form the pattern for that part of the face of the bracket. The upper part of the face of the bracket, shown in the face view by $N^1 U^1$ R M, being a flat surface, as indicated in the side view N U, is obtained by pricking directly from the face view of the bracket. No development of it is necessary. To avoid confusion of lines, the sink piece E F H G is transferred to the right, as shown by $E^1 F^1 H^1 G^1$. The profile of it, as indicated in the side view by U D, is divided into any convenient number of spaces, and through the points lines are drawn crossing it. The stretchout of this profile, as shown by $u^1 d^1$, is laid off at right angles to the lines of the rake, and through the points in it the usual measuring lines are drawn. The T-square

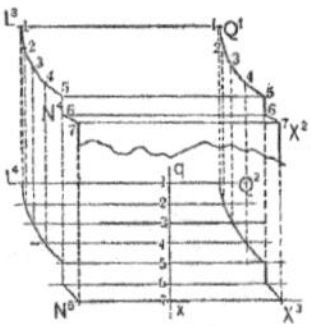

Fig. 462.—Upper Return of Head.
*The Patterns for a Raking Bracket.*

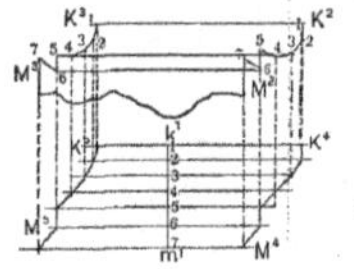

Fig. 463.—Lower Return of Head.
*The Patterns for a Raking Bracket.*

is then placed at right angles to the lines of the rake, and, being brought successively against the points in the sides $E^1 G^1$ and $F^1 H^1$, the corresponding measuring lines are cut. Then lines traced through these points of intersection, as shown by $E^2 G^2 F^2 H^2$, constitute the pattern of the bottom of the sink. Of the strips bounding the panel of the side in the bracket, the piece corresponding to *b c* in the side view is obtained by pricking directly from the face view of the bracket, $A B D^1 C$ being the shape. For the other straight strip bounding

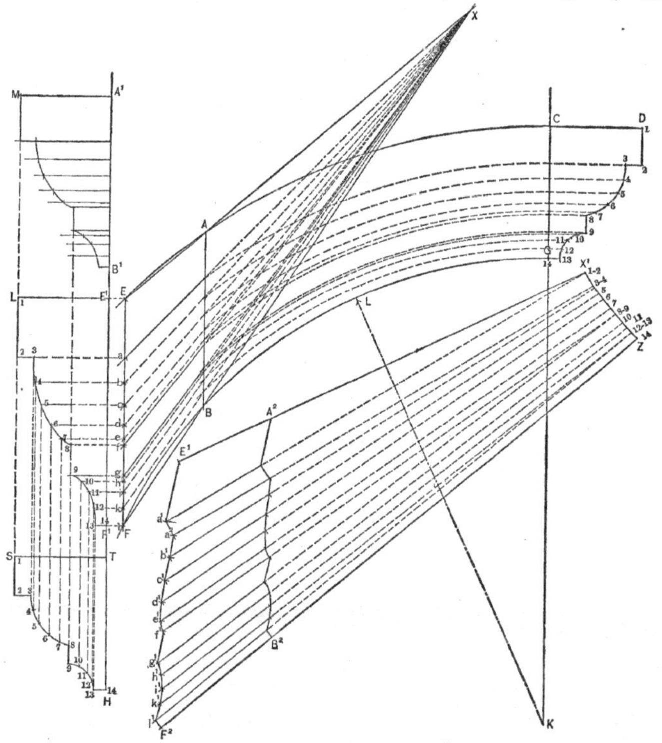


*Fig. 464.—A Raking Bracket in a Curved Pediment.*

this panel, as shown in the side view by *a b*, the length is laid off equal to *a b*, while the width is taken from the face view, equal to the space indicated by A B. For the strip representing the irregular part proceed as follows: Divide the profile *a d c* into any convenient number of parts, from the points in which carry lines crossing the face view of the same part, as indicated by $A B D^1 C$. At right angles to the lines of the rake lay off a stretchout of the profile just named, as indicated by $a^2 c^2$, through the points in which draw the usual measuring lines. Place the T-square at right angles to the lines of rake, and, bringing it against the several points in the line A C and B D', cut the corresponding measuring lines drawn through the stretchout. Then lines traced through the several points of intersection thus formed, as indicated by A' C² and B' D², will be the pattern of the curved strip forming part of the boundary of the panel in the side view of the bracket. For the side of the bracket, including the bottom of the panel last described and the strips forming the sides of the sink in the face of the bracket, we proceed as follows: Through the several points already established in the profile of the bracket, as shown by the side view, and in the profile of the sink and the shape of the panel, likewise shown in the side view, carry lines parallel to the rake, intersecting any vertical line, as X' P². From the points thus obtained in the line X' P², carry lines indefinitely horizontally, as indicated. Upon each of the lines so drawn lay off from the line X' P² a distance or distances equal to the distance or distances upon the

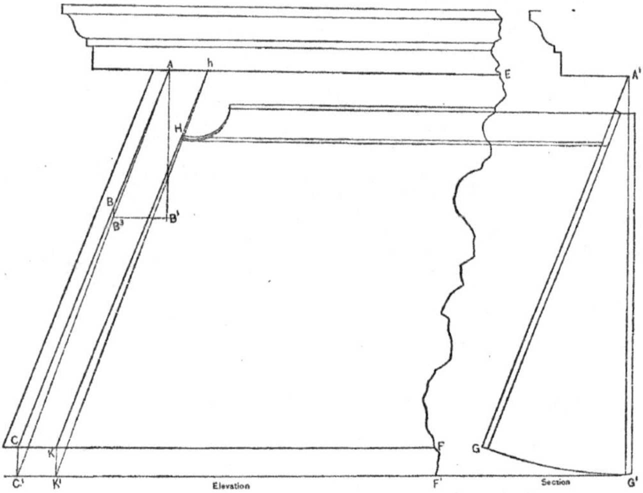


*Fig. 465.—The Principles upon which the Plain Surfaces of a Mansard Finish are Developed.*

corresponding lines drawn across the normal side of the bracket. Through the points thus obtained trace lines, which will give the several shapes in the sides of the brackets corresponding to the shapes shown in the normal side of the bracket. It may be necessary to introduce in the several profiles of the normal bracket other points than those which have been used in developing the patterns described. Use as many points in the several profiles in the normal side of the bracket as may be necessary to determine the points in the side being constructed. Then X' N² P² will be the pattern of the side of the bracket, and U² Z² D² will be the pattern of the strip forming the sides of the sink shown in the face by E F H G, and *b' a' d' c'* will be the shape of the panel in the side of the bracket.

573. *A Raking Bracket in a Curved Pediment.*—Let E A C in Fig. 464 be the arch to which the bracket is to be fitted. C K is the center line of the pediment. Draw the normal profile of the bracket with its back against the center line, as shown by C D G. Divide the face of this profile into any convenient number of parts, as shown by the small figures, and from these points carry lines at right angles to the back of the bracket, cutting the lines C G, as shown. Thence carry lines around the arch from the center K, by which the

same is struck. Let E A B F be the face of the raked bracket as it will appear in elevation. Terminate the arcs corresponding to the points in C G, struck by the center K against the side A B. Draw lines through the points E A and F B, which produce until they intersect in the point X. From X draw lines through each point in A B, crossing the bracket, as shown, continuing them until they cut the side E F in the points *a*, *b*, *c*, *d*, etc. Draw a duplicate of the normal profile below and to one side of the face E A B F, as shown by S T H. Divide the line of the face S H into the same number of parts as used in the division of the face D G, and from these points carry lines upward parallel to the back T H indefinitely. Produce the line of back T H, vertically, upon which to construct the profile of the side E F. Place the T-square at right angles to the side E F, and, bringing it against the several points E, *a*, *b*, *c*, etc., in it, cut corresponding vertical lines drawn from the normal profile S H. Then a line traced through these points, as shown by L F, will be the profile of the lower side of the bracket. Still further produce the line H T, as shown by B' A'. Make B' A' equal to the upper

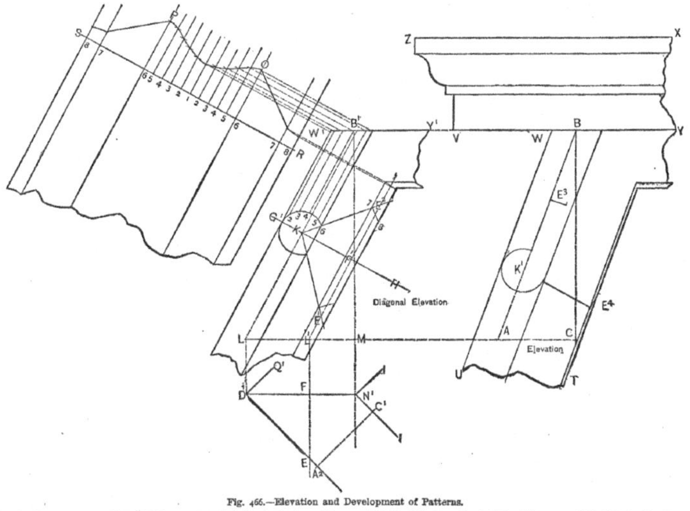


Fig. 466.—Elevation and Development of Patterns.
*The Patterns of a Hip Molding upon a Right Angle in a Mansard Roof, Mitering Against the Planceer of the Deck Cornice.*

side of the bracket B A, as shown in the elevation, and set off in it points corresponding to the points in B A, through which draw lines at right angles to B' A'. Intersect these lines in turn by lines drawn from points in the profile S H, and through these points of intersection draw a line, as shown, from M to B', Then M B' represents the profile of the upper side of the bracket. For the face of the bracket proceed as follows: Lay off E' X' at any convenient place, in length equal to E H. From E' as center, with radius equal to 1 2 of the profile L F', describe an arc, as indicated, and from X' as center, with radius X A, describe an arc intersecting the other in the point *a*'. Draw *a*' X', and to this line, at each extremity, erect perpendiculars, as shown by *a*' *a*² and X' 3, in length equal to the spaces between the points 2 and 3 of the profile L F'. Draw *a*² 3. With *a*² as center, and with radius equal to 3 4 of the profile L F', describe an arc, as shown. From 3, in the line X' Z, as center, with radius X *b*, describe an arc, intersecting it in the point *b*'. From 3, in the line X' Z, erect a perpendicular to the line A² 3, in length equal to the difference in the projection between the points 3 and 4 of the profile C F' as measured upon the line S T, as indicated by 3 4. Draw the line 4 *b*'. Proceed in the same manner from this base, obtaining the point $c^2$, using 4 5 of the profile L $F^1$ and X $c$ as radii for arcs intersecting in the point $C^1$. For the space 4 5 take the difference in the projection between the points of corresponding numbers in the profile L $F^1$, as measured upon S T, setting it off each time perpendicular to the line from which it is drawn, continuing in this manner until all the points are used. Then a line traced through the points $E^1$, $a^1$, $a^2$, $b^1$, $c^1$, etc., to $F^2$ will be the shape of the edge of the face corresponding to the lower side of the bracket. On each of the lines corresponding to these several points, $E^1$, $a^1$, $a^2$, $b^1$, etc., set off a width equal to the width of the face E A B F, measured on corresponding lines. Then a line traced through the points thus obtained, as shown by $A^2$ $B^2$, will be the shape of the face corresponding to the upper line of the bracket.

574. *The Principles upon which the Plain Surfaces of a Mansard Finish are Developed.*—One of the first steps in developing the patterns for trimming the angles of a mansard roof is to obtain a representation of the true face of the roof. In other words, inasmuch as the roof slopes in two ways, the length of the hip is other than is shown in the elevation, and this difference extends in a proportionate degree to the lines of the various parts forming the finish. The true face of a mansard may be obtained by either of the following methods: In Fig. 465, let A E F C be the elevation of a mansard roof as ordinarily drawn, and let A G be the profile or pitch drawn in line with the elevation. Set the dividers to the length $A^1$ G, and from $A^1$ as center, strike the arc G $G^1$, letting $G^1$ fall in a vertical line drawn from $A^1$. From $G^1$ draw a line parallel to the face of the elevation, as shown by $G^1$ $C^1$, and from the several points in the corner finish, as shown by C and K, drop lines vertically, cutting $G^1$ $C^1$ in the points $C^1$ and $K^1$, as shown. From these points carry lines to corresponding points in the upper line of the elevation, as shown by $C^1$ A and $K^1$ $h$. Then A $C^1$ $F^1$ E represents the pattern of the surface shown by A C F E of the elevation. In cases where the whole hight of the roof cannot be put into the drawing for use, as above described, the same result may be accomplished in the following manner: Establish any point, B, in the line of the hip, and from A, in a vertical line, set off A $B^1$, equal to A B. From $B^1$ draw the horizontal line, as shown by $B^1$ $B^2$, and from B drop a vertical line cutting this line, as shown, in the point $B^2$. By inspection of the engraving it will be seen that the point $B^2$ falls in the line A $C^1$ previously obtained, thus demonstrating that the latter method of obtaining the angle by which to proportion the several parts corresponds to the method first described, and therefore may be used when more convenient.

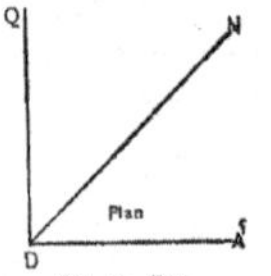


Fig. 467.—Plan.

*The Patterns of a Hip Molding Upon a Right Angle in a Mansard Roof, Mitering Against the Planceer of the Deck Cornice.*

575. *The Patterns of a Hip Molding upon a Right Angle in a Mansard Roof, Mitering Against the Planceer of a Deck Cornice.*—Let Z X Y V in Fig. 466 be the elevation of a deck cornice, against the planceer of which a hip molding, U W Y T, miters. Let the angle of the roof be a right angle, as shown by the plan Q D $A^1$, Fig. 467. The first step in the development of the patterns will be to construct a diagonal elevation of the hip molding. Assume any point, A, in the elevation on a line drawn through the fascia of the profile, as shown by B A. Through A draw a horizontal line indefinitely, as shown by L A C. From B, the point in the line A B against the planceer, drop a vertical line, cutting the horizontal line drawn through A at the point C, all as shown by B C. Produce the line of planceer W Y, as shown by $W^1$ $Y^1$. Draw a duplicate of the plan, Q D $A^1$ in Fig. 467, in such a manner that the diagonal line D N shall lie parallel to the horizontal line drawn through A, all as shown by $Q^1$ $D^1$ $A^2$. At right angles to the line $D^1$ $A^2$, at any convenient point, as $A^2$, draw the line $A^2$ $C^1$, in length equal to the distance A C in elevation, and through $C^1$ draw a line parallel to $D^1$ $A^2$, as shown by I $N^1$, cutting the diagonal line $D^1$ $N^1$ in the point $N^1$. Then $D^1$ $N^1$ represents the diagonal plan of the hip. From $N^1$ erect a perpendicular, $N^1$ M, which produce until it meets the line carried horizontally from the planceer in the point $B^1$. In like manner from $D^1$ erect a perpendicular, which produce until it meets the horizontal line L C in the point L. Connect L and $B^1$, as shown. Then points in L $B^1$ correspond to points in A B of the elevation. Therefore at any convenient point, and at right angles to it, draw the line G H, upon which to construct a profile of the hip molding. Assume any point in the diagonal plan, as E, in the side $D^1$ $A^2$, from which erect a line perpendicular to $D^1$ $N^1$, as shown by E F, which produce until it meets the horizontal line L C in the point $L^1$, and thence carry it upward parallel to L $B^1$, cutting G H in the point $F^1$. On either side lay off a space equal to F E of the diagonal plan, as shown by $F^1$ $E^1$ and $F^1$ $E^2$. Through these points $E^1$ and $E^2$ draw lines to K, being the intersection of the lines L $B^1$ and G H and a point corresponding to $K^1$ of the elevation. Upon these lines K $E^1$ and K $E^2$, at proper distances from K, set off the edges of the hip molding, as shown by $E^3$ and $E^4$ of the elevation. From K as center, with radius corresponding to the radius of the

profile in elevation, describe the shape of the roll, thus completing the profile of the hip molding in the diagonal elevation. Space one-half of this profile, as K E$^{3}$, in the usual manner, through the points in the roll of which carry lines parallel to L B$^{1}$, cutting the line of planceer W$^{1}$ Y$^{1}$, and through the points in the edges of which carry lines, also parallel to L B$^{1}$, until they meet the line of apron of the deck cornice, all as shown in the elevation. At any convenient point at right angles to the line L B$^{1}$ draw the straight line S R, upon which

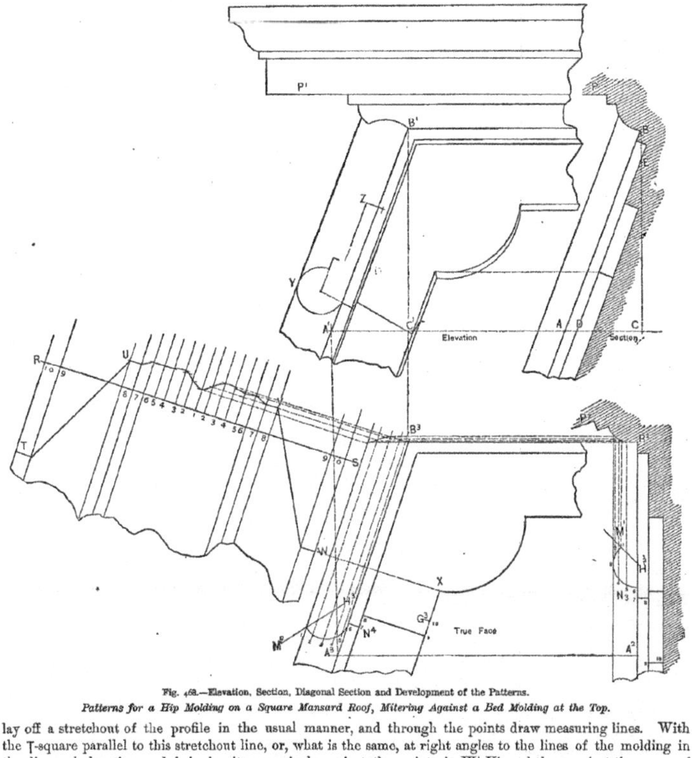


Fig. 468.—Elevation, Section, Diagonal Section and Development of the Patterns.

*Patterns for a Hip Molding on a Square Mansard Roof, Mitering Against a Bed Molding at the Top.*

lay off a stretchout of the profile in the usual manner, and through the points draw measuring lines. With the T-square parallel to this stretchout line, or, what is the same, at right angles to the lines of the molding in the diagonal elevation, and, bringing it successively against the points in W$^{1}$ Y$^{1}$, and then against the apron of the deck cornice, as above explained, cut corresponding measuring lines drawn through the stretchout. Then a line traced through these points, as shown in the engraving, will be the pattern of the hip molding mitering against the horizontal planceer.

576. *Patterns for a Hip Molding on a Square Mansard Roof, Mitering Against a Bed Molding at the Top.*—Let A C B in Fig. 468 be the section of a mansard roof, the elevation of which is shown to the left of the section, and P E be any bed molding, the profile of which does not correspond to the molding used upon the hips. For the pattern of the hip molding to miter against this bed molding we proceed as follows: Since the angle of the roof is a right angle, the elevation may be used by which to construct a true face of the hip. No other section than the original section will be required for that purpose. It is necessary, however, to construct a diagonal section through the hip, in order to get the correct profile of the stay by which to place it in the elevation of the true face. At any convenient place lay off a plan of the roof, as shown by $D^1$ F $D^2$ in Fig. 469, and through this angle draw a plan of the hip, as shown by F K. From $D^1$ erect a perpendicular, $D^1$ $C^2$, in length equal to D C of the section. Through $C^2$, parallel to $D^1$ F, draw $C^2$ K, producing it until it cuts the line representing the plan of the hip. From the points F and K in the lines representing the plan of the hip erect perpendiculars, as shown by F L and K $C^3$. Draw L $C^3$ parallel to F K, as shown. From $C^3$ erect a perpendicular $C^3$ $E^1$, in length equal to C E of the original section. Connect $E^1$ L. Then L $C^3$ $E^1$ will be a diagonal section of a portion of a roof, and L $E^1$ will be the length of the hip through that portion. At right angles to L $E^1$ draw M $H^1$, upon which to construct a section of the hip molding. Take the point G in the line F $D^1$ at convenience, and from it erect a perpendicular to F K, cutting F K in the point H, and produce it also until it cuts the base line of the diagonal section L $C^3$, as shown, and from this carry it parallel to the line L $E^1$, representing the pitch of the hip, until it crosses the line M $H^1$, cutting it in the point $H^1$. Since $D^1$ F $D^2$ represents the angle over which the hip molding is to fit, and since G H is the measurement across that angle, if we set off from $H^1$ in the diagonal section a distance equal to H G, we shall have obtained a point by which the angle contained between the fascias of the hip molding may be determined. From $H^1$ on either side set off the distance H G of the plan, as shown by $G^1$ $G^2$. Through these points draw lines representing the fascias of the hip molding, as shown by O $G^1$ and O $G^2$. Add the fillets and draw the roll, all as shown. In the true face, Fig. 468, draw a half section of the hip molding, as shown. $M^2$ $H^2$ corresponds to M $H^1$ of the diagonal section. Space this profile into any convenient number of parts in the usual manner, and through the points draw lines parallel to the lines of the hip molding indefinitely. Place a corresponding portion of the stay of the hip molding in the vertical section in which $M^1$ $H^2$ also corresponds to $H^1$ M in the diagonal section. Divide this section into the same number of equal parts, and through the points draw lines upward until they intersect with the profile of the bed molding, as shown from $P^2$ to $B^2$. From the points in $P^2$ $B^2$ carry lines horizontally, intersecting the lines drawn from the profile in the true face. Then a line traced through these points of intersection will be the miter line between the hip molding and the bed molding, as seen in elevation. At right angles to the line of the hip molding, as shown in the true face, lay off a stretchout of the hip molding, as shown by S R, through the points in which draw the usual measuring lines. Place the T-square at right angles to the lines of the hip molding, and, bringing it successively against the several points in the miter line, as shown in elevation, cut corresponding measuring lines, which will give that portion of the pattern shown from U to V. In like manner place the T-square against the point X in the true face, which is the point of junction between the flange of the hip molding and the apron of the bed molding corresponding to points 9 and 10 of the profile, and cut the corresponding measuring lines. The pattern is then completed by drawing a line from W to V and T to U.

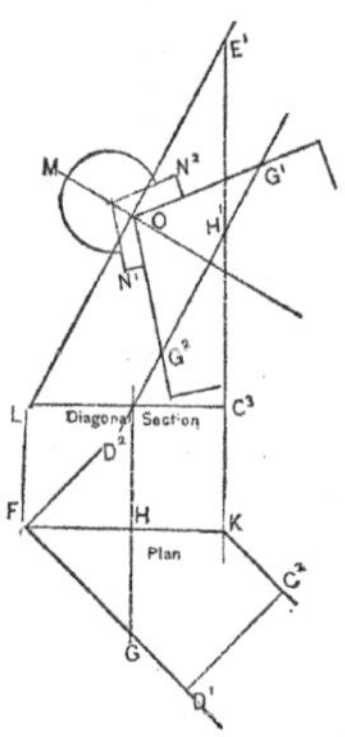


Fig. 469.—Plan of Hip.
*Patterns for a Hip Molding on a Square Mansard Roof, Mitering Against a Bed Molding at the Top.*

577. *The Patterns of a Hip Molding upon an Octagon Angle of a Mansard Roof, Mitering Against an Inclined Wash at the Bottom.*—In Fig. 470, let D B represent the wash surmounting the base molding at the foot of a mansard roof, the inclination of which is shown by B A. Let R S T be the half profile of the hip molding which is required to miter against the wash D B, and let the angle of the roof upon which the hip molding occurs be an octagon angle, as shown by E G H in the plan. Problems of this nature are likely to reach the pattern cutter in various stages of completion, so far as relates to the drawings. They may

present the plan correctly drawn, together with the elevation corresponding thereto, and a section, or nothing but the pitch of the roof, the angle of the miter and the profile of the hip molding may be given. Accordingly, in our description we will start with the smallest number of given parts, and from them develop the several representations of the work, in order to afford the pattern cutter such knowledge as will enable him to start wherever circumstances may require. Assume any point, A, in the pitch of a roof as a starting point by which to measure the angle of inclination. From A drop a vertical line, as shown by A C, and from B, the point of intersection between the roof and the wash, draw a horizontal line cutting the vertical line in the point C. Draw a plan of the wash to an octagon angle, as shown by E G H I K F. Draw the miter line G K in plan. Show a top view of the hip molding as it would appear meeting this wash, by means of lines drawn parallel to the miter line G K, as shown by M L and N O. From the inside line of the wash, at any convenient point, as B', set off B' C', in length equal to B C of the section. Then the point C' in the plan corresponds to both points A and C in the section. From the point C' carry a line horizontally, or parallel to the line of plan, meeting the hip molding in any point, as P. From P and O draw vertical lines indefinitely, which intersect by horizontal lines drawn from the points A and B in the section. Connect the points of intersection between corresponding lines, as shown by the line P' O'. Then P' O' will represent the inclination of the hip molding as seen in elevation. The elevation may be completed by drawing the other lines, as shown. The elevation of hip thus obtained may be used in the following steps, or

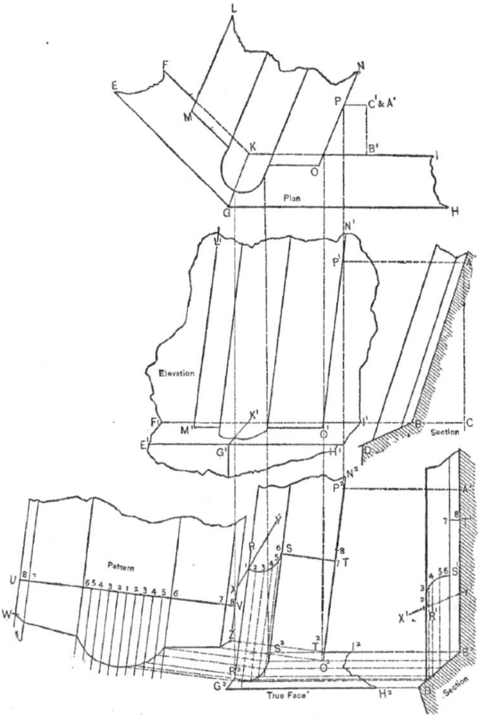


*Fig. 470.—The Patterns of a Hip Molding upon an Octagon Angle of a Mansard Roof, Mitering Against an Inclined Wash at the Bottom.*

the plan itself from which the elevation was constructed may be used for that purpose. So far as cutting the pattern is concerned, it is not necessary to construct an elevation, or if the elevation be correctly given in the original drawings the patterns may be cut by it independent of the plan. Construct a section of the roof and wash, as though the roof were placed in a vertical position. Make A² B² equal to A B of the original section, and let the angle A² B² D² equal A B D of the original section. From the points A² and B² draw horizontal lines, which intersect by points dropped from $P^1$ and $O^1$ in the elevation, or from P and O in the plan, according to whichever is being used for the purpose. Through these points of intersection draw a line, as shown by $P^2$ $O^2$, which will represent the pitch of the hip, as seen in the plan of the roof, and which is to be used for measurement in the patterns. Complete the view of the hip by inserting one-half of the profile of the molding, as shown by R S T. Complete a corresponding view of the wash at the bottom by drawing lines from the point $B^2$ $D^1$, all as shown. Divide the profile R S T into spaces in the usual manner, and from the points carry lines parallel to the lines of the hip molding on to the wash indefinitely. Draw a duplicate profile in connection with the corresponding section, as shown by $R^1$ $S^1$ $T^1$, which divide into the same number of parts, and from the points in it drop lines against the line of the wash, as shown by $D^1$ $B^2$, and from the points in $D^1$ $B^2$ carry lines horizontally intersecting the lines dropped from the profile R S T. Then a line traced through these points of intersection, as shown by $R^2$ $S^2$ $T^2$, will be the miter line formed by the junction of the hip molding with the wash. At right angles

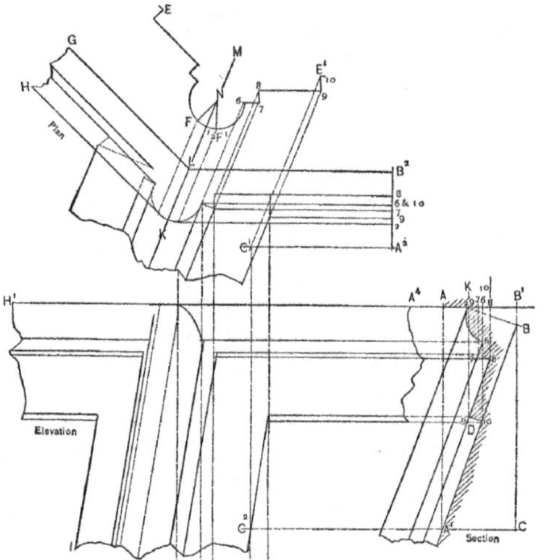


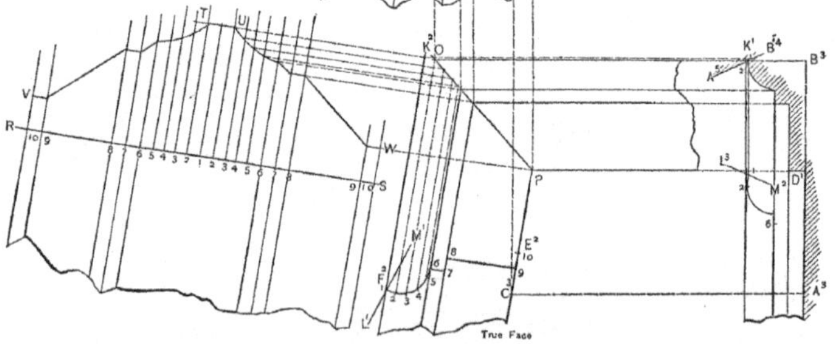


*Fig. 471.—The Patterns of a Hip Molding upon an Octagon Angle in a Mansard Roof, Mitering Against a Bed Molding of Corresponding Profile.*

to the line of the hip molding in the true face lay off a stretchout of the hip molding, as shown by U V. Through the points in it draw measuring lines in the usual manner. Place the T-square parallel to this stretchout, or, what is the same, at right angles to the line of the hip molding, as shown in true face, and, bringing it successively against the points in the miter line $R^2$ $S^2$ $T^2$, cut the corresponding measuring lines. Then a line traced through these points of intersection, as shown from W to Z, will be the cut to fit the bottom of the hip molding.

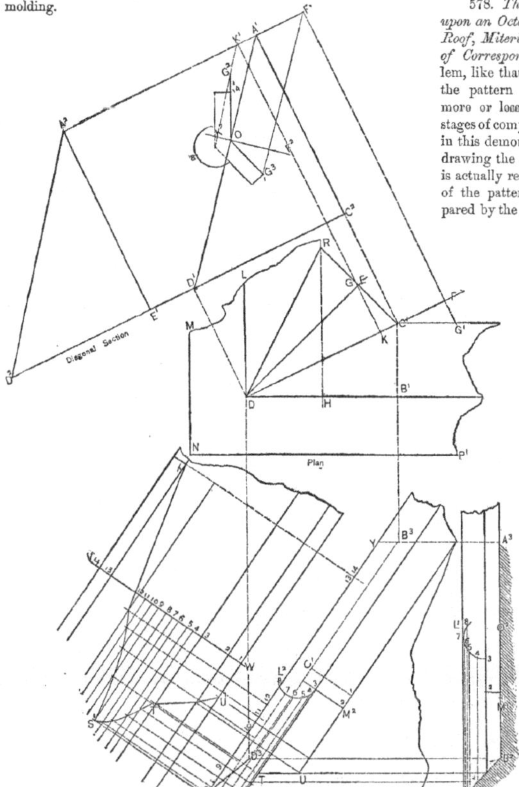


**Fig. 472.—Plan, True Face and Diagonal Section.**
***The Patterns for the Miter at the Bottom of a Hip Molding on a Mansard Roof which is Octagon at the Top and Square at the Bottom.***

578. *The Patterns of a Hip Mold-upon an Octagon Angle in a Mansard Roof, Mitering Against a Bed Molding of Corresponding Profile.*—This problem, like that in Section 577, may reach the pattern cutter in drawings either more or less accurate, and in different stages of completion. Accordingly we give in this demonstration, so far as concerns drawing the elevation, a little more than is actually required for the development of the patterns. The drawings, as prepared by the architect or draftsman, may contain everything necessary to be used and ready for the development of the pattern, or they may contain the elements from which the pattern cutter must construct such views as are necessary for him to use in the latter operation. In Fig. 471, let A B C represent the angle of the pitch of the roof, and let A D B be a section of the bed molding and apron finishing the mansard roof at the top. Let A $B^1$ be a continuation of the line of the planceer. Let G L $B^1$ be an octagon angle representing the plan of the hip over which the molding fits. Let E F $F^1$ $E^1$ be a profile of the hip molding, of which the portions E F and $F^1$ $E^1$ correspond to the bed molding and apron, as shown from A to D. The pattern to be developed is that of the hip molding mitering against the bed molding and apron A D. Commence by constructing a section of the roof, as shown by A $A^2$ C $B^1$, in which draw a section of the bed molding and apron. From the several points in the profile of the bed molding and

apron carry lines vertically cutting the horizontal line A B'. Duplicate this line in plan in the same relative position, as shown by $A^2$ $B^2$, making the several points in it correspond to the several points in A B'. Carry these lines horizontally indefinitely. Draw the miter line K L. Prolong the miter line K L, as shown by L M, upon which, in position corresponding to the position the molding is to occupy when upon the building, draw a profile of the hip molding, in which set off the points corresponding to the points in the profile of the bed molding and apron. From these points carry lines parallel to the miter line intersecting the lines drawn from the points in $A^2$ $B^2$. By this means we have obtained a correct plan of the intersection between the hip molding and the bed molding and apron. From the points in the plan obtained as just described drop lines vertically, and in turn intersect them by lines drawn from the several points in the section, as indicated by the dotted lines in the engraving. By this means we produce an elevation of the junction between the hip molding and bed molding corresponding to the plan already constructed. Next construct a section, placing it in a vertical position, instead of in an inclined position. Set off $B^3$ $A^3$ equal to B A'. Draw a section of the bed molding and apron, as shown by $K^1$ $D^1$, corresponding to K D of the original section. From the points in the section $K^1$ $D^1$ carry lines horizontally, and intersect them by lines drawn from corresponding points in either plan or elevation, according to which one is used, as indicated in the diagram. By this means is produced a representation of the true face of one half of the hip molding. In this true face insert a half profile, as shown by $F^2$ $E^2$, which divide into any convenient number of spaces in the usual manner. Draw the miter line O P, representing the line of junction between the hip molding and the bed molding and apron. Inasmuch as the profile of the hip molding and that of the apron and bed molding correspond, this line O P is a straight line. Were they different in profile it would be other than a straight line. From the points in the profile $F^2$ $E^2$ carry lines parallel to the line of the hip molding, cutting the miter line O P, as shown. Inasmuch as point No. 1 in this profile falls outside of the miter line O P, a separate operation must be performed in order to obtain a measurement for its intersection. Draw as much of the profile of the hip molding as may be necessary in the vertical section, and from point No. 1 carry a line parallel to the lines of the section, intersecting the planceer line $A^4$ $B^4$, which is here inserted, being transferred from the other section for this purpose. From these points of intersection carry a line intersecting a corresponding line drawn from the same point in the profile $F^2$ $E^2$, as shown by the point $K^2$. Having thus obtained all the points in the miter line, for the pattern itself we proceed as follows: Lay off a stretchout of the complete molding, as shown by R S, placing the same at right angles to the lines of the hip, as shown in the true face, and through the points in the same draw miter lines in the usual manner. Place the T-square parallel to this stretchout line, or, what is the same, at right angles to the lines of the hip molding, and, bringing it successively against the points in the miter line, and also against the point $K^2$, cut the corresponding measuring lines, as shown. Then a line traced through these points of intersection, as shown by V T U W, will be the pattern sought.

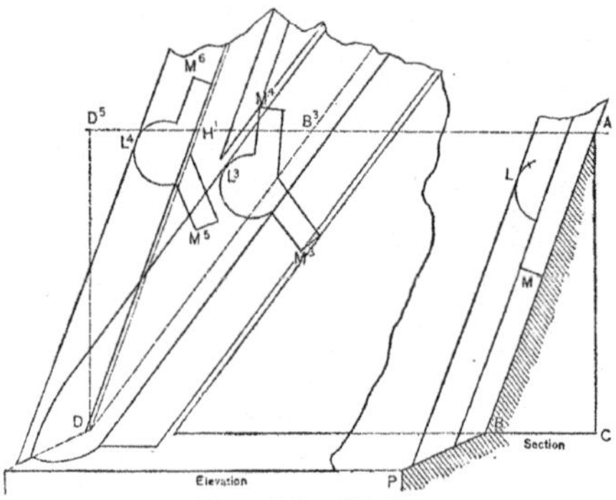


Fig. 473.—Section and Elevation.

*The Patterns for the Miter at the Bottom of a Hip Molding on a Mansard Roof which is Octagon at the Top and Square at the Bottom.*

579. *The Patterns for the Miter at the Bottom of a Hip Molding on a Mansard Roof which is Octagon at the Top and Square at the Bottom.*—Let L D B' in Fig. 472 be the plan of the roof at the base, and R C' G the plan at the top of the portion here made use of for the purpose of demonstration. Let A B C in Fig. 473 of the section indicate the pitch of the roof. Then, since it is square at the base and octagonal at the top, we have two converging hips, represented by R D and C' D, which unite and become a single profile at D. Let B P

of the section represent a wash, the plan of which is shown by M N $P^1$ of Fig. 472. Then the pattern required will be the shape of the hip molding to miter against this wash. But, since the two hip moldings join before the wash is reached, the pattern will be modified to the extent of fitting the inner edge of one against the corresponding edge of the other. This condition of things is shown in the elevation which is here introduced, not for any use it may be in the operation of cutting the patterns, but for more clearly showing the principle. The elevation is drawn by means of intersecting points from the section and the plan. We are compelled to place the cut representing the elevation away from the plan in this instance, and, therefore, the connection between the two is not so clearly represented as it would otherwise be. $D^5$ $H^1$ $B^3$ corresponds to D H $B^1$ in the plan. Horizontal lines from the points A B in the section are drawn, intersecting lines corresponding to the points already named. Let M L of the section represent one half of the profile of the molding which is required to be fitted to the converging hips. Our first step in the development of the patterns is in the construction of a section corresponding to the line of one of these hips as it appears in plan. Lay off $D^1$ $C^2$ equal to D $C^1$ of the plan, and from $C^2$ erect a perpendicular, $C^2$ $A^1$, in length equal to A C of the original section. Connect $A^1$ $D^1$. Then $A^1$ $C^2$ $D^1$ is a section of the roof as it would appear if cut through on the line D $C^1$ of the plan, and $A^1$ $D^1$ is the pitch of the hip. In order to locate the profile of the hip molding upon this section in correct position, take any point in the line R $C^1$, as G. Also lay off a corresponding point on the other arm, as $G^1$. From G carry a line parallel to $C^1$ $A^1$, producing it until it cuts the horizontal line drawn through $A^1$ at the top of the section, as shown by the point $K^1$. From $K^1$ draw a line parallel to the pitch line $A^1$ $D^1$. At any convenient place in $A^1$ $D^1$ establish the point O. From the point O draw a line parallel to $A^1$ $D^1$ of convenient length. From the intersection of the line just drawn through O with the line from $K^1$, set off the distance K G in the plan. In like manner from the point $G^1$ draw a line parallel to $A^1$ $C^2$, cutting the line and the top in $F^1$. From $F^1$ draw a line parallel to $A^1$ $D^1$, intersecting the line at O in the point $F^2$. From $F^2$, on a continuation of the line $F^1$ $F^2$, set off a distance equal to F $G^1$ in the plan, as shown at G. Connect the point O with $G^2$ and $G^3$. Then the point O of the profile will represent the corner of the sheeting boards over the hip. Construct a vertical section of the roof, placing the wash at proper angle with the same. In other words, make $A^3$ $B^2$ $P^2$ equal to A B P of the original section. By means of intersecting points from the vertical section just described and the plan, construct a true face of one of the hip moldings, as shown by Y S. Place a portion of the stay in this true face, locating it so that the point $O^1$, which corresponds to O of the hip section, shall fall upon the angle of the roof. Divide it into any convenient number of spaces, numbering them in the usual manner. From these points drop lines indefinitely through the face of the wash of the vertical section. Place also a part of the profile of the hip molding (greater than one-half) in proper position. From the points in this profile drop lines cutting the wash $P^2$ $B^2$. From the points thus obtained carry lines horizontally crossing the true face, intersecting them with lines of corresponding numbers previously drawn. A line traced through the intersection of these points will give the pattern of the miter in the true face, all as shown by S T U. Note the points where this miter line intersects the miter line of the wash $P^2$ $D^3$, which intersection carry back upon the profile $L^2$ $M^2$, which in this case will correspond to the point 8. Locate the point 8 on the first section of the hip obtained at O, and use the remainder of profile 8 14 for the other operation. Lay off a stretchout of the entire profile of the hip molding, as shown by W V, through the points in which draw the usual measuring lines. With the T-square placed at right angles to the lines of the hip, as shown in the true face, and brought against the points in the miter line S T U, cut so many of the measuring lines drawn through the stretchout W V as correspond to those points. By this means that portion of the pattern shown by $S^1$ $T^1$ $U^1$ will be obtained. For the portion of the pattern corresponding to the part of the hip which miters against the other hip, we have first to construct a true face of the octagon side of the roof. To do this we require a diagonal section of the roof corresponding to the line D E in the plan. Lay off $D^2$ $E^1$ equal to D E of the plan, and from $E^1$ erect a perpendicular, $E^1$ $A^2$, equal to C A of the section in Fig. 473. Connect $A^2$ and $D^2$. Then $A^2$ $D^2$ is the length of the diagonal face of the roof measured on the line D E of the plan. Upon any convenient straight line lay off $D^4$ $A^4$ in Fig. 474, in length equal to $D^2$ $A^2$, and

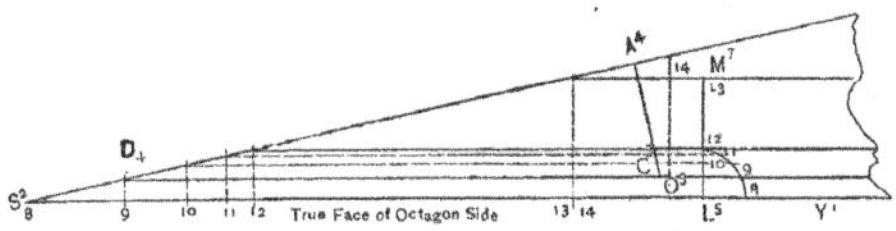


Fig. 474.—Pattern of Miter between Hip Moldings near Base.

*The Patterns for the Miter at the Bottom of a Hip Molding on a Mansard Roof which is Octagon at the Top and Square at the Bottom.*

from A⁴ set off, at right angles to it, A⁴ C³, in length equal to E C of the plan. Then D⁴ A⁴ C³ shows in the flat one-half of the diagonal face of the roof, or what is represented by D E C¹ in the plan. At right angles

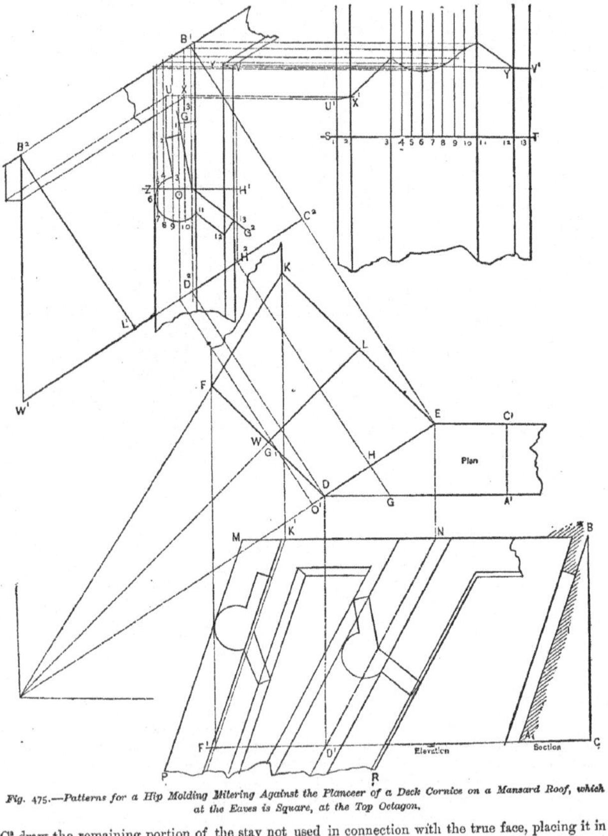


*Fig. 475.—Patterns for a Hip Molding Mitering Against the Planceer of a Deck Cornice on a Mansard Roof, which at the Eaves is Square, at the Top Octagon.*

to D⁴ C³ draw the remaining portion of the stay not used in connection with the true face, placing it in such a manner that the point O³, corresponding to O of the hip section, shall fall upon the line D⁴ C³, which represents the angle of the sheeting board. Through the point 8 of the section L³ M³, corresponding to 8 of the section

L' M', draw a line parallel to D' C', as shown by S' Y'. Then S' Y' corresponds to S Y of the true face. Space the profile L' M' into the same parts as used in laying off the stretchout W V, and through the points draw lines parallel to D' C', cutting the line D' A'. From the points of intersection in the line D' A', at right angles to S' W', draw lines cutting S' Y', giving the points marked 8, 9, 10, 11, 12, 13 and 14. For convenience in using one stretchout for the entire pattern, transfer these points to the line S Y of the true face, and thence, at right angles to S Y, draw lines cutting the corresponding measuring lines of the stretchout. Then a line traced through these points of intersection, as shown from S' to X, will be the remainder of the pattern.

580. *Patterns for a Hip Molding Mitering Against the Planceer of a Deck Cornice on a Mansard Roof, which at the Eaves is Square, at the Top Octagon.*—In Fig. 475 is shown the method of obtaining the miter against the planceer of a deck cornice formed by the molding covering a hip, which occurs between the main roof and that part which forms the transition from a square at the base to an octagon shape at the top. The roof is of the character sometimes employed upon towers which are square in a portion of their hight and octagon in another portion, the transition from square to octagon occurring in the roof. The hip molding with which we have to deal covers what may be called a transition hip, being a diagonal line starting from one of the corners of the square part and ending at one of the corners of the octagon above. In the plan, F D indicates a line across the face of the transition part of the roof at a point somewhere between the top and bottom. D A' indicates a corresponding line through one of the adjacent sides of the roof. C A B is the angle of the pitch of the roof taken at right angles to one of the sides. A' C' of the plan corresponds to A C of the section. D E of the plan represents the line of one of the hip moldings, and W L of the plan is the line through the transition part of the roof corresponding to A' C' of the principal parts of the roof. By means of intersection of lines drawn from corresponding points in the plan and the section already described, an elevation may be constructed, as shown by M N R P, if the same is desired. It is introduced here not for any service which it performs in connection with cutting the patterns, but to better explain the relationship of the several parts with which we have to deal. The first step in the development of the pattern is to construct a section of the roof as it would appear if cut through on one of the hip lines. In other words, to construct a section of the roof corresponding to D E of the plan. To do this proceed as follows: At any convenient place outside of the plan draw D' C', in length equal to D E, and parallel to it. Erect a perpendicular, C' B', in length equal to C D of the elevation. Connect B' D', as shown. Then B' D' will be the length of the hip through that portion of the roof represented by the section constructed, and as shown by D E in the plan. The next step is to construct in connection with this hip section of the roof a true stay of the hip molding. To do this proceed as follows: Take any point, G, in the plan at a convenient distance from the angle W D A. Set off at the same distance from the angle on the opposite side G'. From G carry a line at right angles to and cutting D' C' in the point H', and from this point carry it parallel with the line D' B' indefinitely. At right angles to D' B' draw a line, as shown by Z H', intersecting with the line last drawn from H' in the point H'. From H', along the line H' H', set off a distance equal to H G of the plan. And from O in the line Z H', corresponding to O' of the plan, set off a distance equal to O' G' of the plan, as shown by O G'. Having by these points determined the angle of the hip molding finish, a representation of it is indicated in the drawing by adding the flanges in the roll. Since the miter required is the junction between the hip molding, the profile of which has just been drawn, against a horizontal planceer, the remaining step in the development of the pattern consists simply in dividing the profile into any convenient number of parts, and carrying points against the line of the planceer, as shown at B', and thence carrying them across to the stretchout, as indicated. It is evident, however, upon inspection of the elevation, that the apron or fascia strip in connection with the planceer which miters with the flange of the hip molding, will form a different joint upon the side corresponding to the transition piece of the roof, than upon the side corresponding to the normal pitch of the roof. To obtain the lines for this miter an additional section must be constructed, corresponding to a center line through the transition piece, as shown by W L in plan. Prolong C' D', as indicated, in the direction of W', and lay off W' L', equal to W L of the plan. From L' erect a perpendicular, as shown by L' B', equal to C B of the original section. Connect W' and B', against the face of which draw a section of the apron or fascia strip belonging to the planceer, as shown, and from the points in it carry lines parallel to B' B' until they intersect lines drawn vertically from the flange of the hip molding lying against that side of the roof, all as indicated by U X. From these points carry lines, cutting corresponding lines in the stretchout. Having obtained these points we then proceed. At right angles to the lines of the molding in the diagonal section lay off the stretchout of the hip molding S T, and through the points draw the usual measuring lines, as shown. Place the T-square at right

angles to the lines of the molding, or, what is the same, parallel to the stretchout line, and, bringing it successively against the points formed by the intersection of the lines drawn from the hip molding and the planceer line B', cut the corresponding measuring lines, as shown. In like manner bring the T-square against the points U and X, above described, and W and V, points corresponding with the opposite side of the hip molding, and cut corresponding lines. Then a line traced through these several points of intersection, as shown by U' X' Y' V', will be the pattern sought.

581. *Patterns for a Hip Molding Mitering Against the Bed Molding of a Deck Cornice on a Mansard Roof, which is Square at the Base and Octagonal at the Top.*—The problem presented in Fig. 476 is similar to that described, with the difference that a bed molding is introduced in connection with the planceer against which the hip molding is to be mitered. M E M' represents a plan of the roof at the top, while L D M² represents a horizontal line at some point between the top and the bottom. The intersection of the lines M L and E D prolonged would indicate the corner of the building at the bottom of the roof, the structure being square at the base and octagonal at the top. The first step in the development of the pattern is to obtain a correct representation of the roof as it would appear if cut on the line D E. It is not necessary to take the entire length of the rafter, and therefore we construct a section of the roof corresponding to only so much of it as is indicated in the plan. At any convenient point lay off E² D², Fig. 477, equal to D E of the plan. From the point E² erect a

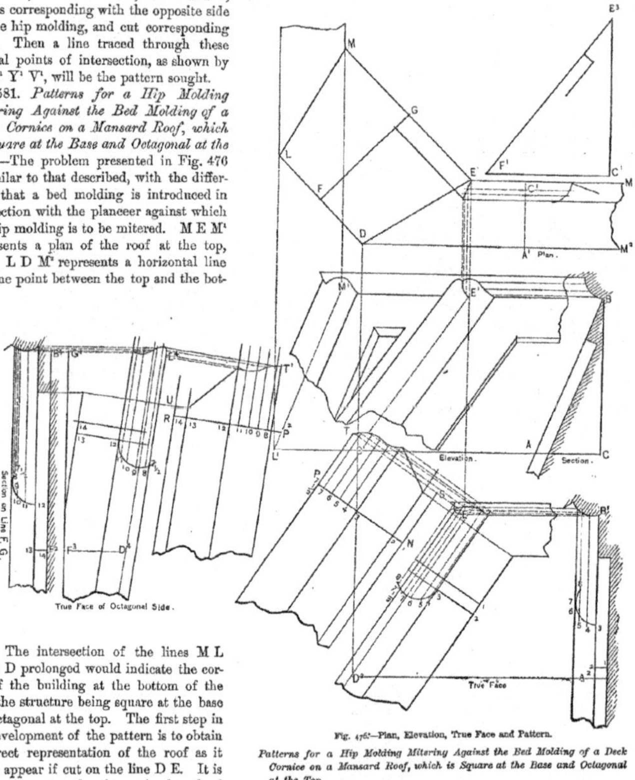


Fig. 476.—Plan, Elevation, True Face and Pattern.

*Patterns for a Hip Molding Mitering Against the Bed Molding of a Deck Cornice on a Mansard Roof, which is Square at the Base and Octagonal at the Top.*

perpendicular, $E^3$ $B^2$, in length equal to C B of the section of the roof. Connect $B^2$ and $D^3$, which will be the pitch of the hip corresponding to the line D E of the plan. Since we have constructed the section $D^3$ $E^3$ $B^2$ away from and out of line with the plan, it is necessary to draw a portion of the plan in immediate connection with the section. Lay off the angle $I^1$ H $A^2$ equal to the angle F D $A^1$ of the plan, and let $A^2$ $C^2$ equal $A^1$ $C^1$ of the plan. Draw H $C^2$, which corresponds to the hip line in plan. From the point H in the plan thus constructed lay off on either arm the points I and $I^1$, equally distant from it and conveniently located for use in constructing the profile of the hip molding. From $I^1$ carry a line parallel to $D^3$ $B^2$ indefinitely. From the point I erect a perpendicular to H $C^2$, cutting it in the point K, which prolong until it meets the base $D^3$ $E^3$ of the diagonal section, from which point carry it parallel to the inclined line $D^3$ $B^2$ indefinitely in the direction of $K^1$. At right angles to the inclined line $D^3$ $B^2$ draw a straight line, $O^1$ $K^1$, cutting the line last described in the point $K^1$. From $K^1$, measuring back on this line, set off the point $I^2$, making the distance from $K^1$ to $I^2$ the same as from K to I of the plan. From $O^1$ in the line $I^1$ $I^3$ set off the distance $O^1$ $I^3$ equal to O $I^1$ of the plan. From these points $I^3$ and $I^2$ draw lines meeting the line $O^1$ $K^1$ at the point of its intersection with the line $D^3$ $B^2$.

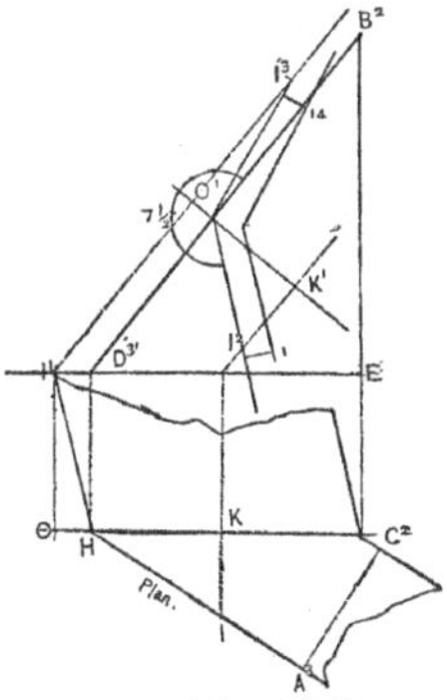

Fig. 477.—Section and Profile.

*Patterns for a Hip Molding Mitering Against the Bed Molding of a Deck Cornice on a Mansard Roof, which is Square at the Base and Octagonal at the Top.*

Complete the profile of the hip molding, as indicated, laying off the width of the fascias from $O^1$ on these lines, adding the roll and edges. The next step in the development of the pattern is to draw a true face of the hip molding, which is done by transferring the section A B to a vertical position, as indicated by $A^2$ $B^2$, Fig. 476, in connection with which the bed molding against which the hip molding is to miter is also drawn, as shown. From the several points in this vertical section draw horizontal lines, which intersect by vertical lines dropped from corresponding points in plan. Then the line $D^2$ $E^2$ is the true face of one-half the hip corresponding to D E of the plan. In connection with the vertical section just described, place a half profile of the hip molding, a true section of which we have obtained by the process already explained, and place a duplicate of this portion of the profile in connection with the true face. Space both of these profiles into the same number of parts, and from the several points in each carry lines upward parallel to the two sections in which they appear, the lines from the profile in the vertical section cutting the bed molding, and the lines from the profile in the true face being continued indefinitely. From the points thus obtained in the bed molding carry lines horizontally, intersecting those drawn from the profile in connection with the true face, producing the miter line, as shown by $E^2$. By inspection of the plan and elevation it will be seen that the miter of the bed molding around the octagon at E is irregular. That is, its miter line does not coincide with the line of the hip D E. If we divide the profile of the bed molding in the vertical section and also the profile of the bed molding, as shown in the plan, into any convenient number of parts, dropping points in the profile of the plan on to the miter line, thence carrying them downward and intersecting them with horizontal lines from the corresponding points of the bed molding in section, as shown at $E^1$, we will have the appearance of the bed molding miter in elevation. By a similar operation the appearance of this miter in the true face could be obtained, but it has here been performed in the elevation, instead of in the true face, in order to avoid confusion of lines. Having obtained this line in the true face, its intersection with the miter line previously obtained at $E^2$ must be noted. A line from this point of intersection must then be carried parallel to the line of the molding back to the profile of the hip, and there marked, as shown by the figure 7½. The position of the point 7½ should now be marked upon the section of the hip molding previously obtained at $O^1$. So much of the profile as exists between 1 and 7½ in the true face is used in obtaining the stretchout of this part of the pattern. The remaining portion of the stay, namely, from 7½ to 14, is afterward used for the true face of the octagonal side for the remainder of the pattern. At right angles to the line of the molding in the true face lay off a stretchout equal to that portion of the profile thus used, as shown by D N, through the points in which draw measuring lines in the usual manner. Place the T-square at right angles to the lines of the molding in the true face, and, bringing it against the several points in the miter line between the hip and bed molding at $E^2$, cut corresponding measuring lines drawn through the stretchout. Then a line traced through these points, as shown by S T, will be the miter line for that portion of the pattern corresponding to the part of the profile thus used. For

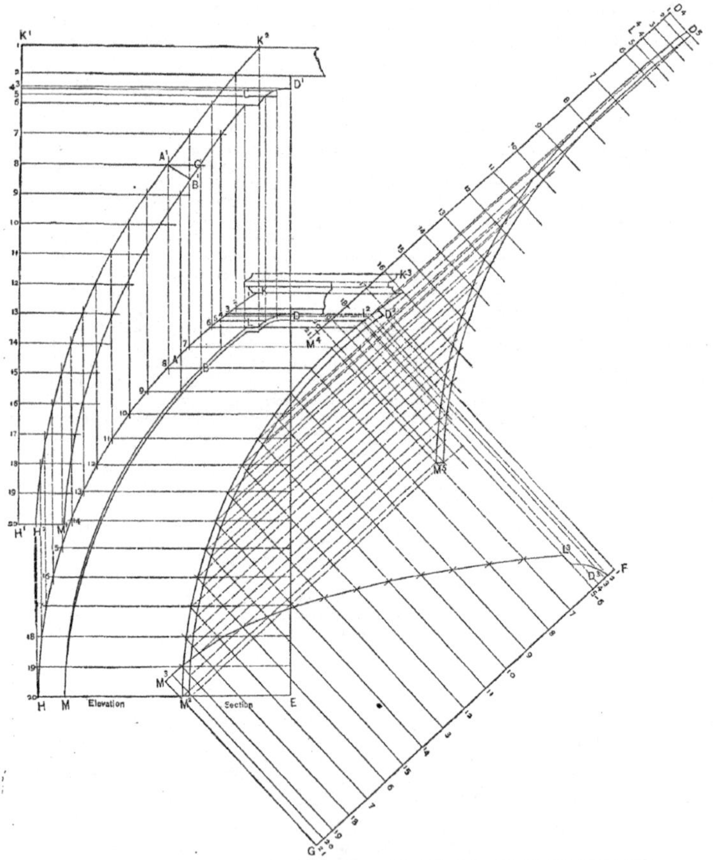


*Fig. 478.—The Pattern for a Hip Finish in a Curved Mansard Roof, the Angle of the Hip being a Right Angle.*

the other half of the hip molding, being that portion which lies on the face of the transition piece, another operation must be gone through. Construct a section of the roof corresponding to the line F G in the plan. At any convenient point lay off $F^1$ $C^1$ in Fig. 477, equal in length to F G. From the point $C^1$ erect a perpendicular, $C^1$ $B^3$, in length equal to C B of the vertical section. Connect $F^1$ and $B^3$. Then $F^1$ $B^3$ is the length of the transition side of the roof through that portion corresponding to F G of the plan. By means of this section in the plan lay off an elevation of one-half of the transition side of the roof, by which to obtain the proper measurement of that flange of the hip molding lying against it. At any convenient point set off $G^1$ $F^3$, in length equal to $B^3$ $F^1$. At right angles to it set off $G^1$ $E^4$, in length equal to G E of the plan, and $F^3$ $D^4$, in length equal to F D of the plan. Connect $D^4$ and $E^4$. Then $G^1$ $E^4$ $D^4$ $F^3$ is an elevation of that portion of the roof represented by G E D F in the plan. In connection with this elevation of the transition face of the roof, construct a vertical section of the roof as it would appear if cut on the line F G. In connection with the vertical section just described, place so much of the stay as was not used for the pattern already delineated, and in the representation of the elevation of the transitional face of the roof place a corresponding portion of the profile, each of which divide into the same number of spaces. From the points thus obtained carry lines parallel to the lines of the respective representations of the part, those in the vertical section cutting the bed molding, and those in the elevation being produced indefinitely. From the points in the bed molding of the vertical section thus defined carry lines horizontally intersecting those drawn from the profile in the elevation, thus establishing the miter line, as indicated at $E^4$. At right angles to the line $D^4$ $E^4$ set off a stretchout of the profile, as shown by R $P^2$, through the points in which draw the usual measuring lines. With the T-square placed parallel to this stretchout line, or, what is the same, at right angles to the line $D^4$ $E^4$, and, being brought successively against the points in the miter line, cut corresponding measuring lines, as shown. Points also are to be carried across, in the same manner as described, corresponding to the bottom of the apron or fascia strip in connection with the bed molding. Then a line traced through these points, as indicated by the line drawn from U to $T^1$, will be the pattern of the other half of the hip molding. By joining the two patterns thus obtained upon the center line of the stay corresponding to P T of the first piece or $P^2$ $T^1$ of the second piece, the pattern will be contained in one piece.

582. *The Pattern for a Hip Finish in a Curved Mansard Roof, the Angle of the Hip being a Right Angle.*—The general features presented in the problem shown in Fig. 478 are similar to some of those already described. The parts requiring special attention are the flange strips, sometimes called sink strips, bounding the fascia of the hip molding, which in curved work must be cut in a separate piece, it being impracticable to turn them from the edges of the fascia. H K represents an elevation of a curved hip molding occurring in a roof, of which E D is the vertical hight and $M^2$ $K^3$ is a section. The first step to be described is the method of obtaining the pattern of the fascias of the hip molding. For this purpose we have shown in the drawing such a representation of it as would appear if the two fascias formed a close joint upon the angle of the roof, and we have supposed that the hip molding or the bead is to be added afterward on the outside over this joint. We therefore consider the part to be dealt with the same as though it were the section of a molding, instead of a section of a roof, and the operations performed are identical with those employed in cutting a square miter. Space the profile into any convenient number of parts, introducing lines in the upper part in connection with the ornamental corner piece, shown by L D, at such intervals as will make it possible to take measurements required to describe the shape of it in the pattern. From this profile, by means of the points just indicated, lay off a stretchout, as shown by $H^1$ $K^1$, and through the points draw the usual measuring lines. Bring the T-square against the several points in H K, and cut the corresponding lines drawn through the stretchout just described. Then a line traced through these points, as shown by $H^2$ $K^2$, will be the outside line of the fascia. For the inside line take the given width of the fascia and set it off from this line, measuring at right angles to it, as indicated by $A^1$ $B^1$, and not along the measuring lines of the stretchout, as would be indicated by $A^1$ C. Then a line traced through these points, as shown from $M^1$ to $L^1$, will be the inside line of the fascia strip. The points in the ornamental corner piece from $L^1$ to $D^1$ are to be obtained from the elevation, in case an elevation is furnished the pattern cutter, by measurement along the lines drawn horizontally through the several points in L D, and which are indicated in the stretchout line already referred to. Or the shape from $L^1$ to $D^1$ may be described arbitrarily at this stage of the operation, according to the finish required upon the roof. The latter plan is the correct one in principle. The method of constructing the elevation, working back from the profile thus established, is clearly indicated by the dotted lines in the engraving. Through the several points in the profile H K horizontal lines are drawn, as shown, and from the inside line of the pattern of the fascia piece, as above described, lines are dropped, cutting these horizontal lines of corresponding numbers. Then a line traced through these points, shown from M

to L, will be the inside line of fascia piece in elevation. For the flange strip bounding the fascia piece, commonly called the sink strip, an elevation of which is shown in the section from $M^2$ to $D^2$, proceed as follows: Draw the line G F approximately parallel to the upper part of the section $M^2$ $D^2$, making it indefinite in length, which cut by lines drawn from the several points in $M^2$ $D^2$, at right angles to it, as shown. From F G, upon the several lines drawn at right angles to it, set off spaces equal to the distance upon lines of corresponding number from D E to the line M L of the elevation. Then a line traced through these points will represent the profile of this flange strip, as indicated by $M^3$ $L^3$. In like manner set off in continuation of it the length measured upon the ornamental corner piece, all as shown by $L^3$ $D^3$ F. From this profile lay off a stretchout parallel to G F, as shown by $M^4$ $D^4$, through the points in which draw measuring lines in the usual manner. Place the T-square parallel to this stretchout line, and, bringing it successively against points in both the inner and the outer lines of the elevation of the flange strip, as shown from $M^2$ $D^2$, cut the measuring lines of correspond-

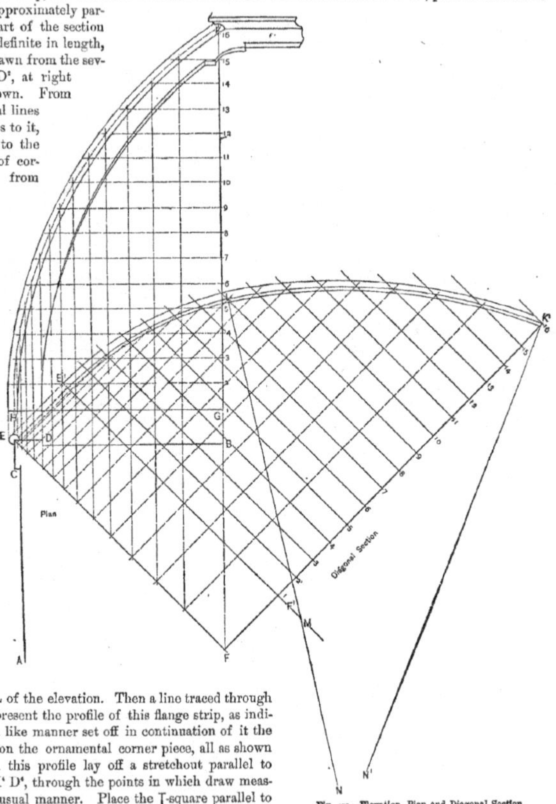


Fig. 479.—Elevation, Plan and Diagonal Section.

*The Patterns for the Bead Capping a Hip Finish in a Curved Mansard Roof, the Angle of the Hip being a Right Angle.*

ing number. Then lines traced through these points of intersection, as shown from $M^4$ to $D^6$, will be the pattern of the flange strip bounding the edge of the fascia.

583. *The Patterns for the Bead Capping a Hip Finish in a Curved Mansard Roof, the Angle of the Hip being a Right Angle.*—Let A E B in Fig. 479 represent the plan of a mansard roof or tower, the elevation of which is shown by H E K, over the hip of which a molding of any given profile is to be fitted, in this case a three-quarter bead. Then the diagonal line E F in the plan represents the hip as it would appear if viewed from the top. At any convenient point parallel to E F, and equal to it, draw $E^1$ $F^1$, and from $F^1$ erect a perpendicular, F $K^1$, in length equal to the vertical line in elevation E K. Divide E K and $F^1$ $K^1$ into the same number of equal spaces. From the points in E K draw lines cutting the profile H K, as shown, and from the points thus obtained in H K drop lines vertically, producing them until they cut the diagonal line E F, as shown. Through the points in $F^1$ $K^1$ draw measuring lines in the usual manner, and intersect them by lines erected perpendicularly to E F. Then a line traced through these points of intersection, as shown by $E^1$ $K^1$, will be the profile to which the molding covering the hip is to be raised. Inasmuch as the usual process of raising the curved molding requires for the adjustment of the machine, as well as for the description of the pattern, a knowledge of the center from which the curve is struck, divide the profile $E^1$ $K^1$ into such parts as will correspond to segments of circles. In this case the section from $E^1$ to L corresponds to an arc struck from the center M, and the section from L to $K^1$ corresponds to an arc struck from a center not shown in the engraving, but which will be found by the intersection of the lines L N and $K^1$ $N^1$ produced. In Fig. 480 we show an enlarged section of the hip molding, including flanges and roll as it would appear at the bottom of the hip, and also another section as it would appear at the top. Upon inspection it is evident that the distortion to which these profiles is subjected is altogether owing to the change of direction in the hip molding. In other words, they are sections taken at right angles to the hip at different points, and therefore the angle in the one is a right angle corresponding to the base of the roof, while the other is an obtuse angle corresponding to a section at right angles through the molding at the top of the roof. The sections will be the same at all points if taken upon horizontal planes. The method of obtaining these several sections from the plan has been clearly described in connection with problems relating to hip finish upon straight mansard roofs, and therefore needs no further description at this time.

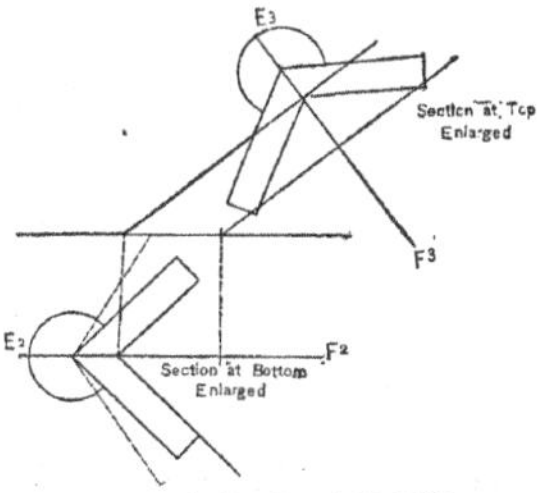


Fig. 480.—Sections through Hip Finish.

*The Patterns for the Bead Capping a Hip Molding in a Curved Mansard Roof, the Angle of the Hip being a Right Angle.*

584. *The Patterns of the Hip Molding Finishing a Curved Mansard Roof which is Square at the Eaves and Octagonal at the Top.*—The problem illustrated in Fig. 481 may be described as a combination of some of features of the last three problems presented. It is ordinarily presented, however, to the pattern cutter in a manner which requires the use of still other principles than those we have explained, in order to develop the several shapes. C D E F represents the plan of the building at the base of the roof, while V G H W represents the plan of the roof at the top. It will be seen that the roof is square at the foot of the rafters and octagonal at the top. The same conditions may arise where the corners of the roof are chamfered, the chamfer being of unequal width, starting at nothing at the bottom and increasing to a considerable space at the top. D G H E in the plan represents a chamfer of this kind, or a transition piece in the construction of a roof which, as above described, is square at the base and octagonal at the top. The same features are represented in elevation by $D^2$ $G^2$ $H^2$ $E^2$. The elevation is introduced here not for any use in pattern cutting, but simply to show the relation of parts. O A B represents a section of the roof, showing the inclination and curve of the rafter. Space the profile O B into any convenient number of parts, and from the points thus obtained draw horizontal lines indefinitely. Draw a duplicate section placed in a horizontal position, as shown by $O^1$ $A^2$ $B^2$, which divide into like spaces, and draw lines from that horizontally cutting the hip line E H in plan, which becomes a miter line so far as the patterns are concerned. The intersections of lines drawn vertically from the miter line E H with those drawn horizontally from the profile O B, give the line of the hip in elevation, as indicated by $E^2$ $H^2$. Take a stretchout of the profile O B and lay it out at right angles to the horizontal lines drawn through the points in it, as shown by $O^3$ $B^4$, through the points in which draw the usual

measuring lines. Cut these measuring lines by lines drawn vertically from the points in E H. Then a line traced through these lines of intersection, as shown by $E^3$ $H^3$, will be the line of the pattern corresponding to

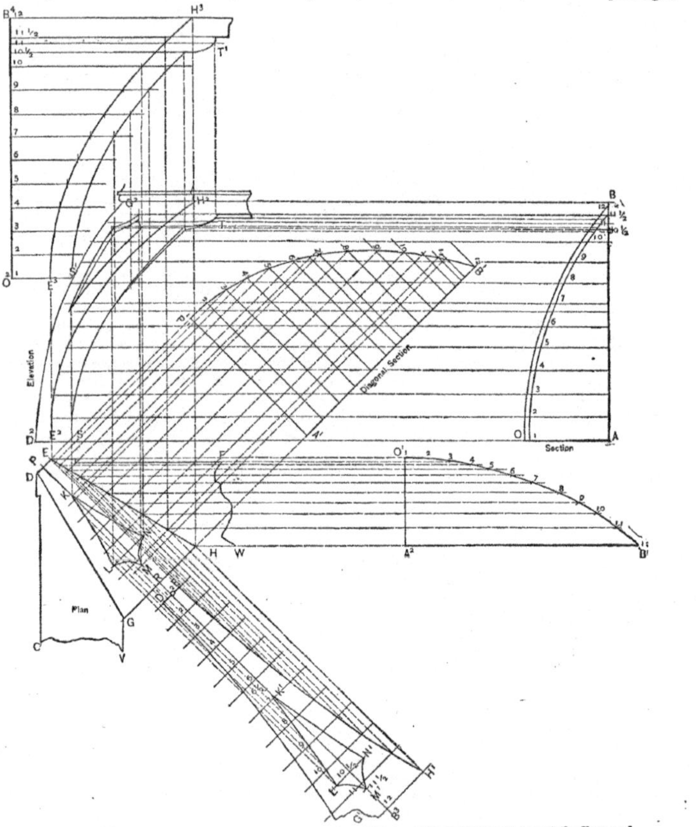


*Fig. 481.—The Patterns of the Hip Molding Finishing a Curved Mansard Roof which is Square at the Eaves and Octagonal at the Top.*

the line E H in the plan. For the width of the flange or fascia piece forming the hip molding, proceed as described in Section 582. For the pattern of the transition piece we proceed as follows: Through the center of the transition piece, as shown in plan, draw the line P R. At any convenient place outside of the plan of the transition piece draw a duplicate of P R parallel to it, as shown by $P^1$ $A^1$, and from the point $A^1$ erect a perpendicular, $A^1$ $B^1$, in length equal to A B of the original section. In $A^1$ $B^1$ set off points corresponding to the points in A B, and through them draw horizontal lines, as shown. Place the T-square parallel to $A^1$ $B^1$, and, bringing it against the points in E A, cut corresponding measuring lines. Then a line traced through these points of intersection, as shown by $B^1$ $P^1$, will complete the diagonal section corresponding to P R in the plan. Of this diagonal section take a stretchout, $B^1$ $P^1$, which lay off on the straight line corresponding to P R produced, all as shown by $P^2$ $B^2$. Through the points in $P^2$ $B^2$ draw the usual measuring lines. With the T-square placed parallel to this stretchout line, and brought successively against the points in E H, cut the measuring lines, as shown. Then a line traced through these points of intersection, as shown by $E^1$ to $H^1$, will be one side of the required pattern. In like manner, having transferred points from E H to the corresponding line D G, cut the measuring lines from it, which will give the other side of the required pattern. By the same general means the shape of the panel occurring in the transition piece is described in the pattern. In its original design it may have been drawn either in the plan or elevation, or it may have been designed in connection with both. Carry lines across it corresponding to the points in E H, and by this means obtain measurement of its width upon lines corresponding to the lines drawn through the stretchout. Use the T-square for cutting these measuring lines in the usual manner. Then lines traced through the points of intersection, as shown by $K^1$ $N^1$ $M^1$ $L^1$, will be the pattern for the panel piece. The flange strips or sink strips are to be obtained by the same general method as described in Section 582. It may be observed in this connection that in the construction of

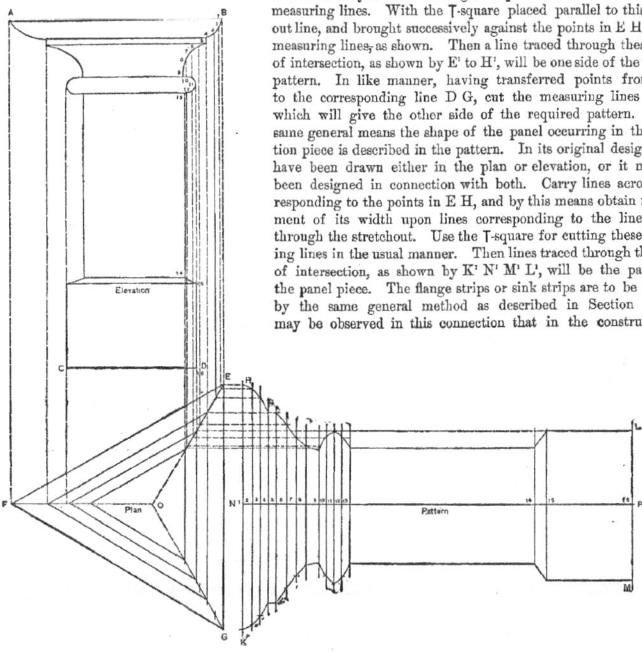


*Fig. 482.—The Patterns for a Pedestal of which the Plan is an Equilateral Triangle.*

the elevation by means of the intersection of lines drawn from corresponding points in the section and plan, it is occasionally necessary to introduce other points than those first inserted, in order to obtain corresponding points of measurement in other representations of the parts. For instance, 6½ of the section O A B corresponds to the lower point of the panel piece, as shown in elevation. A point is necessary to be inserted to locate this part. The same may be said of 10½ and 11½, also shown in the same section. The reader will readily understand that in all profiles spaced in the manner employed in the roof here described, and, in fact, in almost all cases, additional points may be inserted at any time when found necessary. In cases where greater accuracy is required in certain parts of the work than in others, the same end may be accomplished by inserting additional points in this general manner.

585. *The Patterns for a Pedestal of which the Plan is an Equilateral Triangle.*—Let A B D C in Fig. 482 be the elevation of a pedestal or other article of which the plan is an equilateral triangle, as shown by F E G. Construct the elevation so as to show one side in profile, and place the plan to correspond with it. Draw the miter lines E O and G O. Divide the profile B D into spaces of convenient size in the usual manner, and number them as shown in the diagram. From the points thus obtained drop lines, cutting E O and G O, as shown. Lay off the stretchout N P at right angles to the side E G, and through the points in it draw measuring lines. Place the T-square at right angles to E G, and, bringing it successively against the points in the miter lines E O and G O, cut the corresponding measuring lines. A line traced through these points will be the pattern, as shown by H L M K.

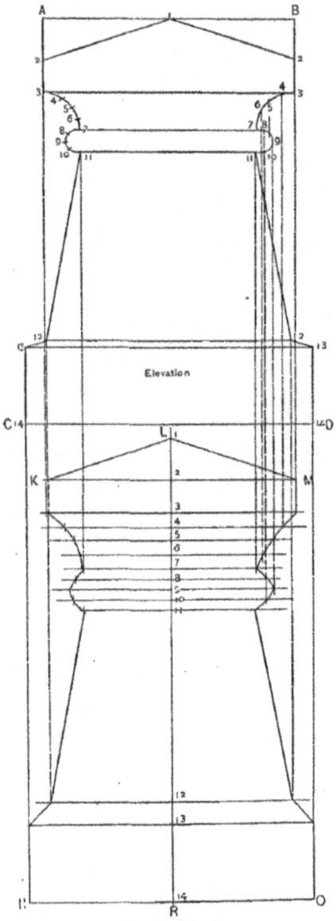


Fig. 483.—Elevation.
*The Pattern for a Pedestal, Square in Plan.*

586. *The Pattern for a Pedestal, Square in Plan.*—In Fig. 483, let A B D C be the elevation of a pedestal, the four sides of which are alike, being in plan as shown by E H G F, Fig. 484. The miters involved are what are called square miters, or miters forming a joint at 90 degrees. A square miter admits of certain abbreviations in the operation of cutting it, which makes it peculiar as compared with others. In the case of miters for all other angles the points must be first dropped from the elevation on to the plan, cutting the miter line, and then in turn transferred to the stretchout, which is laid off at right angles to the side of the plan. This is illustrated in the triangular pedestal just described, and also in the several polygonal shapes following this. A square miter may be cut in the same way, as is shown in Section 440, in which miters for several plans are obtained from the same profile. In practice, however, whether in the case of a four-sided article, as shown in the accompanying diagram, or in the case of a simple miter in a cornice or a gutter, the abbreviated method which is here illustrated is always used. This method, as will be seen, dispenses with the plan entirely. The plan E H G F, Fig. 484, is introduced only to show the shape of the article, and is not employed at all in cutting the pattern. Space the profiles, shown in the elevation by A C and B D, in the usual manner, numbering the points as shown. Set off a stretchout line, L R, at right angles to the base line C D of the pedestal, through the points in which draw measuring lines. Place the T-square parallel to the stretchout lines, and, bringing it successively against the points in the two profiles, cut the corresponding lines drawn through the stretchout. A line traced through these points, as shown by L M O N K, will be the pattern of a side.

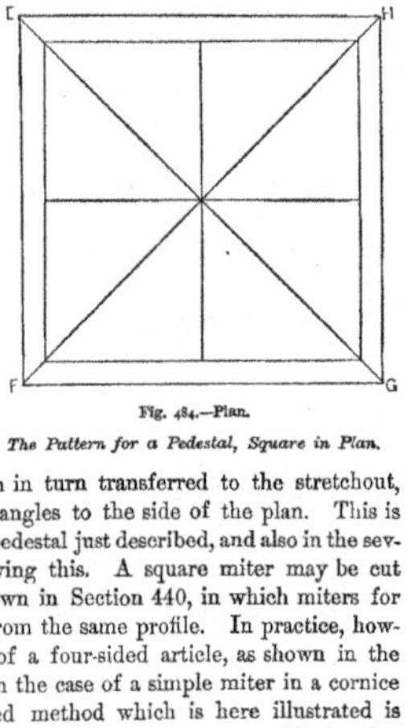

Fig. 484.—Plan.
*The Pattern for a Pedestal, Square in Plan.*

587. *The Patterns for a Vase, the Plan of which is a Pentagon.*—In Fig. 485, let S C K T be the eleva-

tion of a vase, the plan of which is a pentagon, as shown by O C[2] C[3] R P. Construct the elevation in such a manner that one of the sides will be shown in profile. Draw the plan in line and in correspondence with it. Divide the profile into spaces of convenient size in the usual manner and number them. Draw the miter lines C[1] H[1] and C[2] H[2] in the plan, and, bringing the T-square successively against the points in the profile, drop lines across these miter lines, as shown by the dotted lines in the engraving. Lay off the stretchout M N at right angles to the piece in the plan which corresponds to the side shown in profile in the elevation. Through the points in it draw the usual measuring lines. Place the T-square parallel to the stretchout line, and, bringing it against the several points in the miter lines which were dropped from the elevation upon them, cut the corresponding measuring lines drawn through the stretchout. A line traced through the points thus obtained will describe the pattern. In the case of a complicated profile, or one of many different members, to drop all the points across one section of the plan C[1] H[1] H[2] C[2] would result in confusion. Therefore it is customary, in practice, to treat the pattern in sections, describing each of the several pieces of which it is composed independently of the others. In the illustration given we have divided the pattern at the point H, describing the upper portion from the profile and plan, as above, while the lower part is redrawn in connection with a section of the plan, as shown in Fig. 486. Corresponding letters in each of the views represent the same parts, so that the reader will have no trouble in perceiving just what has been done. Instead of redrawing a portion of the elevation and plan, as we have done in this case, various other methods are sometimes resorted to by pattern cutters. It is considered best to work from one profile rather than to redraw a portion of it, as that always results in more or less inaccuracy. Therefore, after using the plan and describing a part of the pattern, as shown in the operation explained above, a piece of clean paper is pinned on the board, covering this plan and pattern, upon which a duplicate plan is drawn, from which the second section of the pattern is obtained. This operation is repeated for each of the several sections of which the pattern is composed. As this method necessitates redrawing the plan each time, which also leads to inaccuracies, some mechanics prefer, after getting one section of the pattern, to erase the points on the miter lines dropped from the elevations, thus adapting them for use again, and employ a fresh piece of paper only for the pattern. This method has the sanction of usage upon the part of some of the best pattern cutters in the country, and is probably quite as accurate as any.

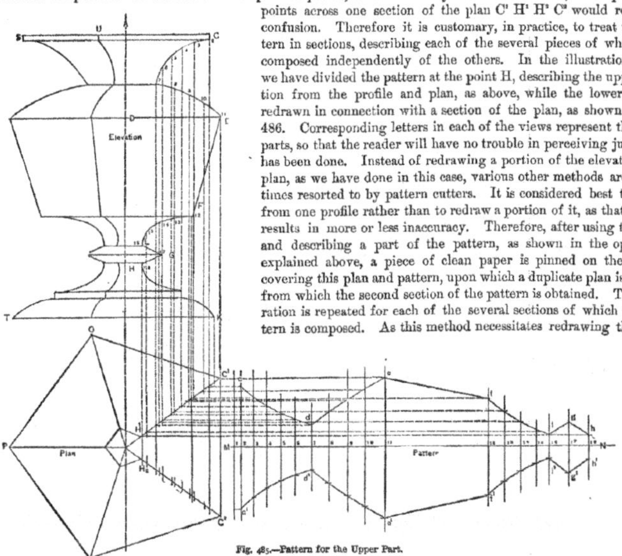


**Fig. 485.—Pattern for the Upper Part.**

*The Patterns for a Vase, the Plan of which is a Pentagon.*

588. *The Pattern for a Pedestal, the Plan of which is a Hexagon.*—In Fig. 487, let C D F E be the elevation of a pedestal which it is desired to construct of six equal sides. Draw the elevation so that one of the sides will be shown in profile. Place the plan below it and corresponding with it. Divide the profile shown by the elevation into any convenient number of spaces in the usual manner, and, to facilitate reference to them,

number them as shown. Bring the T-square against the points in the profile and drop lines across one section of the plan, as shown by H X M. At right angles to this section of the plan lay off the stretchout line N O, through the points in which draw the usual measuring lines. Place the T-square parallel to the stretchout line, and, bringing it successively against the points in the miter lines H X and M X, cut the corresponding measuring lines, as indicated by the dotted lines. Then a line traced through the points thus obtained will be the required pattern, as shown by P S T R.

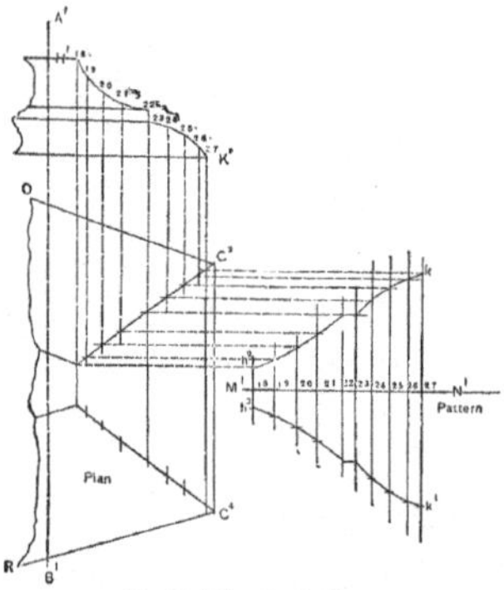


Fig. 486.—Pattern for the Base.
*The Pattern for a Vase, the Plan of which is a Pentagon.*

589. *The Pattern for a Vase, the Plan of which is a Heptagon.*—In Fig. 488, let E L P G be the elevation of the vase. Construct it in such a manner that one of its sides will be shown in profile. In line with it draw the plan, placing it so that it shall correspond with the elevation. Space the profile L P in the usual manner, and from the points in it drop lines crossing one section of the plan, cutting the miter lines R S and H V, as shown. Lay off a stretchout, A B, at right angles to the side of the plan corresponding to the side of the vase shown in profile in the elevation. Through the points in it draw the usual measuring lines. Place the T-square parallel to this stretchout line, and, bringing it successively against the points in the miter lines, cut the corresponding measuring lines, as shown. A line traced through these points, as shown by K O W U, will be the pattern of one of the sides of the vase.

590. *The Pattern for an Octagonal Pedestal.*—Let K H G W L in Fig. 489 be the elevation of a pedestal octagon in plan, of which

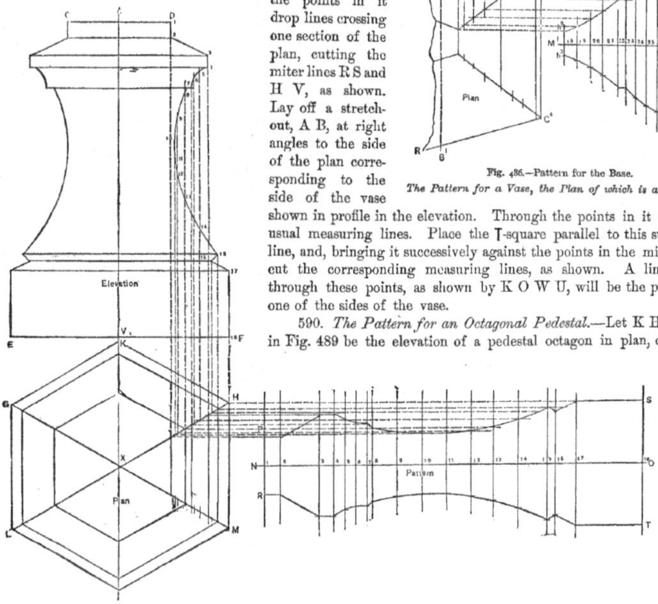


*Fig. 487.—The Pattern for a Pedestal, the Plan of which is a Hexagon.*

the pattern of a section is required. Draw the elevation in such a manner that one side will appear in profile in the elevation. Place the plan so as to correspond in all respects with it. Divide the profile G W in the usual manner, and from the points in it drop points upon each of the miter lines F T and P U in the plan.

Lay off a stretchout, B E, at right angles to the side of the plan corresponding to the side of the article shown in profile in the elevation, and through the points in it draw the usual measuring lines. Place the T-square parallel to the stretchout line, and, bringing it successively against the points dropped upon the miter lines from the elevation, cut the corresponding measuring lines. A line traced through the points thus obtained will describe the pattern of one of the sides of which the article is composed. In cases where the profile is complicated, consisting of many members, and where it is very long, confusion will arise if all the points are dropped across one section of the plan, as above described. It is also quite desirable in many cases to construct the pattern of several pieces. In such cases various methods are resorted to, several of which are fully described in connection with the problem showing a pentagon plan (Section 587). In the present case the pattern is constructed of two pieces, being divided at the point 8 of the profile. The lower part of the pattern is cut from the plan drawn below the elevation, while the upper part of the pattern is cut by means of a part of the plan redrawn above the elevation, thus allowing the use of the same profile for both. The same letters refer to similar parts, so that the reader will have no difficulty in tracing out the relationship between the different views.

591. *The Patterns of a Finial, the Plan of which is Octagon with Alternate Long and Short Sides.*—In Fig. 490, let A L M N O P R S T be the elevation of the finial corresponding to the plan which is shown immediately below it. The elevation is so drawn as to show the profile of one of the long sides, for the pattern of which proceed in the usual manner. Divide the profile A L M N O P into any number of convenient spaces, as shown by the small figures, and from the points thus obtained drop lines across the corresponding section in the plan, cutting the miter lines D E C and $D^2$ F C, as shown. To save space, a duplicate section of the plan is shown below by by $E^1$ $C^2$ $F^1$, and in the demonstration $C^2$ $E^1$ and $C^2$ $F^1$ are to be con-

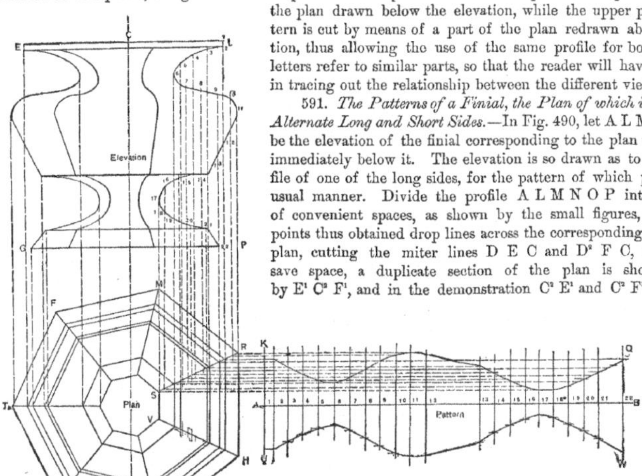


*Fig. 488.—The Pattern for a Vase, the Plan of which is a Heptagon.*

sidered the same in all respects as C E and C F. The same applies to the stretchout lines, which are indicated by the same letters. Perpendicular to D $D^2$ lay off a stretchout, as shown by G H, through the points in which draw measuring lines in the usual manner. Place the T-square parallel to the stretchout line, and, bringing it against each of the several points in D E C and $D^2$ F C, cut the corresponding measuring lines. Then a line traced through these points of intersection will be the pattern sought. For the pattern of the short sides a somewhat different course is to be pursued. A profile of the piece as it would appear if cut on the line C D must first be obtained. To do this proceed as follows: From the points in C E dropped from the profile carry lines parallel to E K across C D, cutting C K, as shown. At any convenient place lay off $B^1$ $P^1$, Fig. 491, in length equal to C D of the plan. On $B^1$ erect the perpendicular $B^1$ $A^1$, equal to B A of the elevation. On $B^1$ $P^1$ lay off points corresponding to the points obtained in C D of the plan, as above explained, and for convenience in the succeeding operations number them to correspond with the numbers in the profile from which they are derived. From the several points in the profile of the elevation draw horizontal lines, cutting the central vertical line A B, as shown. Set off points in $A^1$ $B^1$ in Fig. 491 to correspond, and through these points draw horizontal lines, which number, for convenience of identification, in the following steps. From the several points in B' P' carry lines vertically, intersecting corresponding horizontal lines. Then a line traced through these points, as shown by A' L' M' N' O' P', will be the profile of the short side on the line C D of the plan. After obtaining the profile as here described, for the pattern of the short side proceed as follows: Perpendicular to K E of the

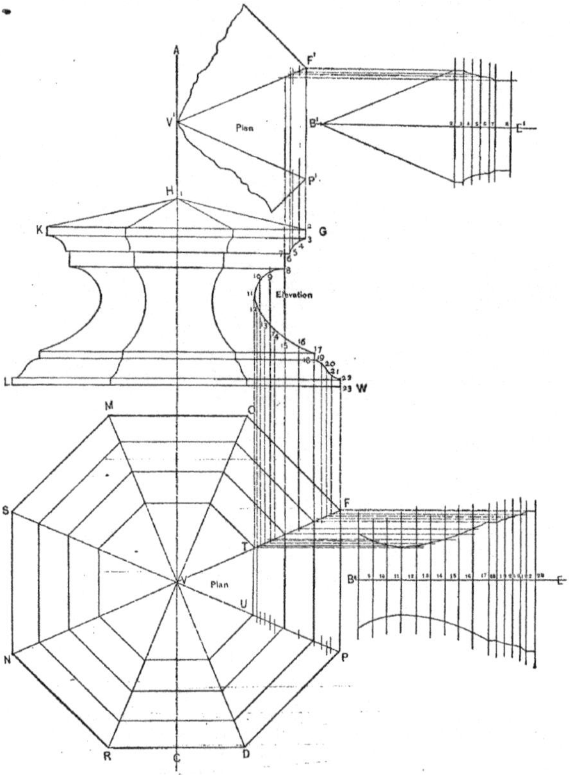


*Fig. 489.—The Pattern for an Octagonal Pedestal.*

short side lay off a stretchout of the diagonal profile, as shown by C' D', through the points in which draw measuring lines in the usual manner. Place the T-square parallel to the stretchout line, and, bringing it against the several points in the miter lines D K C and D E C bounding the short side in the plan, cut the corresponding measuring lines. Then a line traced through these points, as shown in the diagram, will be the required pattern.

592. *The Pattern for a Newel Post, the Plan of which is a Decagon.*—In Fig. 492, let V W U S P O R T be the elevation of a newel post which is required to be constructed in ten parts. Draw the plan below the

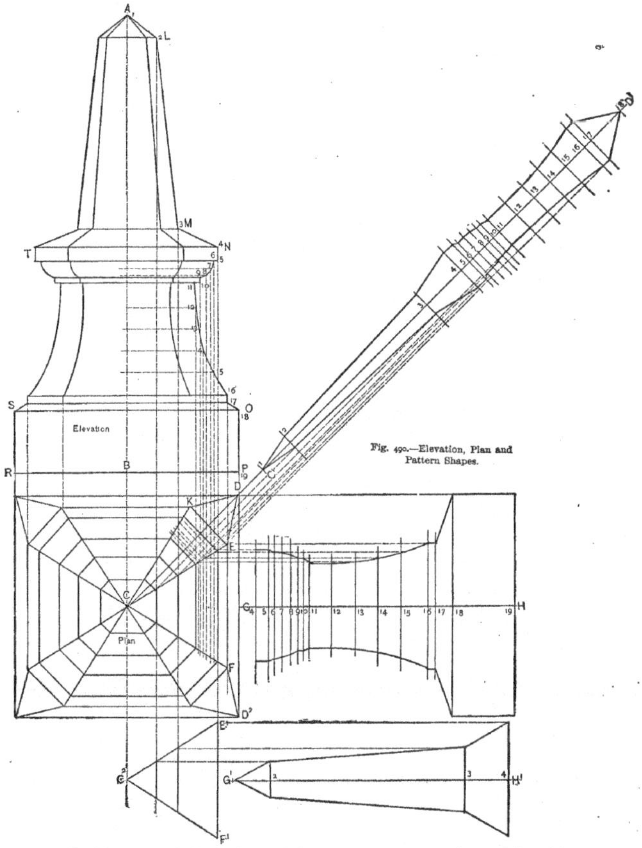


Fig. 490.—Elevation, Plan and Pattern Shapes.

*The Patterns of a Finial, the Plan of which is Octagon with Alternate Long and Short Sides.*

elevation, as shown. The elevation must show one of the sections or sides in profile, and the plan must be placed to correspond with the elevation. Space the molded parts of the profile in the usual manner, and from the points in them drop lines crossing the corresponding section of the plan, as shown by G X H, and cutting the two miter lines G X and H X. Lay off the stretchout line C D at right angles to G H, and through it draw the customary measuring lines. Place the T-square parallel to the stretchout, and, bringing it against the several points in the miter lines G X and H X, cut the corresponding measuring lines. A line traced through the points thus obtained will describe the pattern. In order to avoid confusion of lines, which would result from dropping points from the entire profile across one section of the plan, a duplicate of the cap A' W' is drawn in Fig. 493 in connection with a section of the plan, as shown by G' X' H', which are employed in precisely the same manner as above described, thus completing the pattern in two pieces, the joint being formed at the point numbered 11 of the profile and the stretchout.

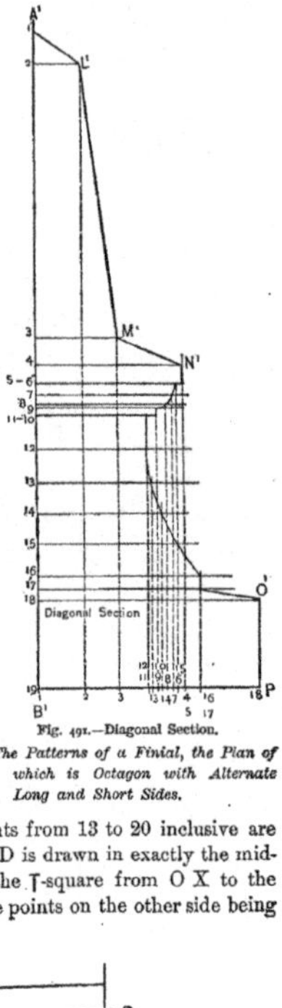


Fig. 491.—Diagonal Section.

*The Patterns of a Finial, the Plan of which is Octagon with Alternate Long and Short Sides.*

593. *The Pattern for an Urn, the Plan of which is a Dodecagon.*—In Fig. 494, let X A G H be the elevation of an urn to be constructed in twelve pieces. The elevation must be drawn so as to show one side in profile. Construct the plan, as shown, to correspond with it and draw the miter lines. Divide the profile A S G into spaces in the usual manner, and from the points thus obtained drop lines across one section, N X O of the plan. Lay off the stretchout C D at right angles to the side N O of the plan. Place the T-square parallel to the stretchout, and, bringing it successively against the several points in the miter lines N X and O X, cut the corresponding measuring lines. A line traced through the points thus obtained will describe the pattern sought. In this illustration we have shown a method sometimes resorted to by pattern cutters to avoid the confusion resulting from dropping all the points across one section of the plan. The points from 13 to 20 inclusive are dropped upon the line O X. The stretchout C D is drawn in exactly the middle of the pattern. Points are transferred by the T-square from O X to the measuring lines on one side of the stretchout, the points on the other side being

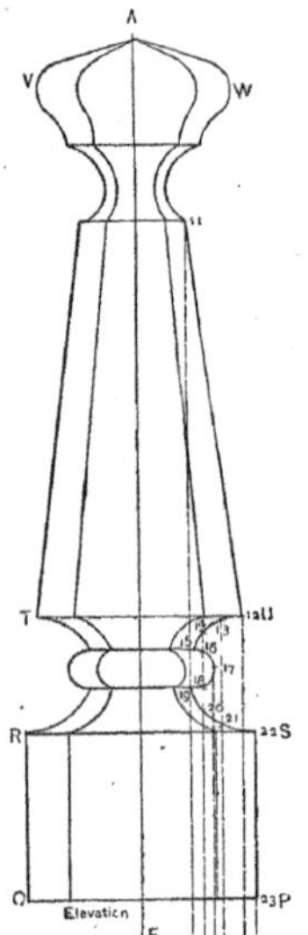


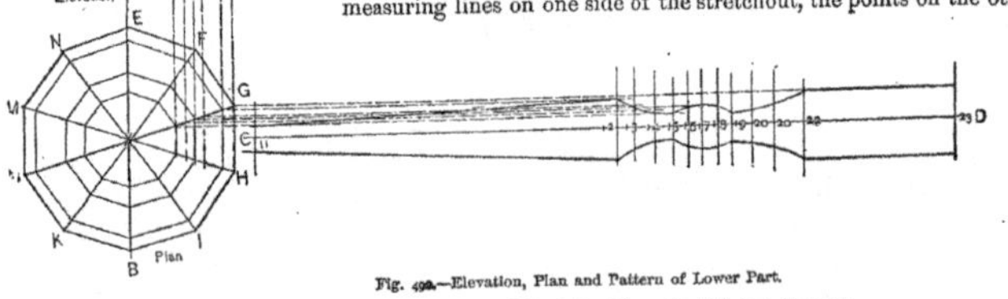


Fig. 492.—Elevation, Plan and Pattern of Lower Part.

*The Pattern for a Newel Post, the Plan of which is a Decagon.*

obtained by duplicating distances from C D on the several lines. The points 1 to 13 are dropped on N X only. The stretchout E F is laid off at right angles to M N and directly in the middle of the pattern, and the T-square being set parallel to E F, the points are transferred to the measuring lines on one side of E F, while the distances on the opposite side are set off by measurement, as described above in the first instance. This plan will be found advantageous in complicated and very extended profiles.

594. *The Patterns for an Elliptical Vase Constructed in Twelve Pieces.*—The first step is to draw an ellipse, by whatever rule is most convenient, of the length and breadth which the vase is required to have. Draw the sides of the vase about the curve, as shown in Fig. 495, in such a manner that all the points X, Y, Z, etc., shall have the same projection from the curve. Complete at least one-fourth of the plan by drawing miter lines, as shown by P C, M C, O C, U C and K C. Above the plan construct an elevation of the article, or over one end draw a profile simply, as shown by H V W L. Only the profile of the elevation is needed for the purpose of pattern cutting, but the other lines are desirable in process of designing, in order that the effect may be considered before the article is constructed.

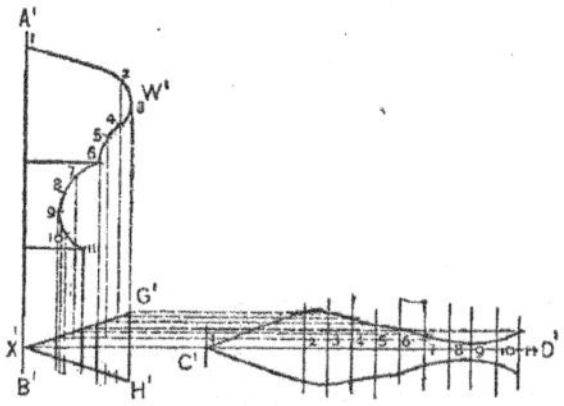


Fig. 493.—Pattern of Cap.

*The Pattern for a Newel Post, the Plan of which is a Decagon.*

Divide the profile H V W L in the usual manner, and from the several points in it drop lines across the corresponding section (No. 1) of the plan. Take the stretchout of H V W L and lay it off at right angles to the side of section No. 1 of the plan, as shown by E F. Through the points in it draw the customary measuring lines. Place the T-square parallel to this stretchout line, and, bringing it against the several points dropped upon the miter lines N C and U C bounding No. 1 of the plan, cut the corresponding measuring lines. Then a line traced through the points thus obtained will be the pattern of section No. 1. Across the second section in the plan, from the points already obtained in U C, draw lines parallel to O N, the side of it, and produce them until they meet A C, which is a line drawn from C at right angles to U O produced. Then the points in A C serve to obtain a profile of the section numbered 2. In like manner continue the points from C O across the third section in the plan, also parallel to O M, the side of it, and produce them until they cut C B, which is a line drawn from C at right angles to O M produced. Then C B contains the points requisite in obtaining a profile of the third section. Continue the points in C M across the fourth section, cutting its other miter line C P. From C draw C D at right angles to the side P M of the section. Then upon C D

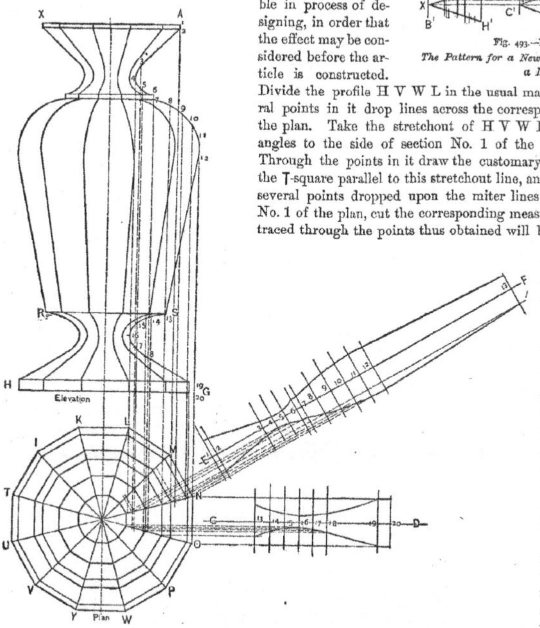


*Fig. 494.—The Pattern for an Urn, the Plan of which is a Dodecagon.*

being cut by the lines drawn across the section, will be found the points necessary to determine the profile of the fourth pattern. Produce the line of the base of the elevation indefinitely, as shown by C' C² C³, and also the line of the top A' B² D³. From the several points in the profile H V W L draw lines indefinitely, parallel to the

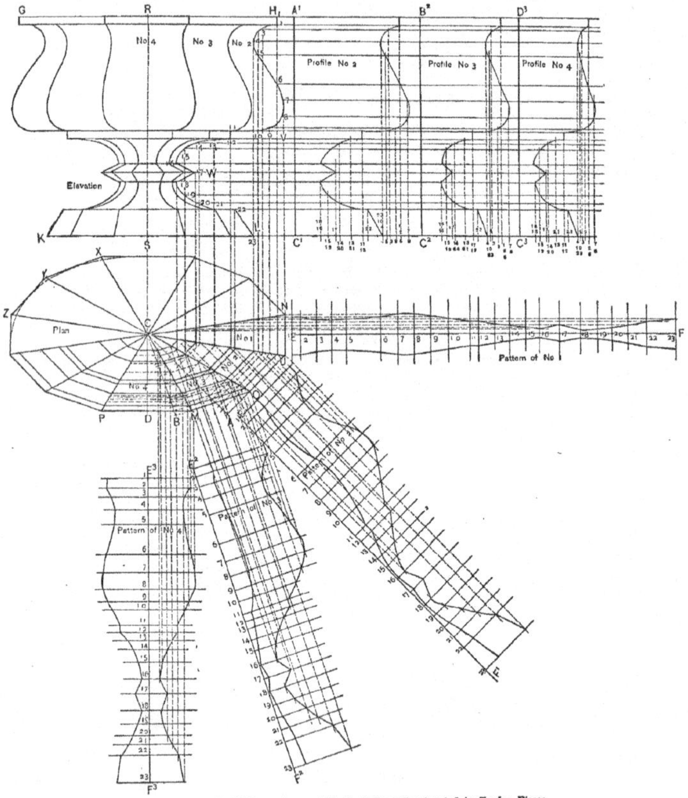


*Fig.* 495.—*The Patterns for an Elliptical Vase Constructed in Twelve Pieces.*

lines just described and as shown in the diagram. From C', upon the base line produced, set off points corresponding to the points in C A of the plan, making the distance from C' in each instance the same as the distance from C in the plan. Number the points to correspond with the numbers given to the points in the profile

H V W L, from which they were derived. In like manner from $C^2$ set off points corresponding to the points in C B of the plan, numbering them as above described. From $C^3$ set off points corresponding to those in C D of the plan, likewise identifying them by figures in order to facilitate the next operation. From $C^1$ erect the perpendicular $C^1$ $A^1$; likewise from $C^2$ and $C^3$ erect the perpendiculars $C^2$ $B^2$ and $C^3$ $D^3$. From each of the points laid off from $C^1$, and also from each of those laid off from $C^2$ and $C^3$, erect a perpendicular, producing it until it meets the horizontal line drawn from the profile H V W L of corresponding number. Then lines traced through these several intersections will complete the profiles, as shown. Perpendicular to the side of each section in the plan, lay off a stretchout taken from the profile corresponding to it, just described, and through the points in the stretchout draw measuring lines in the usual manner, all as shown by $E^1$ $F^1$, $E^2$ $F^2$ and $E^3$ $F^3$. Place the T-square parallel to each of these stretchout lines in turn, and, bringing it against the several points in the miter lines bounding the sections of the plan to which they correspond, cut the measuring lines in the usual manner. Then lines traced through the points of intersection thus obtained, all as shown in the diagram, will complete the patterns.

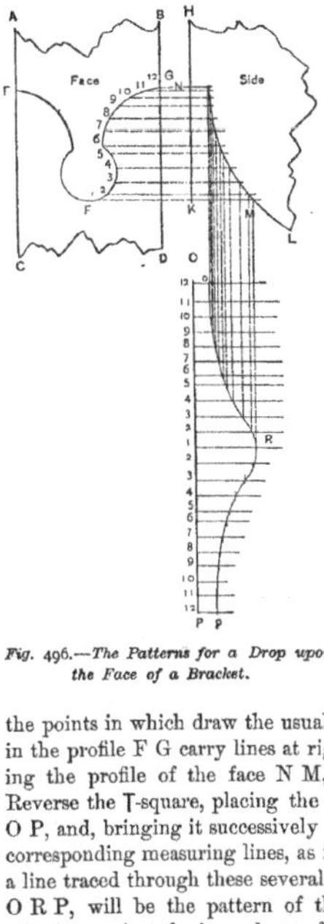


*Fig. 496.—The Patterns for a Drop upon the Face of a Bracket.*

595. *The Patterns for a Drop upon the Face of a Bracket.*—In Figs. 496 and 497 methods of obtaining the return strip fitting around a drop and mitering against the face of a bracket, are shown. Similar letters in the two figures represent similar parts, and the following demonstration may be considered as applying to both. Let A B D C be the elevation of a part of the face of the bracket, and H K L a portion of the side, showing the connection between the side strip of the drop E F G and the face of the bracket. Divide the profile F G into any convenient number of parts in the usual manner, as shown by the small figures. Produce N K, as shown by O P, and on O P lay off a stretchout, through the points in which draw the usual measuring lines. From the points in the profile F G carry lines at right angles to the bracket, intersecting the profile of the face N M, against which the drop is to miter. Reverse the T-square, placing the blade parallel to the stretchout line O P, and, bringing it successively against the points in N M, cut the corresponding measuring lines, as indicated by the dotted lines. Then a line traced through these several points of intersection, as shown by O R P, will be the pattern of the strip fitting around E F G and mitering against the irregular surface N M of the bracket face.

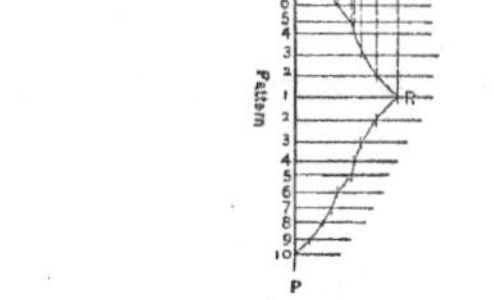


*Fig. 497.—The Patterns for a Drop upon the Face of a Bracket.*

596. *The Patterns of a Boss Fitting over a Miter in a Molding.*—Let A B C in Fig. 498 be the part elevation of a pediment, as in a cornice or window cap, over the miter in which, and against the molding and fascia, a boss, F K G H, is required to be fitted, all as shown by A D E. For the patterns we proceed as follows: Divide so much of the profile of the boss K F H G as comes against the molding, shown from K to F, into any convenient number of parts, and from these points draw lines parallel to the lines of the molding until they intersect the profile of the molding, as shown from N to O. Also draw a line from the point H until it intersects the fascia in the point E. Then the points from N to O and the point E are the points by which measurements are to be taken in laying out the pattern on the stretchout line. In line with the side elevation lay off a stretchout of the boss, as shown by $K^1$ $K^2$, dividing the portion $K^1$ $F^1$, which corresponds to K F of the elevation, into like spaces, through which draw the usual measuring lines.

In the same manner divide the space $G^1 K^1$, which corresponds to G K of the elevation, into the same number of spaces as employed in the portion F K, to obtain points in the profile of the molding N O, and also draw measuring lines, as shown. Place the T-square parallel to the stretchout line $K^1 K^2$, and, bringing it against the several points in N O, cut corresponding measuring lines, as shown. Then lines traced through these points of intersection, as shown by $K^1$ L and M $K^2$, will be the required pattern.

597. *The Patterns of an Octagonal Shaft, the Profile of which is Curved, Mitering upon the Ridge of a Roof.*—In Figs. 499 and 500 are shown the elevation and patterns of a finial of a character somewhat common in cornice work. The shaft is octagon in shape. Four crockets and a point constitute the flower surmounting the same. The neck molding immediately below the flower consists of eight simple octagon miters, the patterns for which are cut by the ordinary rule, and need not be described in this connection. The shaft below the neck molding miters over the ridge of the roof. It is also curved in its profile, and by reason of these several combined features presents conditions differing from other problems of a similar character already demonstrated. For the patterns proceed as follows: Construct a plan of the shaft at its largest section, as shown by A, B, C, etc., from the center of which to two of the angles draw miter lines, as shown by G H and G H. Divide the profile of the side of the shaft J L into any number of parts in the usual manner, and from these points carry lines vertically crossing the miter lines G H and G H. Bisect the section bounded by the miter line, as shown by $E^1 F^1$, upon which line lay off a stretchout of the profile J L, drawing measuring lines through the points. Place the T-square parallel to the stretchout line, and, bringing it successively against the points in G H and G H, cut corresponding measuring lines, as shown, and through the points thus obtained trace lines, all as indicated in the drawing. This gives the general shape of the pattern for the sections of the shaft. By inspection of the plan and elevation together, it will be seen that to fit the shaft over the roof some of the sections composing it will require different cuts at their lower extremities. Two of the sections will be cut the same as the pattern already described. They correspond to the side marked A in the plan and the one opposite. Two others, one of which is indicated in the plan by C, and which is also shown in the elevation by *n m n*, will be cut to fit over the ridge of the roof. The remaining four pieces will be cut to fit against the pitch of the roof, as shown by *n o* in the elevation, and corresponding to the sides, of which B in the plan is one. For the sections corresponding to the one shown in the center of the elevation proceed as follows: From so many of the points in the profile J L as occur below a point opposite the ridge of the roof *m*, draw lines at right angles to the center line of the shaft, crossing the lines K I, K I, representing the pitch of the roof, all as shown. Thus it will be seen that the line drawn from 4 touches the ridge in the point *m*, while the line drawn from 3 corresponds to the point at which the side terminates against the pitch of the roof. Therefore, in the pattern draw a line from the center of it, on the measuring line 4, to the side of it on the measuring line 3, all as shown by *m' n'* and *m' n'*. Then these are the lines of cut in the pattern corresponding to *m n* and *m n* of the elevation. By inspection of the elevation, for the remaining four sides it will be seen that it is necessary to make a cut in the pattern from one side, in a point corresponding to 3 of the profile, to the other, in a point corresponding to 1 of the profile, all as shown by *n o*. Taking corresponding points, therefore, in the measuring lines of the pattern, draw the lines *n' o'*, as shown. Then the original pattern, modified by cutting upon these lines, will constitute the pattern for the other sides. In this connection we may remark that

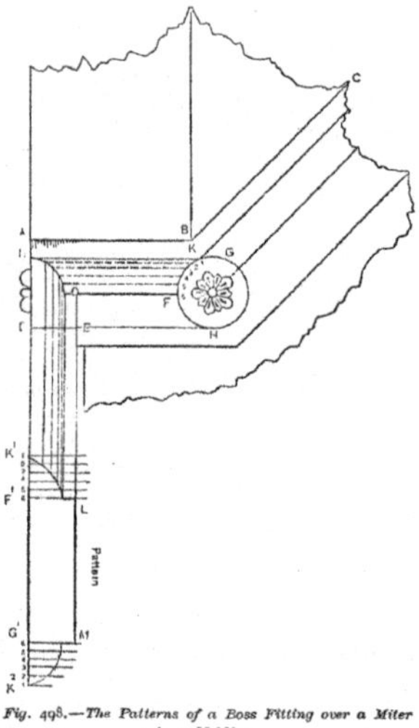


*Fig. 498.—The Patterns of a Boss Fitting over a Miter in a Molding.*

for the crockets and point no pattern can be described. Of course an approximation to the forms shown might be devised, consisting of geometrical shapes, but it is far better in point of construction, wherever possible, to use either pressed work or hand-hammered work instead. Therefore we make no attempt to show patterns for the foliated parts.

598. *To Construct a Ball in any Number of Pieces, of the General Shape of Zones.*—In Fig. 501, let A O G H be the elevation of a ball which it is required to construct in thirteen pieces. Divide the profile into the required sections, as shown by 0, 1, 2, 3, 4, etc., and through the points thus obtained draw parallel horizontal lines, as shown. The divisions in the profile are to be obtained by the following general rule, applicable in all such cases: Divide the whole circumference of the ball into a number of parts equal to two times one less than the number of pieces of which it is to be composed. In convenient proximity to the elevation, the centers being located in the same line, draw a plan of the ball, as shown by K M L N. Draw the diameter K L parallel to the lines of division in the elevation. With the T-square placed at right angles to this diameter, and brought successively against the points in the elevation, drop corresponding points upon it, as shown by 1, 2, 3, 4, etc. Through each of these points, from the center by which the plan is drawn, describe circles. Each of these circles becomes the plan of one edge of the belt in the elevation to which it corresponds, and is to be used in establishing the length of the arc forming the pattern with which it corresponds. Through the elevation, at right angles to the lines of the zones or belts of which the ball is to be composed, draw a diameter, as shown by G A, which produce in the direction of O indefinitely. Construct chords to the several arcs into which the profile is divided by the division lines, which produce until they cut G A O, as shown by 1 2 E, 2 3 D, 3 4 C, 4 5 B and 5 6 A. Then E 2 and E 1 are the radii of parallel arcs which will describe the pattern of the first division above the center zone, and D 3 and D 2 are the radii describing the pattern of the third zone, and so on. From E' in Fig. 502 as center, with E 2 and E 1 as radii, strike the arcs 2 2 and 1 1 indefinitely. Step off the length on the corresponding plan line, and make 1 1 equal to the whole of it, or a part, as may be desired—in this case a half. In like manner describe patterns for the other pieces, as shown, struck from the centers D', Fig. 503; C', Fig. 504; B', Fig. 505, and A', Fig. 506. The pattern for the smallest section, as indicated by F in the plan, may be pricked directly from it, or it may be struck by a radius equal to F 6 in the plan. The center belt or zone, shown in the profile by 1 0, is a flat band, and is therefore bounded by straight parallel lines. The width is taken by 1 0 in the elevation, and the length is measured upon 1 of the plan, all as shown in Fig. 507.

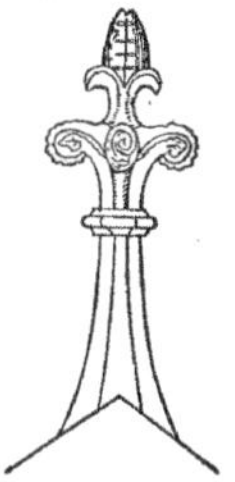

Fig. 499.—Elevation of Finial, of which the Shaft is a Part.

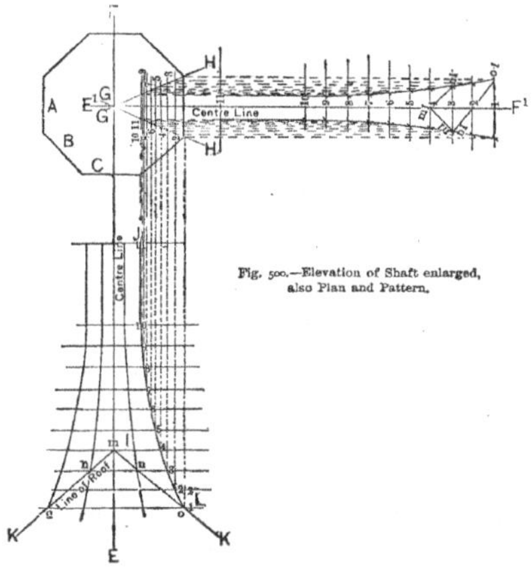


Fig. 500.—Elevation of Shaft enlarged, also Plan and Pattern.

*The Patterns of an Octagonal Shaft, the Profile of which is Curved, Mitering upon the Ridge of a Roof.*

599. *To Construct a Ball in any Number of Pieces, of the General Shape of Gores.*—Draw a circle of a size corresponding to the required ball, as shown in Fig. 508, which divide, by any of the usual methods employed in the construction of polygons, into the number of parts of which it is desired to construct the ball, in this case twelve, all as shown by E, F, G, H, etc. From the center draw miter lines, as represented by R E

and R F. If the polygon is inscribed, as shown in the illustration, it will be observed that the arc of the circle, as, for example, U C, does not form a profile in dimensions corresponding to the middle line of the sections of which the ball is to be constructed. Hence, it is necessary to draw a new profile, which may be done with sufficient accuracy for all practical purposes by taking the radius of the profile, and a point for the center whose distance from the line A V prolonged is equal to the distance from the point U' to U in the plan. Then, from the

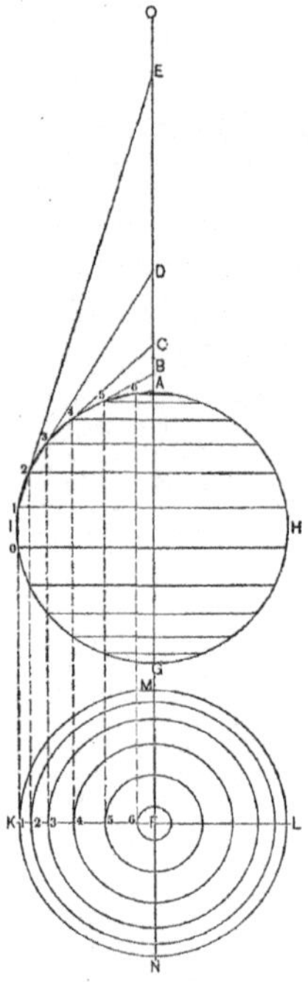


Fig. 501.—Plan and Elevation.

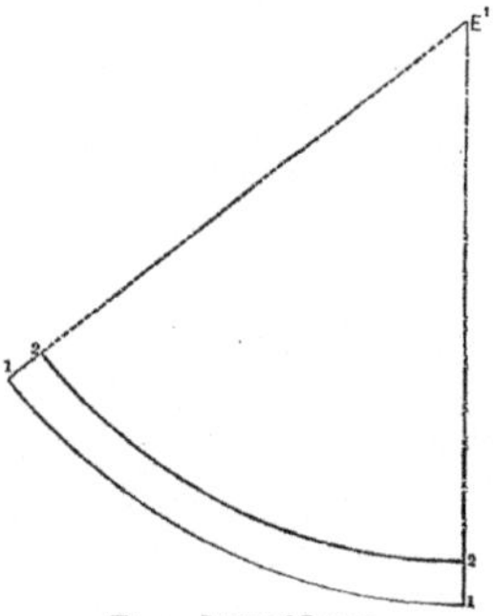


Fig. 502.—Pattern of Zone 1 2.

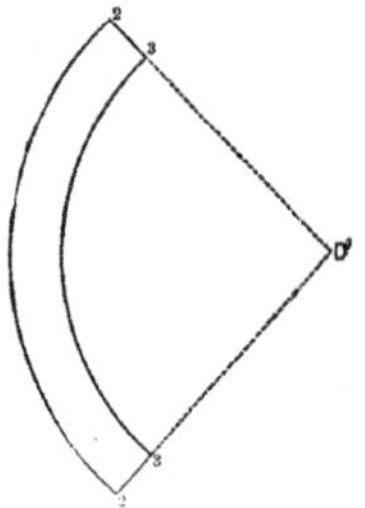


Fig. 503.—Pattern of Zone 2 3.

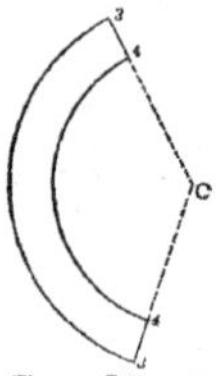


Fig. 504.—Pattern of Zone 3 4.

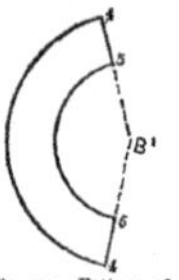


Fig. 505.—Pattern of Zone 4 5.

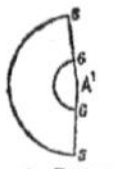


Fig. 506.—Pattern of Zone 5 6.

Fig. 507.—Pattern of Middle Zone.

*To Construct a Ball in any Number of Pieces of the General Shape of Zones.*

point located near V, as above described, as center, and with a radius equal to R U, strike the arc B A, which forms the profile of a section of the ball on its center line. Divide B A into any convenient number of equal parts, and from the divisions thus obtained carry lines across one of the sections at right angles to a line drawn through its center, and cutting its miter lines, all as shown in R E and R F. Prolong the center line R C, as shown by S T, and on it lay off a stretchout obtained from B A, through the points in which draw measuring lines in the usual manner. Place the T-square parallel to the stretchout line, and, bringing it successively against the points in the miter lines R E and R F, cut the corresponding measuring lines, as shown. A line traced through these points will give the pattern of a section. If, on laying out the plan of the ball, the polygon had been drawn about the circle, instead of inscribed, as shown in the engraving, it is quite evident that a quarter of the circle would have answered the purpose of a profile. These points, with reference to the profile, are to be observed in determining the size of the ball. In the illustration presented, the ball produced will correspond in its miter-lines to the diameter of the circle laid down, while if measured on lines drawn through the center of its sections it will be smaller than the circle.

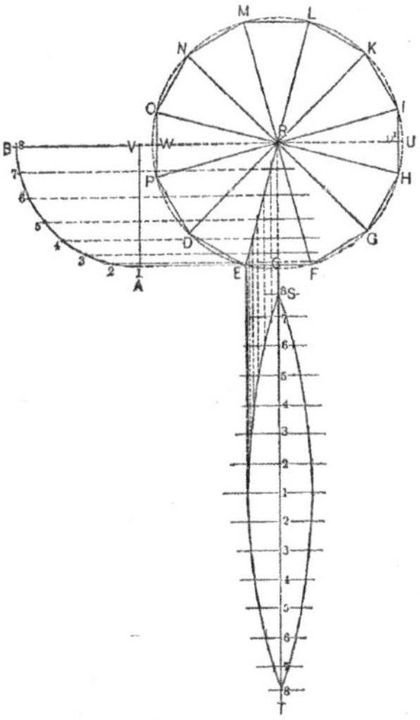

*Fig. 508.—To Construct a Ball in any Number of Pieces, of the General Shape of Gores.*

600. *The Patterns of a Square Shaft to Fit Against a Sphere.*—In Fig. 509, let H A $A^1$ K be the elevation of a square shaft, one end of which is required to fit against the ball D F E. From the center G describe the circle of the ball. Through G draw a vertical line, as shown by F L. At equal distance from either side of this center line F L, draw the sides of the shaft, as shown by H A and K $A^1$, continuing them across the line of the circumference of the ball indefinitely. From the points of intersection between the sides of the shaft and the circumference of the ball, A or $A^1$, draw a line at right angles to the sides of the shaft, across the ball, cutting the center line, as shown at B. Set the dividers to G B as radius, and from G as center, describe the arc C $C^1$. Then H C $C^1$ K will be the pattern of one side of a square shaft to fit against the given ball.

601. *To Describe the Pattern of an Octagon Shaft to Fit Against a Ball.*—Let H F K in Fig. 510 be the given ball, of which G is the center. Let $D^2$ $C^2$ $C^2$ $D^2$ E represent a plan of the octagon shaft which is required to fit against the ball. Draw this plan in line with the center of the ball, as indicated by F E. From the angles of the plan draw lines indefinitely, cutting the circle. From the point A or $A^1$, where the side in profile cuts the circle, draw a line across the center line of the ball F E, cutting it in the point B, as shown. Through B, with the center by which the circle of the ball was struck, describe an arc, cutting the two lines drawn from the inner angles $C^2$ $C^2$ of the plan, as shown by C and $C^1$. Then M C $C^1$ N will be the pattern of one side of an octagon shaft mitering against the given ball H F K. If it be desired to complete the elevation of the shaft meeting the ball, it may be done by carrying lines from C and $C^1$ horizontally until they meet the outer line of the shaft in the points D and $D^1$. Connect $C^1$ and $D^1$, also C and D, by a curved line, the lowest point in which shall touch the horizontal line drawn through B. Then the broken line D C $C^1$ $D^1$ will be the miter line in elevation formed by an octagon shaft meeting the given ball.

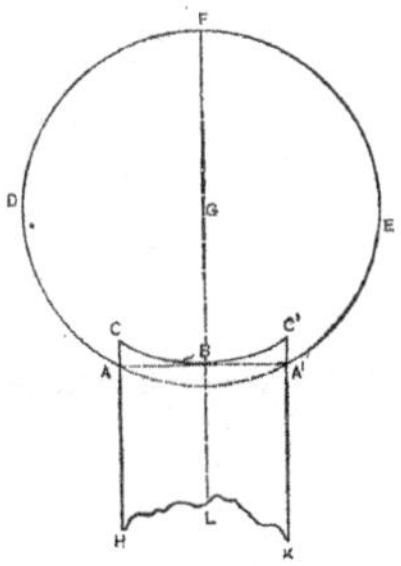

*Fig. 509.—The Patterns of a Square Shaft to Fit Against a Sphere.*

602. *Patterns for the Volute of a Capital.*—Draw an inverted plan of the parts, as shown in Fig. 511, and through the center of one of the volutes draw a line, A B, which shall correspond to the center line of the patterns. Construct the diagonal elevation, as shown, placing it in correspondence with the plan. Divide the volute, as shown in the diagonal elevation, into any convenient number of parts, numbering them for convenience of identification, as shown. From each of the several points in the elevation thus obtained drop lines crossing the plan, as shown. Prolong the line A B, as shown by B C, upon which lay off the stretchout of the several parts of which the volute is composed, drawing the usual measuring lines. Place the T-square parallel to the stretchout line, and, bringing it against the points formed by the lines of the eleva-

tion crossing the plan, cut the corresponding measuring lines drawn through the stretchout line, all as shown. In order to avoid confusion of lines, but one-half of each pattern is shown in the engraving. In ordinary work sufficient accuracy is obtained if the sides of the volute are pricked directly from the diagonal elevation, which saves the long and tedious operation required to develop them. It will be seen, by inspection of the elevation and plan, that the difference in the length of the sides, as shown in the two views, is very slight indeed.

603.—*The Patterns for a Cornucopia in Eight Pieces.*—In Fig. 512 is shown the side and end elevation of a cornucopia which is to be constructed in eight pieces. The first step in the development of the pattern is a correct representation of the article in these two views just named. It is not our purpose in this connection to describe in detail the method of drawing these two views. Certain parts must necessarily be conceived in the mind before they are laid upon the paper. For example, having determined that the article is to be constructed in eight pieces, and that its size at the mouth is to be of given dimensions, draw the section E F Y B D C X A opposite its corresponding line, H' G', in the side elevation. Having determined the length of the article and its general shape, the profiles H' y' of the top, and G' c' of the bottom are drawn. The end elevation is then worked out from these lines by means of corresponding lines carried across, after which the intermediate lines showing the side, which is turned directly toward the sight, are inserted, being derived in the same way from the sectional view. We think this much of a general description will enable the intelligent reader to construct the necessary views of such an article. But, ordinarily, work of this character comes to the pattern cutter already drawn, the labor of delineating it being the work of a draftsman and designer, rather than that of the pattern cutter. Accordingly, we commence our description with the assumption that the side and end elevations have been correctly drawn. By inspec-

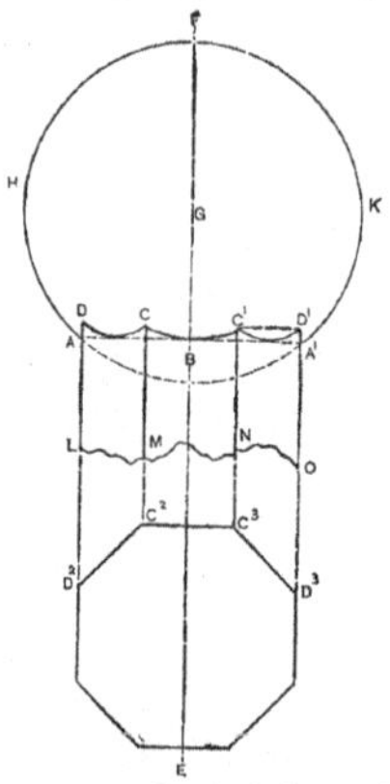


*Fig. 510.—To Describe the Pattern of an Octagon Shaft to Fit Against a Ball.*

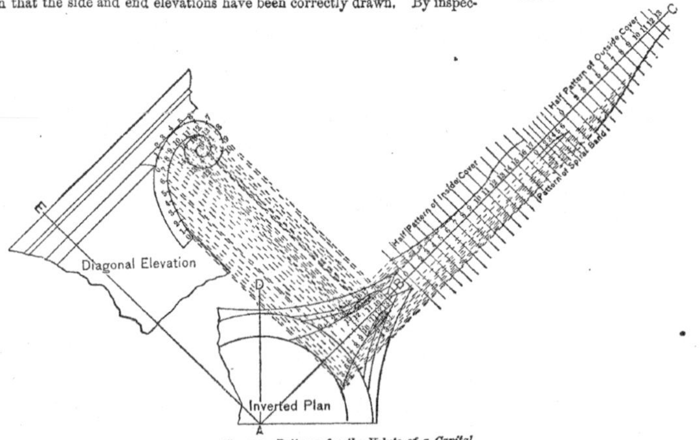


*Fig. 511. Patterns for the Volute of a Capital.*

tion of the side elevation, it will be seen that the profile of piece No. 1 is to be taken directly from the lower line in the side elevation. Therefore, at right angles to X C, forming the side of the plan bounding section No. 1, lay off a stretchout taken from the profile $G^1$ $c^1$, as shown by $G^3$ $C^3$. Through the points used in laying off this stretchout, draw measuring lines in the usual manner. From the corresponding points in the profile $G^1$ $c^1$ carry lines across section No. 1 in the end elevation, cutting the two miter lines, as shown. With the T-square

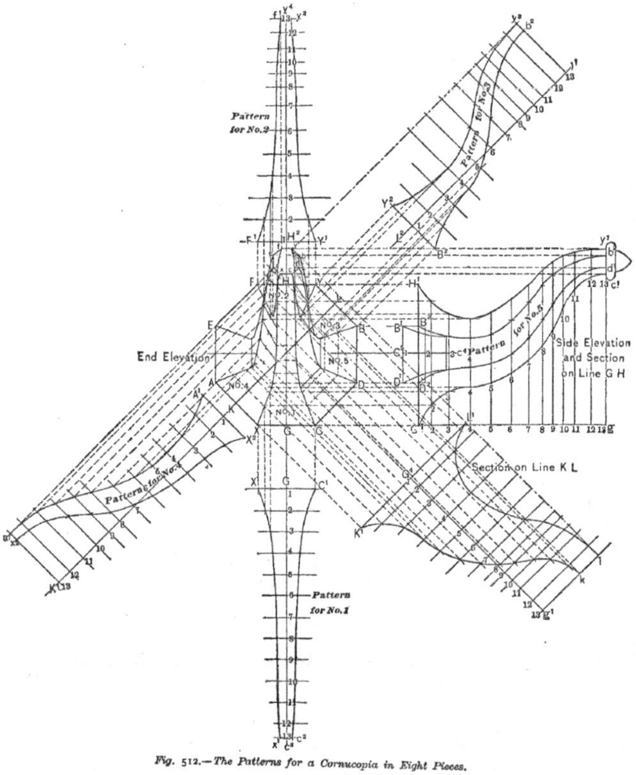


*Fig. 512.—The Patterns for a Cornucopia in Eight Pieces.*

at right angles to the side X C, and brought successively against these points in the two miter lines, cut corresponding measuring lines. Then lines traced through these several points of intersection, as shown by $X^1$ $x^1$ and $C^1$ $c^2$, will give the shape of the pattern of this section. In like manner take a stretchout of the profile of the upper side $H^1$ $y^1$ of the side elevation, and lay it off at right angles to the side of the plan F Y, bounding section No. 2, all as shown by $H^2$ $y^4$, through the points in which draw the usual measuring lines. From the points in the profile H' y' carry lines across the end elevation, cutting the miter lines bounding section No. 2. Then with the T-square at right angles to the side F Y of the plan, bringing it successively against the several points in the miter lines bounding section No. 2, cut the corresponding measuring lines. Then lines traced through these points of intersection, as shown by F' f' and Y² y², will be the pattern for section No. 2. For sections Nos. 3 and 4, situated diagonally to the section by which the stretchouts for the two patterns just described were obtained, another sectional view of the cornucopia must be constructed. To obtain stretchouts for these patterns, a section must be taken through the article at right angles to their respective sides. Our next step, therefore, is to construct a section corresponding to the line K L drawn through the plan, for which we proceed as follows: Through the point G' draw a horizontal line, G' g, upon which drop points from the profile G' c', all as indicated by the small figures. At right angles to K L, at any point convenient for the required section, draw G⁴ g', in length equal to G' g, in which set off points corresponding to the points in G' g. Through these points draw lines after the usual manner of measuring lines. In order to obtain the points in the end elevation from which to draw lines cutting these measuring lines, by which to determine the diagonal section, we proceed as follows: From the points in the line D² d' carry lines cutting the lines A a in end elevation, and in like manner from the line B² b' carry lines cutting the line B b of the end elevation. By this means it will be seen that in the boundary lines of pieces Nos. 4 and 3 the same points have been obtained, both being derived from lines in the side elevation having corresponding divisions. Therefore, if these points be connected by drawing lines across the respective sections, and their middle points be taken, we shall have points of measurement by which to construct the diagonal section. The diagonal line K L cuts a number of these cross lines in the center, but the others, on account of the distortion of the end elevation, will fall at other points than on the line K L. It will be seen, for instance, that the line corresponding to 8 crosses section No. 4 obliquely, but still its center point must be the center point in the section. Therefore, from the center point in the two sections Nos. 4 and 3 thus determined, draw lines cutting the measuring lines drawn through G⁴ g'. Then lines traced through these intersections, as shown by L' l and K' k, will form the diagonal sections of the article corresponding to a line, K L. For the pattern of No. 3 proceed as follows: At right angles to its side, W B, lay off a stretchout corresponding to the side of the section constructed, agreeing with it. In other words, make the stretchout L² l' equal to L' l of the section. Through the points draw measuring lines in the usual manner. Bring the T-square successively against the points in the miter lines bounding No. 3, placing the blade at right angles to the side Y B, and cut corresponding measuring lines. Then lines traced through the points of intersection thus formed, as shown by B² b² and Y² y², will form the pattern of No. 3. In like manner, for the pattern of No. 4 proceed as follows: At right angles to the side A X lay off a stretchout taken from the side of the section K' k, which corresponds with it, all as indicated by K k', through the points in which draw the usual measuring lines. Place the T-square at right angles to the sides A X, and, bringing it successively against the points in the miter lines bounding piece No. 4, cut the corresponding measuring lines. Then lines traced through the points of intersection thus obtained, as shown by X² x² and A' a', will be the pattern for it. The pattern for No. 5 is obtained directly from the side elevation, as shown. That part of it in the smaller portion of the article is bounded by lines so nearly corresponding to the side elevation as to render it impossible in an engraving so small as here represented to distinguish between them. It begins to deviate, however, in a manner that may be shown in points corresponding to line No. 4, and a description of this part will serve to illustrate the principle upon which the development of the pattern is based. Commencing with point 4, lay off a stretchout taken from the corresponding portions of the profile G' C', as indicated by the small figures 3 2 1 in the line G⁵ c⁴. Through these points draw measuring lines in the usual manner, and with the T-square placed at right angles to them, and brought successively against corresponding points in the sides of piece No. 5, as shown in the side elevation intersecting them, trace lines, all as indicated by the lines terminating at B' D'. Then B' b² d' D' will be the pattern of piece No. 5.

604. *The Patterns for a Ship Ventilator, having an Oval Mouth on a Round Pipe.*—In Fig. 513 there are presented the front and side elevations of a style of ship ventilator occasionally employed. It starts from a round pipe, A' B', at the base, and ends in an elliptical shape, as shown by O R P S, at the mouth. The rule which we present for developing the patterns is one allowing the mechanic the largest possible latitude in proportioning the article. It is also one which, with slight modifications, can be made to answer in the patterns of other ventilators of the same general kind which differ in the shape of the mouth. Care must be taken to draw the elliptical lines representing the sections, both in the elevation and in the development of the patterns, by the same means in all cases. For example, if a string and pencil or the trammels are used in drawing

R P S O, the same means should be used in drawing corresponding sections wherever they may be required. The reason for this is very simple. The principle upon which the rule is based is that an oblique section through a cylinder, and also a section through the opposite sides of a cone, is an ellipse. Having established the section through the article at either of the joint lines, both of the pieces which there meet must be based upon that section, so far as their stretchouts and other measurements are concerned, and there should be a correspondence between the several sections in this respect. To draw the section for the edge of one pattern piece with the trammels, and for the other which meets it from centers with the compasses, would hardly produce satisfactory results. It is believed that the method here presented, on account of its brevity, comparing it with other rules which might be used to accomplish the same result, is one that will be found of great service in practical work. If the patterns, as shown in Fig. 515, are not laid out very accurately and carefully, misfits will occur. By the very nature of the operation here described, slight variations in obtaining points of measurements will prove cumulative in character, each succeeding step leading further from the correct line. Hence the necessity of accuracy in applying the rule. Aside from the care necessary to be taken with the sections above mentioned, the parts may be proportioned according to the judgment of the designer and the requirements of the case. Let A' B' be the size of the pipe upon which the ventilator fits at the bottom. Let O P and R S be the dimensions of the elliptical mouth. From these two sections proceed to draw the elevation A B D C. The lines A E G K M C and B F H L N D may be drawn at pleasure. Having determined their form, divide them by points, as shown by E G K M in the one and F H L N in the other, by which to locate the seams between the parts of which the article is to be composed. Connect these two sets of points by lines, as shown by E, F, G, H, etc. The lines R U and S V in the front elevation are to be drawn by eye rather than by any set rule. The only direction that needs to be given is to proportion their sweep to the width suggested by the outlines of the side. Their office in the development of the patterns is to determine the width of the several elliptical sections taken through the article. Therefore, if they are abrupt in their curve at any point, they are likely to produce an unsatisfactory outline in the finished work. By these two elevations the work is laid out as it is to be constructed. As will be evident from inspection of the engraving, a separate pattern will be required for each section. Since all of these, save the lower one, are alike in kind, though differing in size, a single example will be sufficient for showing the principles involved. The pattern of the section E A B F will be the same as that for the corresponding piece in an ordinary elbow, and may be developed as described in Section 511, and therefore need not be specially explained here. The patterns for the other sections will be developed as follows, taking M N D C as an example: This section, for convenience and in order to avoid confusion of lines, is transferred to the opposite side of the front elevation, as shown by W Y Z X. Bisect the several lines of seams between the sections. Thus, bisect the line C D, obtaining the point *n*. Bisect M N, obtaining the point *m*. In like manner locate *k g e*. These points are to be used in determining the width of the several elliptical sections, and for this purpose lines from them are carried

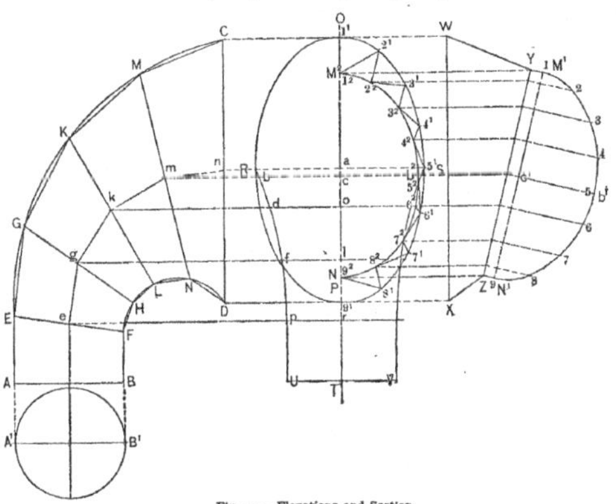


**Fig. 513.—Elevations and Section.**
*A Ship Ventilator, having an Oval Mouth on a Round Pipe.*

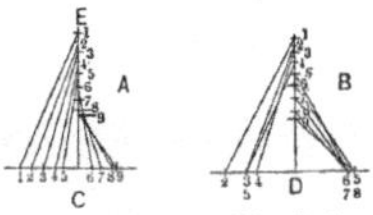


**Fig. 514.—Diagrams of Triangles for Measurements.**
*A Ship Ventilator, having an Oval Mouth on a Round Pipe.*

across the front elevation, cutting the lines R U and S V, as shown. Having drawn the section W Y Z X in line with the front elevation, as already described, drop points from Y and Z perpendicular to the section line O T of the elevation, thus locating the points $M^2$ and $N^2$. Make the distance $b^2$ *c* equal to *b c*. Then draw the ellipse $M^2$ $b^2$ $N^2$, which will be a plan or top view of the section M N of the side elevation. On a line parallel with Y Z construct the section $M^1$ $b^1$ $N^1$, as follows: Let $M^1$ $N^1$ be equal to and opposite Y Z. Let the distance $c^1$ $b^1$ be equal to the distance *c b* of the section. With these points determined, draw the curve $M^1$ $b^1$ $N^1$, which will be a regular ellipse. Divide the sections $M^1$ $b^1$ $N^1$ and O S P into the same number of equal parts, as indicated by the small figures in the engraving. Drop the points 1, 2, 3, 4, etc., on to and perpendicular to the line Y Z; thence carry them perpendicular to the center line O P of the front elevation, cutting the section $M^2$ $b^2$ $N^2$ in the points $1^2$, $2^2$, $3^2$, etc., thus dividing it into the same number of spaces as were given to the original section $M^1$ $b^1$ $N^1$. Next connect the points of the same numbers in the two sections of the front elevation, thus: connect $2^1$ with $2^2$, $3^1$ with $3^2$, $4^1$ with $4^2$, etc.; also connect the points $2^1$ with $1^2$, $3^1$ with $2^2$, $4^1$ with $3^2$, etc., all as shown in the engraving. These lines represent the bases of certain triangles, the vertical hights of which may be measured on the horizontal lines cutting the lines W X and W Z. The next step, therefore, is to construct diagrams of these triangles, as shown by A and B of Fig. 514. Draw any two horizontal lines as bases of the triangles, and erect the perpendiculars E C and F D. On both E C and F D set off the various hights of the triangles, measured as above stated and as indicated by the points 1, 2, 3, 4, etc. Next set off the length of the bases of the triangles as follows: In diagram A, let C 1 equal the distance $1^1$ $2^2$ of Fig. 1, make C 2 equal to $2^1$ $2^2$, make C 3 equal to $3^1$ $3^2$, etc. Connect the points in the vertical lines with the points in the horizontal lines of the same number, thus obtaining hypothenuses of the triangles, or the true distance between the points $1^1$ $1^2$, $2^1$ $2^2$, etc., of the elevation. In diagram B, let the distances D 2, D 3, D 4, etc., represent the distances $1^2$ $2^1$, $2^2$ $3^1$, etc., of the elevation. Having located these points, connect 1 in the vertical line with 2 in the base; also 2 in the vertical line with 3 in the base, and proceed in this manner for the other points. This will give the hypothenuses of the triangles, whose bases are $1^2$ $2^1$, $2^2$ $3^1$, etc., in the elevation. Having thus obtained the true measurements of the various triangles in the envelope of the first section of the ventilator, proceed to develop the pattern for it, as shown in Fig. 515. On any straight line, C M, set off a distance equal to 1 1 in diagram A. From C as center, with radius equal to $1^1$ $2^1$ of the elevation, Fig. 513, draw an arc, which cut by another arc drawn from M as center, with radius equal to 1 2 of diagram B, thus establishing the point 2. From 2 as center, with radius equal to 2 2, diagram A, draw an arc, which cut with another arc drawn from $1^1$, Fig. 515, as center, with radius equal to 1 2 of the elevation, thus establishing the point $2^1$. Proceed in this manner, next locating the point 3, then the point $3^1$; next the point 4, and then $4^1$, etc. It will be noticed that, after passing points 6 and $6^1$, $7^1$ is obtained before 7. This is for the sake of accuracy, as it will be seen by inspection of the elevation, Fig. 513, that the distance $7^2$ $6^1$ is less, and therefore more easily measured in the plan, than the distance from $6^2$ to $7^1$. Having thus located the points 1, 2, 3, etc., $1^1$, $2^1$, $3^1$, etc., draw the lines C D and N M, as indicated in Fig. 515. Connect D and N. Then D N M C will be the pattern for one-half the section M N D C of the elevation.

Fig. 515.—Pattern for First Section.
*A Ship Ventilator, having an Oval Mouth on a Round Pipe.*

605. *The Patterns for a Curved Tapering Horn, Octagonal in Section.*—Let *a b c d k e f l* in Fig. 516 represent a section of the article at the small end. Drop the points *a b c d* vertically to the horizontal line K P. With any given radius, X V, determined by the requirements of the case, and from a center upon the line K P, draw an arc, X Y, which will represent the center line through the article. From the point V in the line K P erect a vertical line, V G. Continue the center line horizontally beyond the vertical V G, as shown by Y Y'. Upon this line construct a section of the required article at the large end, as shown by A B C D W E F. From the points A B C D in this section carry points horizontally until they cut V G, as shown by H T U G. Having thus located the points in the elevation at both large and small ends, complete the figure by drawing the arcs K G, J U, Z T and L H from center, which will be found in the line K P, all as indicated in the engraving. Produce the line G V indefinitely in the direction of $H^1$, upon which locate the points $H^1$ *t u* $G^1$ by duplicating points of corresponding letters in the upper part of the line derived from the larger section. Complete the plan view of the article by connecting the points *u* and *a*, *t* and *l*, $H^1$ and *e*, *f* and $G^1$, and *c* and *b*.

Having thus constructed the several views of the article required for the patterns, proceed as follows: The side J U T Z may be pricked directly from the drawing. The pattern for the upper side, shown by K G in the elevation, may be obtained as follows: Upon any straight line, as K G in Fig. 517, lay off the stretchout of K G of Fig. 516, as indicated. Through the points K and G draw the perpendicular *d k* and D W, making *d k* equal in length to *d k* of Fig. 516, and D W equal in length to D W of the same figure. Connect the points *d* D and *k* W, thus completing the pattern. The pattern for the lower side, shown by L H in elevation, is to be obtained in the same general manner, all as shown in Fig. 518. The pattern for the side Z T H L may be described as follows: Let the line *a u* of the plan view in Fig. 516 be considered the plane in which this face lies. From O, which is the center of the inner arc of this piece, drop a line at right angles to L P, continuing it until it strikes the line of its plane *a u*, as shown by O *o*. Thence carry it at right angles to the line *a u* indefinitely in the direction of R. Continue the line *b a*, which is the profile of this strip, until it intersects the line *o* R in the point R. Then R *a* will be the radius of the arc which will form the inner side of the pattern. In Fig. 519, from R as center, with R *a* as radius, describe the arc *a* H, which in length make equal to the stretchout L H, Fig. 516, all as shown by the small figures. To obtain the line of the outer arc of this piece, from the point O in Fig. 516 draw a line to the point T, cutting the arc L H, as shown in the point 8. In transferring the stretchout this point 8 must be correctly located, as shown by 8 in the arc *a* H, Fig. 519. From the center R, by which the arc *a* H was struck, draw a straight line through the point 8 indefinitely. Take the distance B A in Fig. 516, which is the profile of the wide end of the required piece, as radius, and from H, Fig. 519, as center, describe an arc cutting R B in the point B. Draw H B, which will be the wide end of the pattern. From R, the center by which the inner arc of the pattern was struck, draw a straight line cutting the point *a*, producing it indefinitely in both directions. From *a* set off the distance *a b*, equal to *a b* of Fig. 516, which will be the width of the narrow end of the pattern. The only remaining step necessary is to discover a radius, and a center in the line *b* R produced, by which an arc may be struck which will connect the points *b* and B. This, by experiment, will be found to be R'. For the pattern of the piece K G U J of Fig. 516 the operation to be performed is very similar to that just described. From the point *k* in the side view draw a straight line to the point *t*, which consider the plane in which the outer arc of this piece lies. From the point P draw a line at right angles to K P, which produce until it intersects *k t* produced in the point *p*. Thence at right angles to *k p* draw the line *p* S indefinitely. Produce *k e*, which is the profile of the required piece at the narrow end, until it intersects the line last drawn in the point S. Then S *k* will be the radius of the arc which will form the outer line of the pattern. Transfer the line *k e* S to Fig.

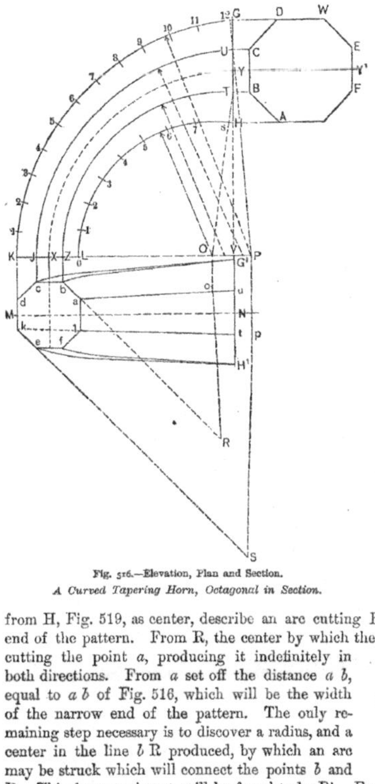

Fig. 516.—Elevation, Plan and Section.
*A Curved Tapering Horn, Octagonal in Section.*

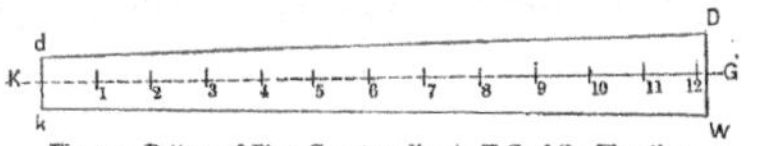


Fig. 517.—Pattern of Piece Corresponding to K G of the Elevation.
*A Curved Tapering Horn, Octagonal in Section.*

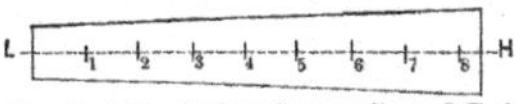


Fig. 518.—Pattern for Piece Corresponding to L H of the Elevation.
*A Curved Tapering Horn, Octagonal in Section.*

520, as shown by *d c* S. From S as center, with radius S *d*, describe the arc *d* D, which in length make equal to the arc K G, Fig. 516. From the point P, Fig. 516, draw a line through the point U, cutting the arc K G in point 12. Carefully locate this point 12 in laying off the stretchout in Fig. 520. From the point 12 in this figure draw a straight line to the center S, as shown. Take the distance D C of the large section, Fig. 516, between the feet of the dividers, and placing one foot on the point D in Fig. 520, swing the other foot around until it cuts the line S 12 in the point C. Then C D will be the wide end of the required pattern. Having now the two ends correctly laid off and the outer arc drawn, it remains to discover a radius, and a center in the line S *c d*, by which an arc may be struck connecting the points *c* C. This is to be determined by experiment, from which it will be found that the center is S' and the radius S' *c*. This method is not to be considered mathematically correct. It is offered on account of its convenience for use and its close approximation to accuracy. It is believed it will be found of greater service in practical work than a rule in which principles are carried to an extreme, resulting in a long and tedious operation.

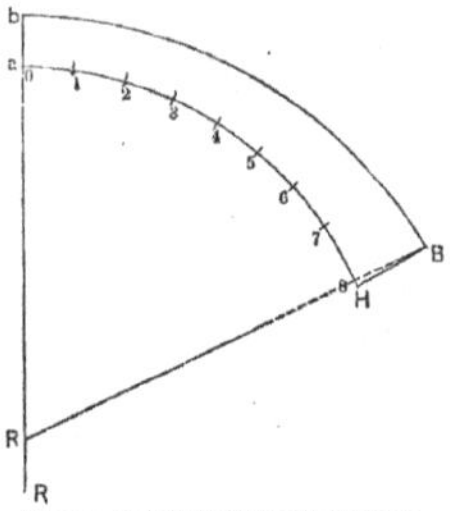


Fig. 519.—Pattern for Inside Flaring Piece.
*A Curved Tapering Horn, Octagonal in Section.*

606. In bringing this work to a close at this point, we do so not because the list of problems which might be presented has been exhausted, but because we think enough has been given to serve every necessary purpose. New problems are continually arising, and the number of combinations which can be made between the various solids known to geometry, which alone can determine the pattern problems that might be enumerated, is almost infinite. The list to which we have given attention in the preceding pages has been gathered during the years in which *The Metal Worker* has been publishing articles upon pattern cutting, and accordingly is believed to embrace all of the more important problems arising in both tin-shop work and cornice making. The fact that many of the demonstrations were prepared in answer to questions propounded by correspondents of that journal, attests the practical bearing upon ordinary workshop practice. In our selection of problems we have been disposed to give preference to those of an elementary character, and which are useful in work of almost daily occurrence, rather than to those of exceptional application, the demonstration of which could not be of interest to any considerable number of mechanics. As elsewhere stated, our aim has been to state principles, with examples of their application, rather than to present arbitrary rules. Rules, when wanted, can be formulated to suit the pattern cutter's requirements, being based upon the principles which it has been our aim to explain.

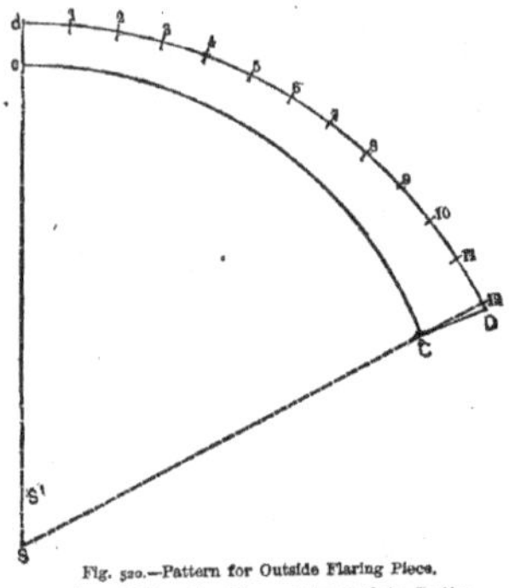


Fig. 520.—Pattern for Outside Flaring Piece.
*A Curved Tapering Horn, Octagonal in Section.*

607. In almost every problem which occurs in practice the mechanic has the choice of several methods. Sometimes these methods differ from each other only in minor particulars and are in reality the same. Still there is in many cases enough difference between them in this respect to warrant a choice. In other instances the difference between methods is radical, making one much more advantageous for employment than the others. The careful pattern cutter will be on the lookout always for points of this kind. He will most carefully avoid falling into ruts or fixed habits, because in the ever-changing conditions of his work set methods often prove a disadvantage. Still other differences in ways of developing patterns may be noted in this connection. The available methods before the mechanic resolve themselves into two general classes. There are what are sometimes called shop rules, and mathematical rules. The former class, while including much that is of no practical value, contains some few methods which, on account of their brevity, as compared with mathematical rules, are really good. Shop rules in general are quite arbitrary in character, at

least upon first sight, but if there is anything in them of merit, upon closer examination an underlying mathematical principle will be found at the bottom of them. This brings us to say that many so-called shop rules may be devised by the intelligent pattern cutter which will be of great use and convenience. Many of the operations in pattern cutting, referring now to the usual mathematical rules, are simply routine in character, and when the mechanic has become sufficiently familiar with the results produced, some of them can be omitted, the net result being laid down arbitrarily by inspection. This is only a suggestion of a way by which shop rules can be devised. Intercourse with mechanics has shown that many of them prefer rules of this kind to those of a purely mathematical character. A few demonstrations properly belonging to this class will be found on the pages preceding, but the majority are mathematical.

# INDEX.

In order to make this work of the greatest usefulness for occasional reference, the names of many articles of ware commonly made in tin shops, but which are not specifically mentioned in the pattern problems, have been incorporated in the index. The sections, figures and pages given in connection with such articles, refer to rules which may be used in developing their patterns.

Spotted something that isn't quite right?

Let us know and we'll fix it for future reprints. And as a thank you for your diligence, we'll send you a free Sothis book of your choice.

publisher@sothis.press

This page intentionally left blank

Except of course for that line above. And this one.
And whilst we have your attention, perhaps you'd
like to follow us on Instagram @sothis.press. Go on.

Made in the USA
Monee, IL
08 November 2026

68106068R00146